The New Economy of the Inner City

Following the restructuring process which swept away the traditional manufacturing economy of the inner city 25 years ago, new industries are transforming these former post-industrial landscapes. These creative, technology-intensive industries include Internet services, computer graphics and imaging, and video game production. The development dynamics of these new sectors are volatile in comparison with those of the classic 'Industrial City'. But these new industries highlight the unique role of the inner city in facilitating creative processes, innovation and social change. Further, they reflect the intensity of interaction between the 'global' and the 'local' in the metropolis, and represent key agencies of urban place-making and re-imaging.

This book addresses the critical intersections between process and place which underpin the formation of creative enterprises in the emergent industrial districts of the 'new inner city'. It contains intensive case studies of industrial restructuring within exemplary sites in prominent world cities such as London, Singapore, San Francisco and Vancouver. The studies demonstrate the global reach of development and innovation across these cities and sites, marked by clustering, rapid firm turnover, and interdependency between production and consumption activity. The evocative case studies, brought to life by interviews, sequential mapping exercises, media narratives, and photography, also disclose the importance of local factors (including urban scale, built form, property markets and policy) which shape both the specific industrial structures and socio-economic impacts.

The New Economy of the Inner City places inner city new industry formation within the development history of the city, and underscores its role in larger processes of urban transformation. The findings inform a critique and synthesis of urban theory which frame the evolving conditions of the 21st century metropolis. This book would be useful to researchers and students of Geography, Urban Studies, Economics and Planning.

Thomas A. Hutton is Professor of Urban Studies and City Planning in the Centre for Human Settlements, School of Community and Regional Planning, University of British Columbia, Canada. Professor Hutton's research interests include new industry formation in the inner city and the role of service industries in urban transformation within the Asia-Pacific.

'With numerous photographs and maps as well as detailed evidence from interviews, and with richly detailed case studies of industrial districts in London, Singapore, San Francisco, and Vancouver, the book documents the emergence of this new urban economy, weaving together industrial restructuring, changing land uses, the city's social composition, public policy, and urban culture. Hutton's approach captures the contingent complexity of metropolitan transformations, though it also unearths striking similarities among cities in quite different nations and cultures. Always informative, this book makes a substantive contribution to the literature on the contemporary space-economy of cities. Highly recommended, for upper division undergraduate, through faculty and research collections'

Robert Beauregard, Columbia University

'The main theme of the book is the restructuring of core urban economic activities and the related regeneration of inner cities in industrialised countries over the past 25 years. . . . the purpose of the book is to link theoretical and empirical debates and findings on industrial restructuring with issues in urban development and urban policy. This aim is achieved in a brilliant way, since the volume is an unconventionally rich source of inspired theoretical discourse as well as carefully conducted field work, based on case studies that fully reflect the changing socioeconomic framework conditions of urban development.'

'[Hutton's] assessments of urban change are careful and thus sympathetic, quite different from the recent hype associated with the apparent general renaissance of cities. Without any doubt, the book sets a standard in contemporary knowledge of the development of inner cities . . . its contribution to the urban discourse will be much appreciated.'

Markus Hesse, University of Luxembourg

'The cases are presented in a systematic way which facilitates comparison. . . . the cases represent cities from different tiers of the world city hierarchy, and, within those cities, locations within the inner city that have experienced a comparable process of industrial restructuring. Since the cases are drawn from four countries in three continents, there is considerable scope for generalization. All of Hutton's cases show an intriguing mix of Fordist, post-Fordist and 'post-post-Fordist' economic activities that are linked in various ways, including collaborations, hybrids, production—consumption interfaces and, of course, also competition. Hutton's monograph offers a useful 'state of the art' of research on the inner-city economy, a welcome critical reflection on over-hyped concepts like the creative class and the creative industries, and several building blocks to bring this research field forward.'

Marco Bontje, University of Amsterdam

'You can find it in any city. It's at most a 20-minute walk from the office towers of the central business district. But it couldn't look or feel more dissimilar. . . . It's a jumble of old workshops, old warehouses, run-down residential streets converted to other uses. It houses thousands of small firms, doing the jobs the downtown needs: printing publicity, offering services, feeding the workers. It's quintessentially messy and extraordinarily lively. It's Clerkenwell in London, Manchester's Northern Quarter, South of Market in San Francisco. Over the last 30 years, such areas have transformed themselves. They've become the seats of the 24-hour city and the creative economy. But what exactly is going on here? Thomas Hutton has the answer. Hutton's answer involves complex, even contradictory processes. Hutton concludes that the inner city is now characterized by a "recombinant economy": industries constantly change track, sometimes competing for space, sometimes collaborating. But within this kaleido-scopic picture, there's a key central process: "new economy" industries, such as com-puter graphics and web design, combine design and technology, service functions and goods production. Great cities are no longer post-industrial: they're something richer and more complex, in which new kinds of industry propel the economy forward. The New Economy of the Inner City is compulsory reading for [those] seeking to understand the contemporary city'

Sir Peter Hall, The Bartlett School, University College London

Routledge Studies in Economic Geography

The Routledge Studies in Economic Geography series provides a broadly based platform for innovative scholarship of the highest quality in economic geography. Rather than emphasizing any particular sub-field of economic geography, we seek to publish work across the breadth of the field and from a variety of theoretical and methodological perspectives.

Published:

Economic Geography
Past, present and future
Edited by Sharmistha Bagchi-Sen and Helen Lawton Smith

Remaking Regional Economies
Firm strategies, labor markets and new uneven development
Susan Christopherson and Jennifer Clark

The New Economy of the Inner City
Restructuring, regeneration and dislocation in the twenty-first-century metropolis
Thomas Hutton

New in Paperback:

Remaking Regional Economies
Firm strategies, labor markets and new uneven development
Susan Christopherson and Jennifer Clark

The New Economy of the Inner City
Restructuring, regeneration and dislocation in the twenty-first-century metropolis
Thomas Hutton

The New Economy of the Inner City

Restructuring, regeneration and dislocation in the twenty-first-century metropolis

Thomas A. Hutton

Routledge
Taylor & Francis Group

LONDON AND NEW YORK

For Lesley and Siena

First published 2010
by Routledge
2 Park Square, Milton Park, Abingdon, Oxon, OX14 4RN

Simultaneously published in the USA and Canada
by Routledge
270 Madison Ave, New York NY 10016

*Routledge is an imprint of the Taylor & Francis Group,
an informa business*

Transferred to Digital Printing 2010

Typeset in Galliard by
RefineCatch Limited, Bungay, Suffolk

British Library Cataloguing in Publication Data
A catalogue record for this book is available from the British Library

Library of Congress Cataloging-in-Publication Data
The new economy of the inner city : restructuring, regeneration and
dislocation in the 21st century metropolis
    p. cm.—(Routledge studies in economic geography)
    Includes bibliographical references and index.
    1. Inner cities—Economic aspects—Case studies.   2. Industrial
organization—Case studies.   3. High technology industries—Case
studies.   4. Urban renewal—Case studies.   5. Urban economics—
Case studies.   I. Title.
    HT156.H88 2010
    307.3′42—dc22   2009038707

ISBN13: 978–0–415–56932–3 (pbk)
ISBN13: 978–0–203–93365–7 (ebk)

ISBN10: 0–415–56932–X (pbk)
ISBN10: 0–203–93365–6 (ebk)

The New Economy of the Inner City

Restructuring, regeneration and dislocation in the twenty-first-century metropolis

Thomas A. Hutton

LONDON AND NEW YORK

For Lesley and Siena

First published 2010
by Routledge
2 Park Square, Milton Park, Abingdon, Oxon, OX14 4RN

Simultaneously published in the USA and Canada
by Routledge
270 Madison Ave, New York NY 10016

*Routledge is an imprint of the Taylor & Francis Group,
an informa business*

Transferred to Digital Printing 2010

Typeset in Galliard by
RefineCatch Limited, Bungay, Suffolk

British Library Cataloguing in Publication Data
A catalogue record for this book is available from the British Library

Library of Congress Cataloging-in-Publication Data
The new economy of the inner city : restructuring, regeneration and
dislocation in the 21st century metropolis
    p. cm.—(Routledge studies in economic geography)
Includes bibliographical references and index.
1. Inner cities—Economic aspects—Case studies.   2. Industrial
organization—Case studies.   3. High technology industries—Case
studies.   4. Urban renewal—Case studies.   5. Urban economics—
Case studies.   I. Title.
HT156.H88 2010
307.3'42—dc22   2009038707

ISBN13: 978–0–415–56932–3 (pbk)
ISBN13: 978–0–203–93365–7 (ebk)

ISBN10: 0–415–56932–X (pbk)
ISBN10: 0–203–93365–6 (ebk)

Contents

Figures

Tables

Preface

This book project developed from a long-standing interest in the evolution of the urban economy, and more particularly the complex production systems, labour markets, and industrial land use of the metropolitan core. Like many of my generation of urban scholars, my research orientation was shaped by the exigent policy crises and theoretical disjuncture generated by the wrenching industrial restructuring of the 1970s and 1980s. Typically urban scholars (largely but not exclusively economic geographers) elected to study the implications of the expansion of service industries, concentrated within the modernist built environment of the Central Business District (CBD); alternatively, the landscapes of disinvestment and decline within the inner city (a domain of urban geographers); or the social upgrading processes in the core for which these first two trajectories represented preconditions: a central research motif for social geographers and sociologists. These shaped the contours of the *postindustrial city*, a term recurrently contested by neo-Marxists, and assuredly a term laden with multiple meanings, but one that conveys something of the traumatic force of the restructuring experience of the period: an indispensable heuristic for describing something surely well beyond the normal, periodic crises of capitalism that form the thematic matrices of the critical studies agenda.

Each of these domains of urban research on the postindustrial city appealed to me as crucially important and interesting. But I elected to study the specialized intermediate services and segmented labour force of the CBD, stimulated by Peter Daniels' *Office Location: An Urban and Regional Study*, published in 1975. Graduate studies at Oxford under the supervision of Jean Gottmann encouraged an interest in the high-powered service economies of New York, Paris, and especially London, as well as the daunting socioeconomic consequences of restructuring in the city. On my return to Vancouver I embarked upon a research program focusing on the development of the office complex of the CBD (with David Ley), and on trade in services (with Craig Davis). In 1988, a research exercise on the economic significance of arts and design activities for *Design Vancouver*, a biennial festival celebrating the city's leading creative talents, drew my attention to the resurgence of employment in the CBD fringe and inner city. This new (or recovered, in historical terms) design orientation was associated with the aestheticization of postindustrial inner city landscapes, and the revalorization of property markets, analogous to the relentless gentrification of Vancouver's older communities. But by the early 1990s it seemed clear that these ascendant industries of the inner city followed logics of location quite different from those of the white-collar services of the CBD, and warranted further investigation. The congregations of graphic designers, artists, architects, and others situated within certain districts of the core reflected at least in part *preference*, not merely a response to the structure of rent gradients within the central city property market.

The rise of a technology-intensive 'New Economy' in the late 1990s appeared to insert a more insistent development trajectory within the inner city terrains of the metropolis, and potentially a successor to the arts and design-based industries which had pioneered the regeneration of employment (as well as new episodes of social displacement) in these zones. The collapse of the technology boom 2000–2001 abruptly terminated this phase of inner city redevelopment, although there are more enduring legacies of the 1990s which have been absorbed into the economic systems of cities and regions among both 'advanced' and 'transitional' societies. Industries that had established within districts of the inner city since the 2000–2001 crash reflect the more durable advantages of this zone for cultural production, but typically deploy increasingly sophisticated technology for design work, production, and communications, the latter function including outsourcing, marketing, and the recruitment of skilled labour. The economy of the inner city thus takes the form of a 'hybridized' structure of cultural production, creative labour, and technology; comprises a complex mix of 'new' and 'old' economy industrial regimes; and presents a rich and diverse array of production, exchange, circulation, and consumption: a 'recombinant' structure of economic activity in the heart of the twenty-first-century city.

This volume represents an interim statement on the nature, development characteristics, and significance of new industry formation within the inner city, based on an extended engagement with the scholarly and policy literatures, and more particularly on case studies of exemplary sites located in cities which have attracted considerable scholarly attention: London, Singapore, San Francisco, and Vancouver. The empirical insights derived from the experiences of these exemplary sites yield intrinsically rich narratives of industrial innovation and restructuring, but also suggest possibilities of theoretical conjecture. Foundational theories of the late twentieth-century city, postindustrialism and post-Fordism, still have much to say about the contemporary metropolis. But scholars (including Allen Scott and Peter Hall, among others) have proposed new concepts to capture changes in the conditions of twenty-first-century urbanism, and it seems to me that the investigation of new industry formation in the core – a salient zone of urban change now, as before – offers a fruitful entrée to retheorization. There are of course limits to the implications to be drawn from a necessarily limited sample, and so the current research enterprise includes collaborative projects, designed to interrogate more incisively implications of new industry formation in the city. This includes a special theme issue of *Urban Studies* which will incorporate original research on new industry formation within the inner city districts of New York, London, Barcelona, Toronto, and Singapore, among other cities, as well as collaborative research on employment and new economic spaces within, first, Asian metropolitan cities and, second, Canadian city-regions. But for the moment my hope is that this present book might add something of value to the burgeoning research literature on industrial change in the metropolitan core, and to the retheorization process just now in its incipient stage.

Acknowledgements

Acknowledgements of copyright holders

The author and publishers would like to thank the following for granting permission to reproduce material in this book:

Cities (Elsevier) for Figure 3.4
Peter Elliott for Figures 3.5 and 3.6
Economic Geography (Clark University) for Figure 3.7
Sir Peter Hall for Figure 3.8
The Ordnance Survey for Figure 4.4
Town and Country Planning for Figure 4.7
The Times of London for Figure 4.10
Oxford Economic Forecasting and the City of London for Tables 4.1 and 4.2
National Statistics of the United Kingdom (NOMIS) and the Controller of HMSO for Tables 5.1 and 5.2
Hays Davidson and John McLean of the Renzo Piano Building Workshop for Figure 5.7
The Urban Redevelopment Authority of Singapore for Figure 6.1
The Straits Times (Singapore) for Figure 6.8
The Singapore Ministry of Labour, Labour Market Statistics, for Tables 6.1, 6.2, 6.3, and 6.4
The City of San Francisco Planning Department for Figures 7.2, 7.3, 7.4, 7.5, 7.6, and 7.7
Pierre Chatel for Figure 7.13
Urban Studies (Routledge) for Figures 8.2 and 8.4
The City of San Francisco Planning Department for Tables 7.1, 7.2, and 7.3
Markus Moos (UBC Geography) for Table 8.4
The City of Vancouver Planning Department for Figures 8.9, 8.11, and 8.13

The publisher has made strenuous efforts to contact all copyright holders to secure permissions; for those whom we have been unable to contact, please inform the publisher as soon as possible.

Acknowledgements of support and assistance

There isn't room here to acknowledge the contributions of all those individuals who generously lent support to me in various forms over the extended period of this project, but I would be remiss in not citing the following as essential to producing this volume.

For the London case study, the advice of Andy Thornley and Andy Pratt of LSE, and Sir Peter Hall of the Bartlett School at UCL, was instrumental. Graeme Evans and Jo Foord (London Metropolitan University) shared their deep knowledge of the Clerkenwell district, while Tim Butler and Chris Hamnett of King's College London made available their formidable inventory of research on gentrification in London.

For the survey work in Singapore's Chinatown, I benefited greatly from the expert insights of K. C. Ho of the National University of Singapore (Sociology) and Kelvin Ang, of the Urban Redevelopment Authority. For perspectives on creative industries in Telok Ayer I would like to gratefully acknowledge Daen Tay, Sarah Tham, Alan Lim, Cassandra Wong, and Raymond Ng.

During my research on industrial change and land use policy questions in San Francisco's South of Market Area, I received invaluable assistance throughout the extended study period, initially from Dr Amit Ghosh, (former) director of policy, and, over the last decade, from Stephen Shotland, senior policy planner, of the San Francisco Planning Department. For valuable insights on the operation of creative industries in the South Park area, I want to especially thank the following: Larissa Sand, Robert Ahearn, and Victor Petruzzi.

In the Vancouver case, Michael Gordon and Ronda Howard generously shared their extensive knowledge of planning for Vancouver's Central Area.

For the preparation of the book manuscript, Kevin Morgan (Cardiff), David Ley (UBC) and K.C. Ho (NUS) each read chapters and made pertinent suggestions for revision.

In the production process, I relied greatly on the skills and commitment of my colleagues at UBC. Carolyn Bell and Karen Zeller, my colleagues at the Centre for Human Settlements, expertly prepared the tables and several of the figures for the volume. For the preparation of the many maps, which represent a central part of the case study narratives rather than simply adjuncts to the text, Eric Leinberger of the UBC Department of Geography provided expert cartographic advice on an endless series of artistic and technical issues, as he has for the last decade. Karen, Carolyn and Eric all exhibited quite extraordinary levels of patience as well as finely-honed skills. I also benefited from the work of a number of able research assistants at CHS, notably Jason Blackman, Tanis Knowles, and April Lawrence.

At Routledge, I want to acknowledge the support and wisdom, on all matters related to publishing, of Andrew Mould, senior commissioning editor for urban studies, as well as my editorial assistants, Jennifer Page and Zoe Kruze, who provided a felicitous mixture of encouragement, assistance, and zeal. The series editors, David Angel, Amy Glasmeier, and Adam Tickell, offered sage advice based on their examination of the prospectus, including a caution not to over-privilege the 'new' over the 'old' in my depictions of the economy of the inner city. I look forward to their judgement as to whether I have succeeded in achieving this crucial balance of perspectives in the narratives of the evocative sites included in this volume. I would also like to express my appreciation for the expert and patient copy-editing services of Susan Dunsmore in the preparation of this book, and to thank Rebecca Hastie of Swales and Willis for her exceptionally proficient work in preparing the book for publication.

Finally, I want to acknowledge the crucial support of the Social Sciences and Humanities Research Council of Canada for this project, including grants awarded under, first, the SSHRC Initiative on the New Economy (INE) program, then more recently under the National Research Cluster (NRC) program.

Tom Hutton
Centre for Human Settlements
University of British Columbia
Vancouver
June 2007

1 The reassertion of production in the inner city

Introduction: beyond the postindustrial metropolis

Over the 1970s and 1980s, long-established models of urban development and spatial structure derived from the concept of the *industrial city* articulated by the Chicago School's practitioners were comprehensively subverted by far-reaching industrial restructuring processes. The spatial configuration of the postindustrial city incorporated a markedly asymmetrical core, comprising a high-growth central business district (CBD) corporate office complex, and terrains of disinvestment and deindustrialization within the CBD fringe and inner city.[1] The collapse of Fordist production and related employment, and the rise of an urban postindustrial social class (Bell 1973), constituted essential preconditions for gentrification and its dislocations within inner city communities. While industrial restructuring was not confined to the central city, the core served as the defining locus of fundamental change in the metropolis, giving rise to an extended urban policy crisis and a trilogy of influential theories: *postindustrialism, post-Fordism*, and *postmodernism*. These theories have been vigorously contested on polemical, theoretical and empirical grounds, but each served to influence a generation of urban scholars and principal lines of urban research and policy studies.[2]

In the early years of the twenty-first century, urban scholars are presented with conditions of theoretical disjuncture associated with new experiences of industrial restructuring and broader shifts in urban development trajectories. The trilogy of theories cited above remains influential, but over the past decade or so the postindustrial metropolis has been subject to processes of change that require a thorough interrogation and conceptual reformulation. In a number of major metropolitan cities, notably London and New York, the long period of industrial decline which had its provenance in the 1960s had largely run its course by the early 1990s. At the same time, the rising arc of intermediate services growth which led employment formation in the postindustrial period began to subside somewhat in many regions over the 1990s, owing to market, technological, and policy factors, as well as a natural maturation process.[3]

But the 1990s also saw the emergence of new industries and labour among advanced societies, subsumed variously within the rubrics of the New Economy, the cultural economy of the city, and the knowledge-based economy. The theoretical significance of these tendencies remains somewhat inchoate, reflecting the volatile nature of growth trends, the complexities of new industry formation, and differentiated experiences from place to place. It is also the case that the New Economy in its varied industrial manifestations co-exists with elements of the 'old economy', and that episodes of change include restructuring events and cycles, rather than a totalizing break with the past. But there is sufficient evidence of novelty to support the idea that we have entered a stage of urbanism in which development conditions present a departure from those of the classic postindustrial era.

Conditions of urban change and challenges to theory

Accelerating processes of change have served to compromise the integrity of urban development models over the past half-century. To illustrate: the tight residential community formations of the metropolis that gave rise to the social ecology models of the Chicago School in the 1920s and 1930s have been increasingly subjected to the destabilizing influences of socioeconomic restructuring, gentrification, demographic shifts, and migration, including international immigration. Recurrent episodes of industrial restructuring have tended to impart 'greater complexity and instability to the restructured social mosaic' (Soja 2000: 282), a major problematic to be confronted by those with affinities to the urban structure models derived from the Chicago School and its later variants. Suburban areas, treated largely as homogeneous residuals of the 'strong centre' metropolis by the Chicago School, now exhibit increasing social and industrial variegation, shaped by immigration, demographic change, and the locational tendencies of manufacturing and many service industries, as well as comprising increasing shares of the city-region's population and employment.

Second, metropolitan cities among advanced societies no longer function exclusively as 'regional central places' in the Christallerian formulation, but rather are critical base points of globalization, forcefully drawn into international circuits of capital, trade and labour formation. High-order global cities are in many ways influenced more by interactions with other such cities (characterized to some extent by features of complementarity, as well as by relentless competition) than with other centres within the national urban system, as exemplified in Chris Hamnett's (2003) portrayal of 'London in the global arena'. The globalization of city-regions has contributed to the restructuring of industries and property markets and has produced new social cohorts, including highly specialized financial intermediaries, and new gentrifiers typified by high incomes and insensitivity to price in the construction of housing choices. Many of the regional functions formerly performed by central cities, including wholesale and retail services, personal services, education, and public administration, are now largely distributed within suburban and ex-urban areas, following the decentralization of population, households, and manufacturing.

Third, long-established production ensembles of the industrial city were largely swept away by the restructuring processes of the 1970s and 1980s, producing a legacy of disinvestment, decline, and deprivation in many cities, and the erosion (or erasure) of working-class communities which in some cases had survived more than a century of wars, depressions, and other upheavals. For many scholars this wrenching industrial restructuring experience produced the contours of a *post-industrial city*, the constitutive features of which included: (1) a dominant intermediate service sector concentrated within the corporate complex of the central business district, shaped in part by new spatial divisions of production labour and the deregulation and privatization movements of the 1980s; (2) a restructured urban employment base, the growth of which was led by the segmented and hierarchical labour of the corporate office sector; (3) an asymmetrical metropolitan core, comprised of the hegemonic CBD and decline in the CBD fringe and inner city; (4) the ascendancy of the high-rise office tower as the supreme expression of capitalist imperatives and modernism; and (5) the emergence of a 'new middle class' of executives, managers, and professionals, an explicitly urban manifestation of Bell's earlier formulation of a postindustrial society, and a cohort defined by particular residential preferences, linked cultural and political values, behaviours, and lifestyles.

As is well known, this postulation of a postindustrial city and urban economy was bitterly contested by other scholars, notably by the neo-Marxists, a contingent which acknowledged the collapse of orthodox communism and a number of its variants, but nonetheless adhered to the centrality of class both to an understanding of society and to the construction of political values and choices. For a number of these leftist scholars, the collapse of traditional manufacturing and the rise of services followed a standard (albeit much deeper) trajectory of capitalist crises. The restructuring of the 1970s and 1980s was construed as the supplanting of manufacturing and organized production labour in the city by service industries, the elite members of which represented the interests of capital, concerned with directing resources to areas of higher profits, and thus benefiting directly from the surplus accruing from these reallocations. Neo-Marxists, regulationists and other critical studies theorists proposed post-Fordism as a descriptor for the collapse of traditional manufacturing and the blue-collar workforce, a concept encompassing the contraction of Fordist manufacturing plants and the evisceration of Taylorian labour.

The return of production to the inner city

The comprehensive restructuring of the urban economy, employment, social class, and space defined variously as postindustrialism or post-Fordism took place over a three-decade period, from, roughly, the early 1960s to the first years of the 1990s. But this late twentieth-century socioeconomic trajectory, as deeply rooted and wrenching as it was, produced not an 'end-state' city, but, rather, conditions for successive new experiences of industrial innovation and restructuring. These included a putative, technology-driven 'New Economy', a 'knowledge-based economy' characterized by enhancements of human capital as well as by a technological deepening of production and labour, and a 'cultural economy of the city'.

The latter has emerged as perhaps the most durable of the abbreviated restructuring episodes of the past decade as interpreted in the urban-regional development literature, embodying as it does the cultural inflection of new industry and labour formation, and a growing emphasis on 'cultural products' as lead product sectors in the economy of the advanced metropolis (Hall 2000).

These episodes of industrial innovation over the past decade and a half have produced characteristic spatial, structural, social, and land use consequences, as articulated in Stephen Graham and Simon Marvin's (2001) account of the problematic features of 'splintering urbanism'. Suburban (and ex-urban) zones of the metropolis can no longer be considered mere residuals of urban change, as they now typically account for the largest shares of population and employment growth within the city-region, including to a degree representations of the new industries and labour linked to the 'New Economy' of the 1990s. But the persistent saliency of the metropolitan core as a critical terrain of metropolitan transformation can be demonstrated by reference to contemporary processes of growth and change.

Industries of the technology-intensive 'New Economy', and those of the 'cultural economy of the city', have been concentrated within the inner city. These new industries, moulded by a synthesis of cultural values and practices, and advanced production and telecommunications technologies, constitute new forms of post-Fordism, and a departure from the segmented office labour force that comprised the dominant employment sector of the postindustrial city. The investments associated with the emergence of these industries, when combined with new commercial, residential, and consumption development in the inner city, suggest a new phase in the insistent relayering of capital in the city, with implications for the spatial configuration and land use patterns of the core, as well as new experiences of social dislocation.

In the aggregate, these trends are suggestive of a 'new inner city' which presents a marked contrast with defining elements of the late twentieth-century postindustrial metropolis, with defining attributes including the emergence of new social groups as well as recurrent experiences of industrial experimentation, innovation, and restructuring. The metropolitan core is shaped by increasingly complex reproductions of social and economic space, together with recurrent experiences of conflict and dislocation, generated in large part by comprehensive redevelopment encroaching upon the marginal communities of the inner city. This new phase includes residential 'mega-projects' (Olds 2001) as well as new industrial formations and spaces of spectacle and consumption. At the same time, these variegated patterns of development follow logics of location incorporating agglomeration economies, social dynamics of innovation and creativity, and local policy factors, so they can scarcely be construed as following a chaotic form of postmodernism as postulated by some urban scholars in the past decade.[4]

These recent transitional features of the urban economy, labour force, and spatial structure support the idea of theoretical engagement. The production of transcendent theories along the lines of models of earlier phases of urban development is, however, constrained by increasing complexity and volatility of the urban condition, and by highly differentiated vectors of urban development, shaped by distinctive global-local interaction and by contingencies of policy and governance systems. Further, while the roots of postindustrial theory lie in the tumultuous fortunes of the older, mature urban societies of the western world, particularly within the city-regions of the 'Atlantic sphere' such as London and New York, it now seems clear that urban theory needs to consider implications of development within Pacific Asia, increasingly a salient theatre of industrial restructuring and globalization. These experiences to some extent follow the earlier cycles of change within the Atlantic core city-regions, but also present contrasts, not least in the role of the state and developmental policy, and, in some cases, the persistence of developmental dualism (Leaf 2005). With the increasing power of global forces, there may also be a need to consider developments in the 'global south' in the formation of urban theory, including industrialization experiences in India, Brazil and Argentina, among other nations, although contrasts may be more important than commonalities. As the experiences of fast-growing Asian cities – vigorously inserted into international circuits of investment, trade and migration, and subject to restructuring processes which follow in some important ways the transformative patterns of more mature regions – can neither be ignored nor readily assimilated within the trend-lines of 'western' examples, the task of theorization is made commensurately more problematic.[5]

New industry formation as an entrée to retheorization

Restructuring processes were central to the formation of the postindustrial city, with the effects of industrial change evidenced in the comprehensive reshaping of zonal structure and land use, the urban space-economy, housing markets, landscapes and built form, and social class. Similarly, recent processes of industrial innovation and restructuring constitute what Ed Soja (2000) describes as the 'still-evolving discourse' of industrial urbanism in the metropolis. There is a need to come to terms theoretically with the dynamic nature of industrial enterprise in the core, as seen in the rise and fall of the so-called dot.coms, in the growing (but uneven) development of the urban cultural economy, and in the evolving regional divisions of production labour.

Volatility and complexity thus comprise defining features of contemporary industrial formation within the metropolis. Rather than over-emphasizing these episodic or transitory aspects, however, this study places the sequence of experiences in the urban core firmly within an appreciation of the logics of industrial change, linked to larger processes of urban transformation. Twenty years ago, Allen Scott referenced in an influential monograph (*Metropolis: From the Division of Labor to Urban Form*, 1988) a sequence of writers who comprised a tradition of scholarship seeking to establish fundamental interdependencies between industrialization and the development of the metropolis. These exponents of industrial urbanism included (among many others) Weber (1899), Haig (1927), Wise (1949), Hoover and Vernon (1959), Hall (1962a), Sjoberg (1965), and Webber (1984). Scott acknowledged that this 'conceptual lineage represents a sort of submerged tradition of urban studies, never as actively or coherently to the fore as

Chicago School theory and its later permutations, but always co-present through time' (Scott 1988: 6). In endeavouring to reassert the saliency of industrialization as an agency of the urban growth and change, Scott enunciated his broad purpose as an attempt to demonstrate

> how the modern metropolis emerges, at least in part, out of the fundamental logic of industrial production in capitalism, and how its geographical form is composed of interpenetrating production spaces and social spaces locationally dominated by the former. In this sense the modern metropolis is both the creation of the social and property relations of capitalism and a specific condensation of them.
>
> (ibid.)

From a vantage point twenty years on, events have in important respects buttressed the force of Scott's principal arguments, seen both in the spread of capitalism and markets across an ever-wider range of jurisdictions following the collapse of the Soviet Union and the COMECON sphere in 1989, and more specifically in the demonstrated power of industrial change to reshape urban society, space, and place.[6]

At the same time, the contemporary urban economy comprises an increasingly complex mixture of production and consumption activity, interacts with social and cultural systems in ever more intimate ways, and is shaped by a mélange of market and policy factors, even in an era in which neo-liberalism is widely acknowledged as in the ascendancy, all points that Scott has acknowledged in his most recent work. An overly austere ideological viewpoint of industrial development in the city, therefore, may carry the risk of delivering a minimalist understanding of the critical interdependencies between the economy and the broader evolution of the metropolis. This present study thus encompasses allied social, cultural, spatial, and policy factors implicit in processes of industrial innovation and restructuring, rather than a narrow emphasis on purely economistic features, as this broader approach accommodates the organizational complexity of contemporary industrial production in the city.

While studies of the development of industries in all parts of the metropolis can assuredly contribute to our understanding the industrialization-urbanization developmental nexus, the metropolitan core continues as before to present particularly exigent opportunities for research. A multiperspectival study of new industry formation in the urban core, situated within the emergent geographies of specialized industrial production and allied features of urban change, can contribute in significant ways to theoretical enterprise in the early years of the twenty-first century.

Defining features of the New Economy of the inner city

The sequence of new (or reconfigured) industries, production networks, and labour situated within the core areas of advanced societies represents a significant reassertion of production in the inner city, contributing to the comprehensive reproduction of the central city. While these new industry formations are characteristic of advanced societies and industrial production systems, they are increasingly in evidence in transitional cities, notably within Pacific Asia, demonstrating the accelerative effect of globalization on development processes. Salient features of the New Economy of the inner city include the following discussed below.

Industrial restructuring and new development trajectories

The inner city terrains of many advanced and transitional cities include residual Fordist industries and artisanal and craft-scale production, as well as concentrations of mainstream business services, such as legal, accounting, and consulting firms, situated in what Peter Hall terms 'inner edge cities'. But *defining* features of the New Economy of the inner city include ensembles of hybridized, knowledge-intensive firms. These include relatively new, creative, and technology-based industries, such as communications consultants, computer software design, computer graphics and imaging, computer networking, and Internet services. The New Economy of the inner city also incorporates established, increasingly technology-intensive creative industries, exemplified by advertising, architects, fashion design, graphic artists and designers, industrial design, film and video production and postproduction, music, and print media. In general, these industries and firms powerfully exhibit contemporary processes of *convergence* in advanced cities and urban production systems, expressed in: (1) a synthesis of cultural and technological factors in production processes; (2) a more intensive articulation of services and manufacturing in the fabrication of high-value 'cultural products' (after Scott 2000); (3) the marked interaction between production and consumption within the postmodern inner city; (4) the (not unproblematic) interface between the arts, 'high culture', and 'edge cultures' practiced by new social actors in the inner city (Zukin 1995); and (5) the blending of factor inputs derived locally, and from external sources via advanced telecommunications.

The revival of industrial districts in the inner city

The rise of new industry ensembles in the metropolitan core involves a fundamental reorganization of inner city space, including 'primary' new production sites as well as place-based production networks and sets of linked industries. The space-economy of the twenty-first-century metropolitan core incorporates new territorial forms of specialized industrial production which accommodate leading-edge firms, together with distinctive consumption, cultural and environmental amenities, effectively demonstrating the commingling of the 'social' and 'economic' worlds of the inner city, and the role of new industry formation in processes of urban place-making.

New industry sites proliferate within the derelict or obsolescent inner city districts of postindustrial cities in Europe and North America, including London, Glasgow, Hamburg, Berlin, Barcelona, Milan, New York, Montreal, Toronto, San

Francisco, and Vancouver. In these cases, new industry formation can be seen as a revival of inner city industrial sites, or as a new phase of the urban services economy. But we can also identify new industry formation as an increasingly significant process of urban change among the 'growth economies' of Pacific Asia, including Tokyo (Shinjuku, Ropponggi, Shibuya), Seoul (Kangnam), Shanghai (East Bund, Suzhou Creek), and Singapore (Telok Ayer, Far East Square). The emergence of New Economy sites in the inner city can therefore be constructed legitimately as a global phenomenon, albeit one shaped by distinctive aspects of contingency, including policy interventions. There are also features of volatility within the new industrial spaces of the inner city, associated with pressures of market competition and cost, the sensitivity of emergent industries to new technologies and their destabilizing influences, and the operation of local property markets, which may in some cases favour high-end housing over employment-generating land uses.[7]

New divisions of labour in the central city

Emerging divisions of production labour in the twenty-first-century central city represent in several important ways a marked contrast to the labour markets of the postindustrial urban core. During the postindustrial period, circa 1965–1995, the restructuring of the central area's labour market incorporated, first, a calamitous decline in long-established Fordist manufacturing industries and allied industrial labour; and, second, the rapid expansion of a highly segmented office labour force. This office workforce was dominated by an elite cohort of managerial and professional workers, but included within middle and lower echelons of this hierarchy were supervisory personnel, sales staff, clerical and secretarial labour, technical workers, and janitorial and maintenance workers.

The office workforce is still the largest element of the central city labour market, but we can discern significant shifts in the divisions of production labour. Office employment has come under increasing pressure in many cities, with suppressive effects on some of the largest cohorts. Corporate mergers and downsizing have tended to concentrate corporate power among cities at the peak of the global urban hierarchy, and have cut significantly into managerial occupations within many secondary labour markets, while the intensification of capital (notably in the form of new communications technologies) has severely impacted clerical labour. Overall, the generous staffing of offices prevalent during the peak of the postindustrial period has been supplanted by much leaner employment configurations.

At the same time, the comprehensive redevelopment of the CBD fringe and inner city has produced new social, spatial, and technical divisions of production labour. As observed above, there are aspects of both complexity and volatility in the New Economy, and features of continuity as well as discontinuity, but for the purposes of illustration we can describe the restructuring of labour elements in the inner city as follows: (1) emerging *social* divisions of labour in the form of 'cultural product' sectors (goods and services), consistent with Scott's hypothesis concerning the urban cultural economy (1997), and corresponding to the contemporary and historical roles of the inner city as a site of artistic production, creativity, and applied design; (2) a reordered *spatial* division of labour (after Massey 1984), with the revival of production labour within the inner city, presenting a more balanced spatial profile of employment in the core, relative to the spatial asymmetries of the postindustrial urban workforce dominated by the office workforce of the CBD; and (3) new *technical* divisions of production labour, which take the form of 'neo-artisanal' labour (after Norcliffe and Eberts 1999) emphasizing production process synergies between the arts, creativity, technology, and entrepreneurship. These new social, spatial, and technical divisions of production labour in the inner city are in important ways complemented by the growth of employment in consumer industries (for example, in retail businesses, restaurants, and coffee houses) and institutions (design schools, galleries, non-governmental organizations [NGOs], and community-based organizations [CBOs]) which support the operation of the New Economy of the inner city, illustrating one dimension of the relational geographies of specialized production in the twenty-first-century urban core.

Implications: spheres of impact in the city

The principal purpose of this study is to examine the implications of new industry formation in the city's inner districts in a more systematic fashion than has been the case hitherto. But as a means of initially framing this investigation we can on a prima facie basis identify some general areas of economic, social, and environmental impact of new industry formation in the inner city. First, the rise of new industry formations within inner city districts of the metropolis can generate the following economic impacts. New industry formation in the inner city can play a part in the reconfiguration of the metropolitan core's space-economy, redressing to some extent the spatial imbalance of the postindustrial core which heavily favoured the corporate complex of the CBD, and partially offsetting job losses in central city industries and occupations. Second, new production enterprise can contribute to local area regeneration, in the form of business start-ups, infrastructure investments, employment formation, and supply and subcontracting opportunities, as well as injections of entrepreneurship in lagging areas of the inner city likely to be deficient in these attributes. A third category of economic impacts can take the form of regional development linkages and supply chain functions: research has disclosed dense patterns of connection between inner city clusters and other sites within the metropolis; these can include: (1) centripetal (or 'inward') linkages between inner city industries and firms located within the proximate CBD; and (2) centrifugal linkages, as observed in subcontracting relations between small and medium-sized firms in the inner city, and larger corporations in suburban and ex-urban sites.[8] Fourth, the competitive advantage of inner city districts in design, creative, and knowledge-based industries provides a platform for penetrating export, as well as domestic, markets. Finally, the imageries associated with the ascendant industries of the inner city, and allied consumption and spectacle, contribute to the metaphors of urban transformation, as seen in descriptors such as the 'cultural economy of the city', the 'New Economy', and the 'creative city'. These imageries of innovation and enterprise are widely deployed by municipal officials in attempts to attract new investment for regeneration.

Social impacts of new industry formation may include the formation of a putative 'creative class' popularized by Florida and others (Florida 2002), although there is skepticism regarding both the true empirical dimensions of this cohort, as well as its status as an autonomous group distinct from the 'new middle class' which comprises the dominant social class of the central city (Hamnett 1994; Ley 1996).[9] Social benefits of local regeneration underpinned in part by new industries can include expanded local employment and income opportunities, but experiences of dislocation and displacement may also be significant. Dislocation can take the form of direct displacement of residents from communities infiltrated by New Economy firms, as experienced in the South of Market Area of San Francisco in the late 1990s, or inflationary pressures on land prices in areas proximate to New Economy sites.

Environmental impacts of new industry formation include: (1) the preservation of individual heritage buildings via adaptive reuse for new enterprise; (2) the larger-scale reconstruction of inner city landscapes, in which new industries play a modest part, together with new residential development, consumption, and other forms of amenity; and (3) the reterritorialization of the inner city space, with new industry formations contributing to the reconfiguration of established inner city districts. New industry formation over the past two decades has inserted a new force in the processes of urban place-making. To an extent this place-making role has been driven by 'spontaneous' industry formation, but increasingly is guided by government and public agencies, observed in the many examples of 'cultural quarters' now acknowledged as a mainstream instrument of the state's repertoire of programmatic options for regeneration (Bell and Jayne 2004).

The New Economy of the Inner City: purpose and conceptual framework

A burgeoning research literature has disclosed important features of new industry formation in the inner city, emphasizing the importance of specific industries, situated within a range of instructive cities and sites, and offering profiles of innovation and enterprise. There is now a clear need for a deeper study of new industry development in the inner city, including consideration of localized impacts and implications for the larger metropolis, situated more centrally within industrial urbanism as a genre of urban studies and geography, and directed toward theoretical conjecture. There is also scope for incisive comparisons of new industry formation in cities situated within different regions and echelons of the global urban hierarchy, to identify both pervasive and more contingent conditions and experiences. *The New Economy of the Inner City* builds on streams of new industry scholarship; intersects urban and economic geography at the frontiers of redefining change; and contributes to the development of new theory for this contemporary period of complex urban change. Specific objectives of the monograph are as follows:

> To critique, consolidate and synthesize current research orientations on the processes of reindustrialization in the postmodern urban core, as well as the larger implications for the transformation of cities and urban regions.
>
> To develop critical analytical perspectives on generative processes of change in the metropolitan core, emphasizing the saliency of industrial restructuring and new industry formation, but encompassing critical social and policy factors.
>
> To identify new models of advanced industrial production in the metropolitan core, with special emphasis on ascendant industries and new social, spatial, and technical divisions of labour, and aspects of inter-industry complementarity and competition in the New Economy.
>
> To interrogate the role of these specialized industries in the formation of new industrial districts and precincts; in the simultaneous and contingent experiences of regeneration and dislocation; and in the larger processes of multiscalar change in the urban core and metropolis at large.
>
> To propose robust, integrative new models of urban structure and land use in the core, derived from: theoretical engagement; a literature review and synthesis; and a rich program of fieldwork and site visits conducted within influential city case studies.

The central research questions that shape the investigation are as follows: In what ways does new industry formation, together with related social dynamics, contribute to the respatialization of the inner city and the reconstruction of the postindustrial urban landscape? How can we characterize (and model) trajectories and sequences of new industry formation, including the emerging forms of territorial production in the inner city? Can we identify 'global' or universalizing tendencies in new industry formation within the inner city – and what are the expressions of, and limits to, these pervasive tendencies? What are the principal contingencies of difference in process and experience, including scalar considerations, industrial structure, and overall stage of development, as well as the role of the state, at central, regional and local levels? And, more specifically, in what ways do these local contingencies shape outcomes of new industry formation, in terms of *regeneration*, or, alternatively, *dislocation*, observed in the reproduction of space, labour market impacts, and externalities for social groups? Finally, how might an appreciation of these complex processes and outcomes contribute to new typologies of industrial change, developed in an era of recurrent change, and toward a larger retheorization of the twenty-first-century city?

Research orientation and reference points

The potential scope and thematic richness of scholarship on processes and outcomes of new industry formation within the inner city can be demonstrated by manifest connections to several of the most influential contemporary social science research orientations. Within the realm of economic geography, we can readily discern a significant interface between reindustrialization and the region, as seen in the following streams of research: (1) the cultural turn in economic geography (see Barnes 2001), associated with the rise of the urban cultural economy, and the increasing centrality of 'cultural products' (Scott 1997) to the economic base of the advanced city-region; (2) the emphasis on 'relational' processes of industrial innovation situated within urban-regional space (Bathelt and Glückler 2003), configured by complexities of 'actor-structure' relationality, 'scalar relationality', and 'socio-spatial' relationality, replete with 'interconnections and tensions' (Yeung 2005); and (3) explanations of knowledge-intensive industrial innovation and enterprise, notably work on collaboration and the role of proximity, exemplified by Cooke and Morgan's perspective on the 'associational economy' (Cooke and Morgan 1998), Amin's research on globalization and regional development (Amin 1998), Boschma's critique of proximity and innovation in advanced industrial systems (Boschma 2005), Grabher's inquiry into collaboration and the 'ecologies of creativity' (Grabher 2001; 2002), Gertler's explication of proximity, culture and tacit knowledge exchange among advanced economies (Gertler 1995; 2003); and Beyer's research on advanced services and the New Economy (Beyer 2000).

But in addition to this substantial body of research on the workings of specialized production systems, there is now a significant scholarly literature addressing interactions between processes of industrial change, space and place in the city, exemplified in Molotch's work on Los Angeles (Molotch 1996); Helbrecht's examination of the 'creative metropolis' in Munich and Vancouver (Helbrecht 1998); Pratt's ongoing work on London and San Francisco (for example, Pratt 1997a; 2000); Bathelt's examination of clustering processes in Leipzig (Bathelt 2005), and Capello and Faggian's analysis of 'collective learning and relational capital' in local innovation in Milan and Piacenza (Capello and Faggian 2005). The literature now includes case studies of new industry formation and experiences in exemplary cities and sites: see, for example, Indergaard's study (2004) of 'Silicon Alley' in Manhattan, Pedro Costa's analysis of the imposition of new media industries on the old cultural quarter of Lisbon (Costa 2004), and Lloyd's monograph on 'neo-Bohemia' (Lloyd 2006), the intersections of art and commerce in Wicker Park, Chicago. Research on the employment and labour market implications of new industry formation in the inner city include Post's work (Post 1999) on London's 'City Fringe', and Schön, Sanyal, and Mitchell's (1998) examination of high-technology industry impacts on low-income communities. This outline demonstrates a clear trend in research orientation, from a single-minded emphasis on technology and the market as the motive forces of new industry formation, toward richer, multiperspectival approaches that take in the complex social, cultural, physical, and policy factors.

Research model and methodology

The research platform for this study incorporates a sequence of investigations of interdependencies of industrial change and urban transformation, a lineage which includes earlier work on the corporate office complex of the central city (see, for example, Hutton and Ley 1987; Davis and Hutton 1991). But the direct provenance of this study is derived from more recent investigations of the economic structures of the contemporary inner city: the historical and contemporary importance of creative industries in the shaping of production landscapes (Hutton 2000); the synergies of technology, culture, and place in the 'New Economy of the Inner City' (Hutton 2004a); the influence of urban theory on planning interventions designed to reshape inner city production, consumption, and residential landscapes (Hutton 2004b); and the interrelationships between spatiality, built form, and creative industry formation in the inner city (Hutton 2006).

The research model for this study is comprised of (first) theoretical engagement, emphasizing the critical intersections of process and place within the metropolitan core; second, intensive case studies of new industry formation within exemplary cities and sites that enable deeper investigation, disclosing the empirical richness of the contemporary industrialization experience, and, third, the deployment of field research in the service of conceptual enterprise concerning the evolution of the twenty-first-century city. The research methodology (described in greater detail in Appendix A) includes key informant meetings and exchanges concerning the broader experience of new industry formation in the metropolitan core; an extensive literature review, and conceptual critique and synthesis; an ongoing program of fieldwork in principal sites (London, Singapore, San Francisco, and Vancouver) conducted over the past decade (see Appendix B for fieldwork schedule); and presentation of observations and interim findings at international conferences, seminars, and workshops. Beyond the four principal city case studies, site visits to cities such as Cologne, Milan, Florence, Amsterdam, and Seattle have enlarged an understanding of the scope of new industry formation, augmented by an ongoing dialogue with colleagues working primarily in other cities. The research methodology comprised a repertoire of techniques and procedures that have included semi-structured interviews with new industry workers, city planners, and NGOs, and detailed mapping and photographic work, designed to vividly evoke the distinctive spatiality and 'look and feel' of new enterprise in the reconstructed territories of the inner city.

Shaping the book: concept and constituencies

The monograph is intended to place the processes of new industry formation reshaping the 'new inner city' firmly within a narrative of industrial change in the advanced metropolis, acknowledging the thematic density and complexity of new industry formation. This project entails an examination of the cogency of the major foundational theories of late twentieth-century urban change: postindustrialism, post-Fordism, and postmodernism. The starting position is that each still has much to say about the configuration of the metropolis, but is in need of refurbishment.

New industry formation in the inner city also represents a particularly fertile ground for investigating processes of industrial experimentation and innovation among advanced economies. The suitability of the inner city as a testing ground for conceptual innovation in these domains is defined by qualities of scale, built form and spatial intimacy; by the complexity of industrial organization and institutional structure; and by the markedly social nature of production and work in the central city (Evans 2004). The program of fieldwork conducted since 1993 takes in several episodes of industrial change, enabling reflection on the larger significance of these recurrent restructuring experiences.

Rationale for the case studies: cities and sites

The cities selected for the extensive program of fieldwork – London, Singapore, San Francisco, and Vancouver – occupy different echelons of the global urban hierarchy, are distantiated by urban scale and by vast tracts of space, and characterized by contrasting forms of governance and local planning systems (Table 1.1). These contrasts will of course account for a significant measure of difference in new industry structures and experiences, and will form part of the narrative and analysis in each of the case study cities, presented in Chapters 4–8.

But for the purposes of this project, London, Singapore, San Francisco, and Vancouver also exhibit important developmental commonalities and theoretical connecting points, with respect to the following attributes.

Global-local interaction and interdependency

London stands with New York at the apex of the global urban hierarchy, but for each of these four cities experiences of industrial change and community reformation are increasingly shaped by interdependencies between, first, global processes (foreign development investment [FDI], the role of multinationals, or international trade); second, transnational urbanism (international immigration, growing expatriate populations, the role of the city as inter-cultural production and transmission site); and, third, local factors, including governance structure, developmental histories, spatial structure, and environmental factors. The nature of these interactions will form part of the narrative for each of the case studies.

Industrial restructuring trajectories

Each of the four cities in the sample has assumed a distinctly postindustrial trajectory, with manufacturing in a state of secular decline, and with 85–90 per cent of the metropolitan labour force engaged in services industry employment. Specialized services, including higher education and other public sector agencies, as well as a larger platform of intermediate services, represent the dominant suite of industries and labour in each metropolis. In London and Vancouver, postindustrialism

Table 1.1 Comparison of governance structures for London, Singapore, San Francisco and Vancouver

	London	Singapore	San Francisco	Vancouver
1. Area (km²)	1590	693	122 (City) 6408 (MSA)	115 (City) 2880 (CMA)
2. Population (millions)	7.4	4.5	0.7 (City) 4.1 (MSA)	0.5 (City) 2.3 (CMA)
3. National context	Capital of the United Kingdom of Great Britain and Northern Ireland	Sovereign city-state	City and county, California, U.S.A.	Third-layer census metropolitan area (CMA) in Canada
4. Regional governance context	Greater London Authority (GLA) (executive policy and planning with some central government powers)	Republic of Singapore (national government)	Association of Bay Area Governments (ABAG)	Metro Vancouver Regional District
5. Governance leadership	Mayor (elected) and GLA Assembly	• Prime Minister and Cabinet • President (ceremonial)	Mayor and Board of Supervisors	Chair and indirectly elected GVRD Board members
6. Local government	City of London Corporation and 32 London Boroughs, Wards (borough leaders)	n/a	City of San Francisco (11 Districts)	22 Municipalities (mayors and council members)
7. Principal planning bodies	Greater London Authority	Urban Redevelopment Authority (URA)	• City of San Francisco Planning Dept. • City of San Francisco Redevelopment Agency	• Metro Vancouver Planning Dept. • 22 municipal planning departments
8. Economic development bodies	London Development Agency (reports to Mayor's Office)	Economic Development Board	• Bay Area Economic Forum • Mayor's Office of Economic and Workforce Development	Metro Vancouver Economic Council
9. Structure/ organization of metropolitan core	• City of London • City of Westminster • 'City Fringe' wards and districts	• CBD • CBD fringe • Inner city heritage districts (e.g. Chinatown)	• CBD • CBD Fringe • South of Market (SOMA) (e.g. Mission, South Park, South Beach, etc.)	Central Area; new 'Metropolitan Core' used for strategic planning purposes

has constituted a policy value of the state and/or local authorities, while in all four cities agencies have endeavoured to support the emergence of new industries (including creative industries) in the city by means of a suite of policies and regulatory initiatives, although the policy models exhibit significant elements of differentiation.

Proliferation of new industry sites in the inner city

While the experiences have been subject to volatility, each of the cities in the sample represents an important global example of inner city new industry formation, with multiple sites situated within the metropolitan core. These sites vividly exhibit the range of reindustrialization processes observed within the 'new inner city', exemplified by distinctive ensembles of sectors, firms, labour, and production networks, and contribute to the reproduction of space in the metropolis.

Social reconstruction of the inner city

In London, Singapore, San Francisco and Vancouver, the past two decades have seen the social reconstruction of the inner city, strongly associated with the hegemony of the professional, managerial, and entrepreneurial elites of the new middle class, as well as the more recent ascendancy of New Economy workers. These cohorts are powerfully linked to continuing gentrification processes, and to the intermingling of the social and working worlds of the inner city. Further, the formation of new residential districts and industrial production sites represent together important forces in the comprehensive redevelopment of the metropolitan core, although this association of new industry and housing is characterized by conflict and tension as well as by complementarity.

Theoretical significance

Each of the cities in the sample offers a rich social and historical setting for the investigation of new industry formation processes. The developmental experiences of London, Singapore, San Francisco, and Vancouver exhibit wider theoretical significance, and have generated sustained interest among urban studies scholars across a range of disciplines. These theoretical associations will be explicated in the case studies and concluding chapter, but include (in the case of London) Glass (1963) (on the genesis of gentrification), Hall (1962a) (industrialization and its spatial ramifications), Hall *et al.* (1973) (the London region and innovation in planning), Hall (1998) (ideology and global processes in the redevelopment of the city), Sassen ([1991] 2001) (London as global city), and Hamnett (1994 and 2003) (the emergence of the new middle class in the global city); Perry, Kong and Yeoh (1997) (Singapore as an exemplary developmental state), and Ho (1994 and 2005) (occupational change and social class reformation) for Singapore; Brechin (1999) (capital and its impact on city formation and inequality), and Hartman (2003) (capital, redevelopment and displacement in the inner city), for San Francisco; and Ley (1996) (new middle-class formation and the remaking of the central city), Olds (2001) (cultural, capital and mega-projects), and Hutton (2004a) (the influence of theory on planning for urban structure and land use), for Vancouver. These references point to the saliency of each city as a site of contemporary urbanism, framed in theoretical and normative terms.

Structure of the book: process, place, and case studies

Following this introduction, the volume contains eight chapters: two addressing the intersections of 'process' and 'place', five chapters for the presentation of the case studies, and a concluding chapter which draws out the most salient insights and theoretical significance of the cases. Chapter 2, 'Process: Geographies of production in the central city', rehearses the principal interpretations of con-temporary new industry formation within the inner city, encompassing, variously, research that emphasizes the influence of markets, industrial restructuring ten-dencies, the evolution of post-Fordist production regimes, environmental factors, social and cultural attributes, changes in the structure of the regional space econ-omy and divisions of production labour, and institutional factors. A model of new industry formation within inner city settings, synthesizing insights derived from this discussion, is presented as a means of communicating the complexity and interdependency of generative processes.

Chapter 3 ('Place: The revival of inner city industrial districts') entails a con-sideration of both general and more locally contingent experiences of new indus-try formation within inner city districts, drawn from a number of particularly evocative and instructive cases, and comprising a set of 'reference cases' that complement and inform the detailed case studies. The narrative here starts with the historical concept of the inner city production district and its more recent mutations, including the Oltrarno in Florence and the Old Quarter of Hanoi followed by a rehearsal of the industrial district literature, from Marshall to Markusen. Next, the evolution of the inner city industrial district in the global city, and the co-presence of 'old economy' and 'New Economy' industries and firms, are illustrated in the 'triple narratives' of industrial districts in Manhattan. References to the very recent development of New Economy sites in Asian cities, including Tokyo, Seoul and Shanghai, acknowledge the global reach of new industry formation, as well as the sequences of accelerated industrial restructuring which typify the experience of leading economies in the region.

The case studies open with two chapters on the extraordinarily rich London example of new industry formation within the inner city of a first-order global metropolis. Chapter 4 ('Restructuring narratives in the global metropolis: From postindustrial to "new industrial" in London') performs a thematic linking function in the book, drawing together interdependencies of process and place described in Chapters 2 and 3 to elucidate London's historical and contemporary experiences of industrial development. To this end, Chapter 4 offers a recitation of globalization and regional change as they reshape the space-economy of the metropolis, followed by a study of the classic light manufacturing districts of London's East End, and a concise account of the modern post-Fordist industries that cohabit these 'residual' terrains of the 'old economy'. Chapter 4 develops a sketch of the complex structures and systems of industrial production encompassed within London's 'City Fringe', a territory roughly coterminous with the traditional light industrial sector described by Peter Hall, J.E. Martin, and others. This account incorporates the rise of arts and cultural activities within the inner city, and their uneven articulation within industrial production networks of the metropolis. The discussion sets the stage for the presentation of three instructive vignettes of new industry formation in Chapter 5 ('London's inner city in the New Economy'), situated in resonant New Economy/old economy sites in Shoreditch, Clerkenwell, and Bermondsey. These studies demonstrate the importance of localized contingency of new industry formation in its more specific forms and expressions, as well as the uneasy co-existence between creative industries with more potent agencies of change: the commercial-financial sector, and the social reconstruction of inner London. The role of the latter in driving inflationary pressures within the London property market suggests a profile of 'precarious reindustrialization' in the metropolitan core, rather than the deeply entrenched manufacturing ensembles of the nineteenth and early-to-mid-twentieth centuries.

The second case study concerns the evocative experience of new industry formation within the textured landscapes of Singapore's Chinatown (Chapter 6, 'Inscriptions of restructuring in the developmental state: Telok Ayer, Singapore'). The sequence of restructuring episodes in the compact heritage district of Telok Ayer, immediately adjacent to the central business district, represents in scalar terms a modest aspect of the larger story-line of induced restructuring, modernization and globalization in South-East Asia's primary city-region. But imprints of industrial change within Telok Ayer compressed within the period 2000–2006 present vivid evidence of the global sweep of the abbreviated processes of globalization among advanced societies. The experience of Telok Ayer also demonstrates the possibilities of 'spontaneous' innovation within a larger context of induced restructuring in the exemplary 'developmental state', as well as the resilience of high-amenity inner city sites in the face of recurrent economic change. Finally, the observations of inner city new industry formation in Telok Ayer indicate the potential contributions of micro-scale production spaces to the early twenty-first-century 'cultural turn' in economic policy-making, and the constant refinement of competitive advantage as a means of maintaining an edge over regional and more distant rivals.

Chapter 7 ('The New Economy and its dislocations in San Francisco's South of Market Area') presents a case study of the New Economy experience in San Francisco's South of Market Area (SOMA). Although often positioned as a discrete and unique event, the dramatic expansion of the dot.coms in SOMA in the last years of the twentieth century took place in an inner urban setting with a long history of redevelopment and dislocations, and was shaped in part by a pre-existing cultural development trajectory. The latter process incorporated major new cultural institutions as well as congregations of artists, designers and entre-preneurs, each of which comprised elements of the New Economy matrix in this seminal innovation district. Chapter 7 pays special attention to the fortunes of South Park, SOMA's New Economy epicentre, encompassing wild swings of innovation and growth, the 2000–2001 crash, and the slow recovery of the district, as well as discussing the implications for local planning and policy responses to rapid restructuring sequences in the metropolis.

The last case study (Chapter 8, 'New industry formation and the transform-ation of Vancouver's metropolitan core') offers an evaluation of the role of industrial restructuring in the sequence of transformations of Vancouver's metro-politan core, emphasizing the influence of new industry formation since the implementation of the *Central Area Plan* in 1991. The 1991 Plan consolidated the CBD and privileged housing within the larger central area beyond this more compact district. The *Central Area Plan* also facilitated the emergence of New Economy formations in the metropolitan core by means of land use and zoning policies, heritage planning, adjustments to building regulation, and investments in amenity which proved attractive to design-based industries, firms and labour. But the spectacular redevelopment of new residential districts, particularly in the form of high-rise condominiums, has shaped a comprehensive social reconstruc-tion of the core, endorsed in the City's 'Living First' strategy which views residen-tial development as an economic as well as social program. This Living First experience has been widely acclaimed as a bold and progressive approach to a new twenty-first-century urbanism based on livability and sustainability, embodied in an 'eco-density' strategy. But the scale and pace of this residential development may also compromise new industry precincts, as well as the tenure of marginal populations brought within the ambit of a high-growth, high-externality 'new inner city'. Studies of new industry processes in three instructive inner city sites – Victory Square, False Creek Flats, and Yaletown – offer insights into the import-ance of local context and contingency in the shaping of new industry formations within the reconstructed landscapes of the urban core, as well as a window onto the evolving planning styles and policy discourses associated with the reassertion of production in these crucial terrains of the city.

The volume closes with Chapter 9 ('The New Economy of the inner city: an essay in theoretical synthesis'). This concluding chapter comprises: first, a set of reflections on the four principal case studies; second, the distinguishing synergies and interdependencies associated with the 'recombinant economy' of the twenty-first-century inner city; next, the role of these specialized industries in the reshap-ing of the 'new inner city'; succeeded by a foray into theoretical synthesis derived from each of these conceptual iterations, buttressed by the reference cases pre-sented in Chapter 3. A final commentary on some prospectively fruitful themes for scholarly investigations suggested by the observations of this volume is intended as a contribution to the evolving discourse on industrial urbanism and its manifold interrelationships with larger experiences of urban growth and change in the new millennium.

2 Process

Geographies of production in the central city

Introduction: post-Fordism and its antecedents in the inner city

The collapse of Fordist manufacturing industries and labour, characterized by a steep gradient of industrial disinvestment and decline, and visceral socioeconomic consequences for employees and for working-class communities, looms large both in urban studies narratives and in discourses of economic geography (Bluestone and Harrison 1982). The wrenching imageries of firm closures, rationalization and relocation, and attendant contractions of employment and incomes, were greatly exacerbated by both the scale and rapidity of collapse, and also by the locational specificity of the impact, with the inner city bearing the brunt of industrial decline. This was especially the case in Britain, where the restructuring of the 1960s and afterwards swept away many of Europe's oldest industrial districts ('first in, first out'), but was also manifested in other western European societies, as well as in the industrial heartland of North America.

Debates concerning causalities and consequences of industrial decline were sharpened by the larger ideological conflicts over economic policy values of the 1970s and 1980s. The election of Margaret Thatcher and the Conservative Government in Britain in 1979, and Ronald Reagan's accession to the US presidency a year later, signalled to many observers a dismissive posture toward traditional industry on the part of the state, and contributed a bitter polemical edge to the literature on deindustrialization. The meaning of postindustrialism thus assumed a strongly normative tone, as well as retaining its descriptive usage.[1]

The dimensions of the collapse of Fordism justify its central position in the narratives of urban change. But it is important to recall the more extended lineage of the city as site of industrialization and labour formation, in order to contextualize the complex industrial organization of the contemporary city. After all, while Fordism undeniably represented the introduction of a new paradigm of efficient labour organization, forcefully reshaping the systems of industrial production in advanced cities and societies, it held sway as the dominant industrial regime for only the middle of the twentieth century, circa the mid-1920s to the 1970s. Especially in the larger and older metropolitan cities, notably London, Manchester, New York, Chicago, and Montreal, the urban manufacturing sector of the mid-twentieth century comprised a rich admixture of artisanal industries and labour, light manufacturing, craft production, and in some cases a residual heavy industry sector, as well as a substantial and growing intermediate services sector, anticipating the take-off growth of producer services in the 1970s and 1980s. The motive forces for industrialization from the early nineteenth century included not only new technologies and systems of production, but also the growth of consumer markets in the metropolis, the availability of large pools of labour, and the specialized skills of immigrant workers.[2]

The diversity of production regimes and industries persisted in the modern urban economy, although Fordism to be sure constituted the lead modality for advanced manufacturing. The rapid expansion of Fordist production among advanced societies following the Second World War was constructed upon a historically unique moment: a tacit bargain between corporations, labour and the state to maintain with minimum discord a growth-oriented, capitalistic manufacturing sector, subject to regulation, but supported as well by high levels of demand for all sorts of goods from mass consumer markets. New factories were built in suburban areas of the metropolis and in smaller cities and towns to supply this demand, but the inner cities also benefited from this secular expansion of industrial production.[3]

The industrial restructuring of the late twentieth century has thus been typified as the defining crisis of Fordism, measured in firm closures and contractions of labour, particularly skilled and semi-skilled operatives. Fordist industries situated in affluent, high-salaried, and high-cost societies were particularly vulnerable to the emergence of a new international division of labour (Fröbel *et al.* 1980) in which mass-production capacity and employment formation shifted to areas of lower labour and operating costs, principally in East and South-East Asia at first, then to Brazil and other regions.[4] This new international division of labour was driven by the reallocation of capital on the part of multinational corporations (MNCs), but was also accelerated by the deregulation of financial markets and the free trade movement of the 1980s. But the force of industrial restructuring of the 1970s and 1980s served to displace many of the older (i.e. pre-Fordist) production sectors as well, including craft industries and light manufacturing firms, many of which experienced rising cost pressures (for rents, labour, and materials) and changing consumer market preferences (demand for cheaper goods at comparable quality). Heavy industries, confronted by deepening obsolescence, low output values, and shrinking markets, also faced closure or relocation from the inner city.[5] Thus, the industrial restructuring of the late twentieth century was to a large extent one of *general* manufacturing decline in the metropolis, rather than solely a crisis of Fordism.

While the march of restructuring imposed relentless pressures on the urban manufacturing sector in advanced societies, some industries survived. In particular, larger plants requiring more extensive space, but fewer localized inputs, relocated to the suburbs, as recounted in Allen Scott's well-known review of industrial location in *Urban Studies* (Scott 1982), although in London and other cities employment losses accruing from outright closures (or 'factory deaths') were larger than those associated with decentralization. Within the inner city the remnants of the Fordist economy co-existed with residual craft industries and a promiscuous assortment of quasi-industrial activities (such as auto repair and printing), and in some cases office subcentres or commercial strips. But in general the structural and spatial coherence of the old inner urban industrial economy, in terms of long-established sectors, firms, and production networks, together with allied community and social groups, and a host of institutional affiliates, was effectively lost by the end of the 1980s.

Contours of the postindustrial inner city economy

During the protracted period of industrial decline in the late twentieth-century metropolis the dominant story-lines concerned, first, the socioeconomic consequences of pervasive firm closures and massive employment loss within the production districts of the metropolitan core, and, second, the parallel process of social upgrading of residential neighbourhoods within the postindustrial terrains of the inner city. In contrast to these twin narratives of industrial decline and socioeconomic succession, the location of new businesses and activities within the postindustrial terrains of the inner city over the 1980s took the form of an altogether smaller story. In some of the larger cities the redevelopment of the inner city included important secondary office centres (such as Hammersmith in West London, and Shinjuku in Tokyo), but more frequently encompassed a variegated mixture of arts and cultural activity, commercial businesses seeking relief from the steep rent gradients of the CBD, local retail and personal services, and the remnants of the Fordist manufacturing economy.

The scholarly literature on the generative processes of economic growth and development in the metropolis over the last third of the last century tended not unnaturally to focus on the implications of the expansion of the central corporate office complex, and the rapid development of industry (services as well as manufacturing and ancillary activities) in suburban locations. In particular, the central role of the CBD office complex with its constituent head office and specialized financial functions was implicated in the formation of the global city, while the rapid growth of manufacturing in the suburbs and metropolitan periphery represented a major development in the evolving space-economy of the city-region, with each attracting a large constituency of scholars.[6]

The rise of a specialized service economy in the central city

While our focus in this volume is on the processes of industrial change within the old production districts and neighbourhoods of the inner city, a concise rehearsal of growth dynamics for the central office complex is required to place these recent experiences within the larger framework of industrial restructuring in the metropolitan core. The origins of a high-rise office district in the CBD can be traced back to the first skyscrapers, initially in Chicago and then in New York, over the last two decades of the nineteenth century. The expansion of trade and commerce, coupled with the banking and service requirements of urban manufacturing, underpinned growth in office development and employment across a broader range of cities, including major urban centres in Europe as well as in North America in the first decade of the twentieth century and in the interwar period.

The postwar period marked the start of a half century of growth and development in the city's central office complex. A rough periodization of change is shown in Table 2.1, including the take-off era of the 1950s, encompassing growth both in intermediate and final demand services. This initial expansion period included both public sector and consumer services, fed by high levels of population growth and increasing household incomes, allocated in part to service expenditures as well as the purchase of consumer goods, as well as office employment growth driven by a growing commercial business sector. To a considerable extent, however, services (management, clerical, and technical functions) were still *internalized* within industrial corporations and enterprises.

The 1970s saw another growth spurt in services, marked by a rapid increase in the *externalization* of specialized services supply (Gottmann 1970). A portion of this intermediate services employment was located in (or relocated to) suburban and peripheral business centres (see Daniels 1985). But the dramatic growth of the CBD office complex was a defining event of the era, shaped by, first, a growing need on the part of business and industry for specialized service inputs which could be more efficiently provided by outside agencies (including consulting legal, accounting, financial, and IT services), and, second, the tremendous agglomerative power of the CBD (Bourne 1982). During this period the major occupations within the central city office workforce – executives, managers, secretaries, sales, technical and more menial service workers – each experienced rapid growth.

A third phase of urban services development can be traced to the 1980s, which, as observed above, was shaped increasingly by global processes, abetted by the liberalization of finance and consumer markets enacted by neo-liberal governments. The new international division of labour, which deprivileged the Fordist industries and workers of the metropolis, correspondingly favoured the post-Fordist office economy of the CBD, as the knowledge and skills of its professional workforce could not easily be replicated in the burgeoning manufacturing regions of transitional societies. With the secular contraction of manufacturing, and the rise of specialized finance and business services, the emphasis on defining positionality within the international urban hierarchy shifted from Peter Hall's original idea of the 'world city' (Hall 1966), to Saskia Sassen's concept of the 'global city' (Sassen [1991] 2001), with banking, finance, and producer services endorsed as chief measures of primacy.[7] By the last decade of the twentieth century, putative rivals to the global cities in Europe and America had emerged in Pacific Asia, a system of cities which included Hong Kong, Shanghai, Seoul, Taipei, and Singapore, as well as Tokyo, adding another dimension to the relentless pursuit of global competitive advantage.

Table 2.1 Principal phases of service industry development within advanced societies, 1950–2000

Internalization PHASE I *Take-off growth in services c. 1950–1970*	*Externalization* PHASE II *Expansion of producer and consumer services c. 1970–1980*	*Internationalization* PHASE III *Rise of intermediate services and public sector retrenchment c. 1980–1990*	*Globalization* PHASE IV *Sectoral maturation and 'information economy' phase c. 1990–2000*
1. 'Growth industries' include services for final demand: • K-12 education (baby boom) • retail and personal services (demographic trends, household income growth) • public administration and health services (demographics, and expanding role of government and public sector)	1. Continued growth of public and consumer services: • K-12 and university/college education (demographic factors) • tourism (disposable income growth)	1. Rapid growth of intermediate services: • banking and financial institutions • head offices, including MNCs and MNEs • producer services (internationalization trend)	1. Rapid growth in IT, informatics, knowledge based services industries (informationalization of economy)
2. Business (producer) services largely internalized within corporations: • intermediate services as 'lubricants' of the industrial economy	2. Increasing internationalization of business or producer services; rapid growth in: • banking and finance • legal and accounting • consultancies	2. Retrenchment in government and public services (demographic factors and 'neoconservative' political agenda)	2. Globalization of services trade and exchange (market liberalization, interaction)
3. Key service occupations include: managers, teachers, clerks and secretarial staff, health care workers	3. Key service occupations include: managers, professionals, higher education, health care specialists, tourism workers	3. Key occupations include: • banking, finance and securities specialists, computer and information technology personnel	3. Slower growth in some services industries as tertiarization attains 'mature' phase among many OECD economies

4. *'Cities, services and space'*:
- cities as 'regional higher order central places'
- growth of CBD 'corporate complex' in larger business centres

4. *'Cities, services and space'*:
- cities as specialized regional and national service centres
- growth of CBD 'corporate complex' in larger and medium-size business centres

4. *'Cities, services and space'*:
- cities as nodes of international trading and investment networks
- increasing multinucleation and 'edge cities'

4. *'Cities, services and space'*:
- emphasis on 'global competitive advantage' for major service centres
- emergence of 'specialized service clusters' for metropolitan space-economies: knowledge, culture, recreation, and consumption

Selected Benchmark Events

1950	1960	1970	1975	1980		1990	2000
High point of 'Fordist' industrial regime	Decline of 'Fordist' production in major industrial regions	First OPEC price shock (1973)	Expansion of 'flexible production'	Election of: M. Thatcher (1979) R. Reagan (1980)	Deregulation and privatization	increasing globalization	'Asian crisis' (1997–)

Fin-de-siècle *processes of change in the core*

The discussion presented here acknowledges the skeletal features of specialized services as the dominant engine of the metropolitan core economy. The CBD office complex's high growth rates, dramatic physical development in the form of ever-higher point towers, and (at the high end of the employment structure) exalted occupational status and incomes all serve to further diminish the status of industrial production in the core beyond the confines of the CBD. That said, an appreciation of the sociocultural bases of services growth (or *tertiarization*) in the core adds nuance to the basic story-line, and perhaps also anticipates the reassertion of industrial production in the inner city. A basic outline of these sociocultural factors is presented in schematic form in Table 2.2. While economic (i.e. market) forces continue to play lead roles in new industry formation of all kinds, Table 2.2 also recognizes the importance of sociocultural attributes which include: (1) human capital (skills, education, and expertise); (2) social capital embodying civil and community assets and traditions, including international connectivity as well as resilience; (3) sociocultural agglomeration, interpreted here as the density of the urban social ecology, the social content of business interaction, and community information and exchange networks; (4) cultural capital (language, dialect, fine arts and design, public institutions); and (5) identity and image, including the authenticity, legibility, distinctiveness, and durability of urban communities, incorporating symbolic assets and resonance: memory, landmarks, and places of communal activity. These factors are central to the recent development of the urban service sector, but, I am suggesting here, also underpin the development of new production industries and firms in the inner city, as we shall observe in subsequent chapters.

The defining strands of contemporary urbanism – the growth of the CBD office complex, suburban industrial development, and social upgrading in the metropolitan core – continue to represent the principal genres of theoretical and normative scholarship. But the rapid growth of new industries, firms, and employment within the postindustrial districts of the metropolitan inner city in the last decade of the twentieth century inserted a new episode into the narratives of urban growth and change, attracting scholars across a range of disciplines, including urban and economic geographers, sociologists, urban studies researchers, and urban and community planners. The establishment of design-based firms and institutions in 1980s, followed a decade later by the meteoric growth of the dot.coms acknowledged as manifestations of a 'New Economy', suggested a significant recovery of production in the spaces of the metropolitan core. Even for economists, working in what many regard as an essentially aspatial discipline, the apparent emergence of a 'New Economy' associated with the accelerative effects of advances in telecommunications technologies, and their potential implications for productivity growth and labour market enhancement, represented an economic benchmark. Indeed, the more euphoric chroniclers of the New Economy predicted an economic revolution accruing from the introduction of new communications and production technologies, transforming the

Table 2.2 Sociocultural bases of urban tertiarization

SERVICE INDUSTRY LINKAGES	BASIS OF ALL SERVICE INDUSTRY GROWTH	LOCAL / FINAL DEMAND SERVICE INDUSTRIES	CONTACT-INTENSIVE SERVICE INDUSTRIES	CREATIVE AND EXPERIENTIAL SERVICES INDUSTRIES	GLOBAL AND MULTINATIONAL SERVICE INDUSTRIES
	• especially SMEs (small and medium-size enterprises)	• entrepreneurial firms, local retail and personal services • NGOs and CBOs • public services	• all contact-intensive services • intermediate demand firms • knowledge-based services	• applied design services • cultural activities • arts and education • tourism industries	• head office operations • banking and financial industries • international producer service industries
	←	←	←	←	←
SERVICE FIRM EFFECTS	FIRM START-UPS AND GROWTH	COMMUNITY CONTEXT FOR TERTIARIZATION	FACILITATION OF INFORMATION FLOWS	CULTURAL MILIEU	GLOBALIZATION AND TERRITORIALIZATION EFFECTS
	• firm start-ups • productivity • management and leadership • value-added effect	• responsiveness • quality of life factors • trust and transparency	• market intelligence • complement to economic agglomeration • transmission / diffusion of knowledge, ideas, opportunities	• creativity • design inputs to production • quality of life factors • 'experiential' attraction	• power of locality and 'territorialization' • aspects of competitive advantage
	←	←	←	←	←

(*Continued Overleaf*)

Table 2.2 Continued.

URBAN SOCIO-CULTURAL ATTRIBUTES	HUMAN CAPITAL	SOCIAL CAPITAL	SOCIOCULTURAL AGGLOMERATION	CULTURAL CAPITAL	IDENTITY AND IMAGE
	• skills • education • expertise • entrepreneurship • knowledge • adaptability	civil society assets • values, behaviours • traditions • connectivity • resilience • productive diversity	• density of social ecology • social content of business interaction • proximity of firms and institutions • community information networks	• language and dialect • fine arts and applied design • galleries and museums • community preferences, tastes • public forums • restaurants, coffee houses	• distinctiveness of urban / community identity • power and resonance of identity • authenticity • legibility • durability
	Quality of urban labour force	Community assets and traditions	Density and patterns of social interaction and exchange	Urban cultural assets and institutions	Symbolic assets and resonance

workplace, restructuring markets, and enabling quantum leaps in industrial productivity.[8]

In retrospect, the larger claims of the New Economy phenomenon, situated largely within the reconstructed production spaces of the metropolitan core, or among suburban/ex-urban science parks and major research universities, seem more ephemeral than visionary, although it is the case that there is a substantive legacy of technological deepening of the economy which can be ascribed in part to the experimentation and innovations of the 1990s. The market correction (or crash) of 2000–2001 eradicated many of the high-flying corporations of the tech-boom, severely damaged the stock value of others, and generally suppressed the exuberant labour market demand for some of the ascendant New Economy occupations, including, for a time at least, software developers, Internet providers, and multimedia specialists, among others. But the intensification of technology is after all currently observed across a range of industries and institutional applications, including video game production, graphic design, and other creative industries, as well as more generally in business, manufacturing, and higher education, so the 1990s New Economy produced a legacy of transformations as well as firm closures and unemployment.

Following the crash of the turn of the twentieth century, the redevelopment of the metropolitan core has included a new cycle of industrial innovation, business start-ups, and employment growth. But the scale and specific nature of this most recent redevelopment phase vary considerably from place to place, offering possibilities of comparative investigation. Perhaps we now have a temporal vantage point from which to undertake a more reflective assessment of causality and consequence. That said, a survey of the research literature discloses a highly diverse assortment of interpretations of new industry formation in the inner city, reflecting the volatility of processes and trends, as well as the range of disciplines engaged in this field of study.

Interpretations of contemporary new industry formation

Experiences of new industry formation within certain districts of the central city fringe and inner city over the past decade and a half or so have proven problematic in a number of respects. In particular, the volatile tendencies of new industries observed through a series of abbreviated periods of innovation and restructuring have been challenging from the viewpoint of community economic development and 'regeneration'. The displacement effects of the firms in areas of established populations and businesses constitute one problem set, while the transitory nature of industry formations over much of this period compromises regeneration strategies reliant upon reasonably stable patterns of labour demand and other supply relations. This volatility, associated with cycles of innovation and restructuring and other market conditions, including the behaviour of the property market, also places exigent pressures on the firms and workers themselves, manifested in job and income insecurity, and in acute problems in maintaining a semblance of balance between working and home life. These pressures are articulated in the work of Andy Pratt and Helen Jarvis (2002), among others.

Framing generative forces of new industry formation

The diversity, complexity, and volatility of new industry formations within the inner city have also caused difficulties for scholars attempting to position these signifying processes of change within the larger narratives of urbanism, and in theoretical syntheses of industrial restructuring and urban transformation. Further, the breadth of *relational* development characteristics which define in part the nature of the new industry phenomenon within the metropolitan core tends to produce a remarkably diverse array of explanations of generative process, emphasizing, variously, market, social, cultural, institutional, behavioural, and environmental factors. Scalar issues also form part of the discourse, in terms of framing the dimensions of new industry formation within the urban economy, and the interface between global processes and local factors. There is also a regional context for innovation and restructuring which takes in the spatial reassortment of industries and activity at the metropolitan scale. The diversity (and perhaps inchoate state) of analysis and discourse on new industry formation within the inner city can be exemplified by the following interpretations.

The reassertion of production in the inner city

This interpretation of new industry formation in the inner city implies a historical perspective in which an evolving mix of generative processes, including market liberalization and globalization, industrial restructuring sequences, sociocultural influences, and institutional and policy factors, underpin the contemporary revival of the postindustrial landscapes of the metropolitan core. Here we can reference Saskia Sassen's injunction to recall the 'deep economic history' of the metropolis, with specific reference to Chicago, but also with wider applications (Sassen 2006). This deep economic history perspective encompasses the structural advantages of the inner city (centrality, social density, and labour supply) for many forms of specialized production. In this economic history viewpoint, cities and sites inevitably experience major recessions and restructuring episodes, but in time new bundles of activities are reconstituted within the central and inner city. In the contemporary city the metropolitan core is seen as the premier terrain for new phases of industrial experimentation and innovation, underpinned by major assets of the central city which include, notably, skilled labour supply, density of information, the built environment, and amenity.

In this historically informed perspective on the evolutionary sequences of the city, the collapse of Fordist manufacturing represents a particularly savage episode in the long history of the core's industrialization and periodic restructuring experiences, but this is not inevitably a terminal event which closes off renewal opportunities for all time. Indeed, the range of new production and consumption industries may be interpreted as a partial recovery of the functional diversity of the inner city prior to the modern era of industrialization, with its provenance in the early nineteenth century, although there may be problematic aspects of these new industries in terms of displacement.

Centrality of culture to the metropolis

The city has always functioned in part as a crucible of cultural expression and production. This cultural production role is exemplified both by orthogenetic cities such as Paris, Rome, and Beijing, as well as (in both similar and different ways) by heterogenetic cities like Los Angeles, Melbourne, and Singapore. In the latter cases, a diversity of social, cultural, and artistic motifs and symbols may be synthesized and deployed for production purposes (for example, in architecture and industrial design), in contrast to generally older cities possessing (and in some ways constrained by) a foundational or 'master' cultural narrative.[9]

A deeper contemporary association between culture and the city is articulated in Allen Scott's model of the urban cultural economy (Scott 1997; 2000), and is manifested in the rise of an associated urban 'creative class', proposed by Richard Florida (2002). This creative class can be positioned as an extension of the 'new middle class' of elite service workers (Ley 1996; Hamnett 2003), or, alternatively, as an essentially new cohort, with distinctive skill sets, lifestyles, residential preferences, and identities (Florida 2002). There is a lively debate about the extent and durability of the cultural economy (Ley 2003; Scott 2003), and about the efficacy of supporting policy interventions (Evans 2001; Markusen and Schrock 2004). But the notion that the cultural inflection of production favours the metropolitan core, with its unique mix of agglomeration economies, sociocultural diversity, amenities, and heritage landscapes, seems tenable. A key reference point is the social reconstruction of the urban core and its influence on the reshaping of the metropolitan space-economy (Zukin 1998), observed in the emergent geographies of production and consumption in the central city's economy.

Inner city industries as contemporary expressions of post-Fordism

Here the importance of flexible specialization processes, in an era of market segmentation and consumption preferences as markers of identity and class affiliation, is acknowledged, exemplified in the design and fabrication of 'cultural products', and in the deployment of creative labour. These new cultural production processes incorporate synergies between design, technology, and space in the twenty-first-century metropolis. While the placement of creative industries within larger and long-established urban economic development trajectories is very much open to contestation, it does seem clear that some theoretical accommodation of the cultural economy and the city must be essayed in contemporary urban studies and economic geography. Post-Fordism, after all, essentially acknowledges the *absence* of standardized mass-market manufacturing within advanced city-regions following the late twentieth-century industrial restructuring experience, rather than an industrial construct with specific form, while *flexible specialization*, although a useful descriptor, is now perhaps too generic for contemporary applications. In the spirit of this theoretical enterprise Allen Scott has recently proposed the notion of the 'cognitive cultural economy' as a conceptual successor to the more austere descriptor of post-Fordism, a quarter of a century or more since the collapse of large-scale, standardized manufacturing as a centrepiece of the urban economy (Scott 2007).[10]

The industrialization of artistic production

Artists have long been concentrated in major cultural centres such as Paris, Vienna, and Florence over the centuries, and have more recently recolonized, in larger numbers, working-class neighbourhoods (Bianchini and Ghilardi 2004). But here we can acknowledge a contemporary industrial process which structures the expansion of creative enterprises and labour as elements of newly-articulated systems of production incorporating the inner city's artists, designers, and cultural workers. These form in the aggregate a 'cultural production pyramid' (Ho 2007), comprising an extensive base of artists and high-risk creative firm start-ups, and a peak consisting of professionals such as elite artists, fashion designers, architects, graphic designers, and others. Artists also represent a crucial occupational cohort in creative, technology-intensive industries such as video game production, and are engaged in many forms of new media activity. This contemporary regime of artistic-industrial articulation, while undoubtedly of real economic importance in both established and emerging centres of artistic production, is, however, fraught with instability and displacement, owing to the anomalous positionality of artists and designers within the fragmented creative production system in the metropolis, the subsistence level of most artists' incomes, and (as a consequence) the somewhat precarious tenure of artists in the steeply inflating property markets of the inner city.[11]

Creative industries and the development of the urban service sector

A number of scholars have suggested that design functions, artistic labour, and creative task specializations follow the progression of service industry development in the metropolis. In this interpretation the cultural economy is largely comprised of service industries, such as architecture, advertising, graphic arts and design, and industrial and fashion design, among many others. Thus, over the past decade or so, service industry scholars have enlarged this field of research beyond the long-established investigation of banking, finance, and mainstream business (or producer) services to include applied design and creative services (Hutton 2000). The growth of design services responds in large part to the increasing demands of segmented consumer markets for 'culturally differentiated' goods and services. What distinguishes many of these design service industries from other services, notably the mainstream, office-based intermediate services such as legal, accounting, and consulting businesses, is, first, a more intimate contact with physically tangible products and goods, typified by industrial design and fashion design; and, second, a locational tendency that favours inner city locations. Thus design services constitute elements of the regenerative development of the postindustrial inner city, together with new housing, amenities, and public institutions. In this sectoral perspective, the space-economy of the inner city has transited from its original 'factory world' structure of industrial production, to one defined by an emergent service economy situated in offices, studios, lofts, and workshops.

New industries and the urban property market

Critical studies scholars (notably neo-Marxists) have identified the workings of the urban property market as central to the redevelopment of the inner city, driven by a structural 'rent gap' that activates new investments and the insertion of new social actors (Hackworth and Smith 2001). According to this viewpoint, artists and other pioneering creative individuals and industries are attracted by the lower rents associated with obsolescent (or derelict) inner city landscapes, representing the vanguard of social upgrading forces. This pattern has been repeated in many cities over the past three decades, with London and New York as prime examples. Artists and designers comprise a substantial set of economic agents in the inner city, manifested in the growth of studios and galleries, and a retail presence in shops and stores that exhibit the fruits of artistic production. These communities of artists are, however, unstable over the longer run, as the revalorization of the inner city led by members of the new middle class, professional companies, and commercial property firms increasingly displaces artists and other low-income residents, in the classic formulation of urban transition and succession (Bridge 2001).

Spatiality, built form, and creative industries of the inner city

A steady flow of monographs and other studies have demonstrated that cultural industries and creative labour are attracted to the inner city by a combination of factors that include the intimate spatiality of these inner districts (Hutton 2006), the sensuousness of aestheticized landscapes (Ley 2003), and both the 'concrete' (functional) and 'representational' (symbolic) value of historic buildings and their potential for adaptive reuse (Helbrecht 2004). In other words, new industries, firms, and creative workers responded initially not merely to the cheaper rents of the postindustrial inner city, but, rather, evinced a positive affinity toward vernacular building types, and the 'look and feel' of former factories, warehouses, and institutional and residential structures available for adaptive re-use, as well as social and institutional factors. The 'studio workshop' environment is thus acknowledged as an industrial counterpart to the 'loft living' construct popularized in the writing of Sharon Zukin (1989).

An institutional perspective on new industry formation

While the scholarly record discloses numerous examples of 'spontaneous' new industry sites within the inner city, there are also cases where institutions and agencies have played central roles in start-ups and development. Graeme Evans (2001), for example, has shown that the emergence of new industries within the inner city has been induced by a mélange of public policies and programs which include land use and zoning regimes designed to facilitate the entry of new industries; heritage policies which conserve the highly textured buildings and landscapes preferred by many creative and knowledge-based industries; local regeneration and community economic development programs; public investments in amenity and in cultural institutions, and the introduction of new educational and training initiatives, including programs in fine arts, digital design skills training, management, marketing, and entrepreneurship. While the public sector, and more specifically agencies of the local state, is frequently engaged in this mission, there is also scope for involvement on the part of private bodies (including regeneration consultants and property companies), as well as NGOs and community-based organizations, enlarging the institutional capacity for the promotion of new industries and firms in the inner city.

The metropolitan space-economy and spatial division of labour

Another viewpoint on the emergence of creative firms and institutions within the metropolitan core situates the industrial regeneration of the inner city within a regional framework. In this interpretation the emergence of the inner city as a site of industrialization, business start-ups, and labour formation is positioned not as a spatially isolated or discrete phenomenon, but rather as an integral part of the evolution of the metropolitan space-economy, shaped by the specialization of internal production spaces. In this regional context, the inner city exhibits a number of advantages for creative and contact- and knowledge-intensive industries, in relative, if not absolute terms. This idea of intra-regional competitive advantage for specialized sectors and industries also follows in part Scott's model of location of industry within the metropolis (Scott 1982), with smaller, contact-intensive enterprise favouring the agglomerations of the urban core, and larger concerns less reliant on proximate inputs locating in suburban zones.[12] Peter Hall's more recent model of global city structure and function (2006), which includes culture, the arts, higher education, tourism and government as well as corporate control, finance, and producer services, also has some traction in this viewpoint, as it inserts a broader range of industries into the spaces of the metropolitan core.

New industry formation and the 'social nature' of advanced economies

Nigel Thrift and Kris Olds, among others, have written about the 'extraordinarily social nature' (Thrift and Olds 1996: 316) of advanced economies, ostensibly in contrast to the more atomistic world of Fordist production, the industrial shop floor, and even the segregated occupational structure of the office economy. Further, the intensive social basis of working life as well as community function forms part of Jane Jacobs' ideal city, and a part of her critique of modern planning (Jacobs 1961). Creative and knowledge-intensive industries congregating within inner city districts, characterized by an intimate interface between production and consumption activities, and lubricated by the rich amenity base (consumption, recreation, interactive spaces, and social density) of the core, constitute a prime evocation of the social nature of specialized production among advanced societies. Meric Gertler numbers among those who have written about the importance of socio-spatial factors and the exchange of tacit knowledge in the workings of the New Economy (see, for example, Gertler 2003), and here the characteristic intimacy of the inner city may be acknowledged as intrinsically conducive to this function.[13]

The production economy of the 'new inner city': a synthetic model

The diversity of explanations described in this inventory reflects the extraordinary richness of the new industry formation experience within the contemporary inner city. These incorporate a number of pervasive processes and trends which have been documented across a range of cities and sites, both among developing and transitional societies, as well as more locally-contingent influences. A working hypothesis for our purposes is that new industry formation constitutes an exemplary case of the workings of global and local forces, or 'glocalization' of development tendencies in Swyngedouw's terminology. The next task for this present study, then, is to treat this list as an investigative point of departure, and to proceed to a synthetic framework which can be deployed to guide the analysis of new industry formation and its consequences and developmental saliency, giving prominence to the most consequential processes.

Figure 2.1 presents a model of the production economy of the 'new inner city' of the twenty-first century that synthesizes the factors and forces of causality described above. A guiding assumption for this exercise is that there is no single, universal explanation of industrial growth and change in the contemporary inner city, but rather a complex mélange of factors in play, in effect, a range of developmental trajectories; the intertwining of social, cultural, market, and policy factors; a constantly shifting interaction between regional and local forces; all mediated through the lens of space and place in the metropolis. These multiple factors have in turn produced a distinctive production economy in the inner city, which co-exists, at times uneasily, with other key elements of the core, including the financial-commercial sector, new residential neighbourhoods, a large consumption sector, and the marginal residential communities increasingly under pressure from encroaching uses.

At the centre of the model are the defining features of the production economy of the inner city, including elements of multiple production regimes (artistic, artisanal, Fordist, and post-Fordist); diverse industries, including those representative

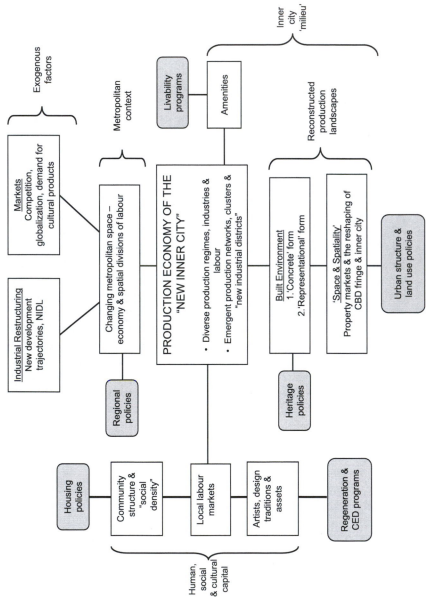

Figure 2.1 Factors shaping the production economy of the 'new inner city'.

of the New Economy, the cultural economy of the city, and the knowledge-based economy, as well as financial and commercial firms situated in sub-centres or inner-edge cities; 'new industrial districts', comprising consumption, amenity and housing as well as production sectors (described in Chapter 3); and emergent production networks, variously structured within the confines of clusters which exhibit familiar agglomerative tendencies, or within more extended production networks, incorporating localized inter-firm linkage patterns, as well as increasingly extensive outsourcing.

Arranged schematically around this specialized production economy of the inner city are clusters of factors, and bundles of influences, involved in the shaping of this new industrial construct. Viewing matters initially 'from the ground up' we can identify a cluster comprising the 'reconstructed production landscapes' of the inner city, taking the form of the distinctive spaces and built environment of the inner city as fundamental to the formation of small-scale, specialized new industries. 'Space and spatiality' are shaped by the property market, and by urban structure and land use policies, while the defining elements of built form are represented by the heritage buildings of the old industrial inner city – factories, warehouses, workshops, stations, and so on, made available for adaptive re-use, and preserved by local preservation policies, heritage societies, and other agencies. The existence of consumption and other forms of local amenity, increasingly supported by public agencies, also forms an essential feature of the inner city milieu for cultural industries.

The reconfiguration of the economy, spaces, and landscapes of the central city is shaped in large part by the relayering of capital, observed at the urban zonal level as a shift in investment from the CBD to the CBD fringe and inner city. This spatial reorientation of capital (public as well as private) has endowed the inner city with a 'New Economy' comprised of creative, increasingly technology-intensive sectors and industries, and the 'revival' of inner city industrial districts and production landscapes (Hutton 2000). The relayering of capital has also generated new housing (urban mega-projects, see [Olds 2001] as well as adaptive reuse and live-work studios), and consumption amenities, in the form of museums, stadia, galleries, restaurants, and the like. Historically, the relayering of capital in the city has produced social dislocation and intensified class conflict, as powerfully articulated in David Harvey's (2003) evocation of redevelopment, displacement, and communitarian action in Paris in the Second Empire. The reconstruction of the contemporary inner city is replete with analogous dislocative outcomes, as Don Mitchell has testified (Mitchell 2003). For this study, the investigation will examine the competition for capital between new industries and high-end housing in signifying inner city precincts, notably in Clerkenwell (in London) and Yaletown (in Vancouver).

As our earlier review demonstrated, however, new industry formation is shaped not only by local agencies and actors, as important as they are, but also by powerful structural forces. As Figure 2.1 shows, larger industrial restructuring tendencies and market shifts comprise a set of 'top-down' influences in the emergence of new industry formations in the inner city, including demand for cultural products and services of all kinds. These exogenous forces interact with local factors to produce new industry constructs among a growing number and range of cities and sites, not only in the 'West' but also in cities of Pacific Asia, including Singapore, Shanghai and Tokyo. The increasing pervasiveness of these sites among an expanding number of cities certainly suggests that the New Economy of the inner city is at least in part a global phenomenon.

Between the global and more intensely localized spheres of (inter)action, the city-region represents a salient intermediary scale for understanding the processes of new industry formation (Figure 2.1). Again, as observed in the preceding review, global economic forces are in part mediated by (and reflected in the reconfiguration of) the metropolitan space-economy, which incorporates local-ized competitive advantage for various forms of specialized production. The most spectacular expressions of global spaces within the space-economy of the metropolis include of course the new financial mega-districts, such as Lujiazui in Shanghai's Pudong redevelopment zone, and Canary Wharf in London's Dock-lands; Olympic Games and international expositions that have transformed urban sites in Los Angeles, Melbourne, Paris, and many other cities; and the new cul-tural mega-spaces, such as Bankside in London, and numerous other sites in Barcelona, Berlin, and elsewhere. These expressions of globalization, replete with big visions, spaces, budgets, architects, and impacts, exemplify the practice of urban place-making writ large (Peck 2001; C. Hall 2006). But the proliferation of new industry sites within the inner city districts of both advanced and transitional cities, though calibrated at a much smaller scale, and routinely blended into local regeneration schemes and urban design programs, nonetheless exemplifies another expression of global forces in the reproduction of the city (Hubbard 1996; Evans 2003).

It seems clear, then, that new industry formation in the inner city constitutes a quintessential multiscalar process, including interactions of the global, regional, and local. As we saw in our earlier discussion of the processes underpinning the growth of the urban service sector (Table 2.1), there is also a complex set of human, social, and cultural capital factors that shape the location, industrial mix, occupational structure, and development characteristics of these new industries. These include the presence of artists and designers, as well as localized design traditions which may underpin and inform contemporary cultural production. The nature of local labour markets in the city will also influence the moulding of the employment pool available for specialized new industries, although the degree of 'fit' is often problematic for local regeneration purposes, and efforts to improve this labour–enterprise match lie at the core of many community development programs. Finally, the overall structure of the local community in which these new industry sites are embedded (or proximate to) can also influence development, as the level of 'social density' may have the effect of increasing interaction possibilities, as well as potential labour supply and customer access. Here, local planning for residential development (including social and other non-market housing) may represent a salient policy field.

Conclusion: processes in the reconstruction of economic space

This chapter has offered a necessarily concise overview of processes seen as influential in the formation of new industry sites within the inner city districts of metropolitan areas. An attempt was made to present an appreciation of the structural forces reshaping the production economy of the metropolitan core in the postwar era, including the rise of a corporate office complex in the CBD, which emerged as the largest industrial agglomeration and densest employment district in the city-region, as well as the collapse of Fordist production and labour in the inner city. These remain dominant story-lines within the narratives of urban growth and change among advanced city-regions in the 'West', and are increasingly in evidence within certain Pacific Asia cities, although the larger role of the state in the latter cases makes for some important distinctions in outcome.

The second overarching purpose of Chapter 2 has been to acknowledge the range of explanations for the rise of a 'New Economy' of specialized, creative, and knowledge-intensive enterprises within the former postindustrial terrains of the metropolis. This new industry formation experience has been fraught with disjuncture, and characterized by apparently abbreviated restructuring cycles, but represents both a significant reassertion of production firms and labour in the inner city, and an influence on the larger reconstruction of the metropolitan core. The multiple influences shaping this new industrialization sequence, the complexity of industrial organization in the core, the problematic nature of outcomes, and the appeal of this study area for a wide range of disciplinary scholars, each contribute to this marked variegation of explanation. A conceptual model of forces underpinning the production economy of the 'new inner city' was presented as a means of articulating a synthesis of these explanations. This discussion leads naturally to a more incisive examination of space and place in the New Economy of the inner city, a task for Chapter 3.

3 Place

The revival of inner city industrial districts

Introduction: new production spaces in the metropolitan core

In the previous chapter we addressed the complex mélange of factors – market, sociocultural, physical, and policy – that have been identified as underpinning the reassertion of industrial production within inner city districts of the modern metropolis. From this survey a synthetic model of principal motive forces was constructed as a means of interpreting the processes of new industry formation within the metropolitan core, including the rich interactions that comprise the relational geographies of production within advanced economic systems. The basic dimensions and operational characteristics of this model can be identified, with some variation, in new industry sites across a diverse range of cities.

Where this discussion of interactions between space and process naturally leads to is a more grounded consideration of the significance of *place*, that is, an examination of the experiences of new industry development in specific cities and sites. As a guiding principle we can acknowledge that the formation and operating characteristics of new industry sites in the inner city exhibit features of contemporary innovation and change, including the deployment of advanced production and communications technologies; but also embody attributes of older production sites and systems. This theme is addressed in Graeme Evans' treatment of 'cultural industry quarters: from pre-industrial to post-industrial production', in Bell and Jayne (2004), an approach that does justice to the extended lineage of localized, high-value, and specialized production within the city. As Evans affirms, the promotion of cultural industry quarters internationally tends 'to neglect both the historic precedents and the symbolic importance and value of place and space' (Evans 2004: 91). While certain basic features of industrial sites situated within cities (and extended regional territories) have proven fairly durable, however, new forces, both exogenous and internal to the metropolis, are putting pressures on these districts and their constituent industries and firms. An appreciation of these influences is essential to placing experiences of new industry formation within both historical and contemporary development contexts.

To particularize, the new industrial formations of the inner city exhibit characteristics of several distinctive territorial forms of production, associated with certain benchmarks in the history of industrialization, while at the same time displaying 'New Economy' features. Indeed, one of the heuristic values of studying the inner city's contemporary industrialization experience lies in its capacity to demonstrate some of these important historical continuities. Other salient characteristics of the 'New Economy of the inner city' include the co-presence of 'old' economy and 'New Economy' industrial regimes, sectors and firms, as noted in the previous chapter, as well as the differentiation of industrial production between sites *within* the metropolis, reflecting the internal specialization of the urban economy.[1]

As a means of drawing out the evolving intersections between processes of industrial change and 'place', an essential preliminary to the presentation of the intensive case studies to follow in this volume, Chapter 3 will address the following themes. First on the agenda is a perspective on the progression of the artisanal district, a precursor both to the urban industrial district, and also to the formation of contemporary new industry sites and cultural production quarters. References include the Oltrarno artisanal district of the Santo Spirito Quarter in Florence, and the Ancient Quarter of Hanoi, each of which exemplifies resilience over the course of history's upheavals, as well as the reshaping power of globalization processes. Second, the industrial district as a concept for investigating formative processes in the spatial organization of production will be explored. Bennett Harrison's influential survey of the industrial district discourse over the last century will be deployed as a framing device for this treatment. The next section will more firmly place the industrial district within the city, with a description of three prominent sites in Manhattan which reflect principal phases in the economic life of the metropolis: (1) the garment district, an evocation of the 'old/industrial' economy; (2) the corporate office complex, the 'global financial-services economy'; and, finally, (3) the saga of Silicon Alley, an exemplar of the 'New Economy'. The conclusion will link the themes disclosed in the industrial districts of one first-order global city, New York, to the narratives of industrialization and restructuring in another: London, presented in Chapters 4 and 5 of this volume.

Changing fortunes of the artisanal district

The historic artisanal and fine crafts district, although later supplanted by the true industrial district of the late eighteenth and early nineteenth centuries, represents an instructive point of departure for our investigation of contemporary reindustrialization processes in the metropolis. We can reference examples of artisanal districts which underscore the traditional role of the inner city as perdurable site of fine arts, applied design, and specialized production, serving initially wealthy patrons among royalty and the aristocracy (Braudel 1982), and now increasingly recast as sites of globalization and spectacle in the early years of the twenty-first century.

Global spectacle in the Oltrarno Artisanal District, Florence

Florence's Oltrarno ('beyond the Arno') artisanal district was formed when the Medicis moved from the Palazzo Vecchio on the north side of the Arno to the Palazzo Pitti on the south side (or left bank), and can therefore claim a history of production in fine arts and crafts of four and a half centuries.[2] Artisanal production is concentrated within a triangular zone, bounded on the east by the Palazzo Pitti and the via Guicciardini, on the west by the via Maggio, and the north by the Borgo San Jacopo. The Palazzo Pitti constitutes a principal reference on the eastern boundary of the area, while the Santo Spirito basilica (Brunelleschi) is a major landmark to the west. Until relatively recently the area was just outside the principal tourist areas clustered on the northern bank of the Arno, which include the Uffizi, Palazzo Vecchio, the Duomo, San Lorenzo, Santa Maria Novella, and Santa Croce.

The spatiality of the Oltrarno takes the form of an intimate, semi-enclosed precinct with a central north–south street, the via Toscanella, several shorter cross-streets, and the Piazza della Passera which serves as a space of social interaction, including as it does several restaurants and a prominent espresso bar-ristorante serving both locals and tourists (Figure 3.1). The area's built form includes two- and three-storey nineteenth-century buildings, as well as some older structures, with workspace on the ground level, and residential uses on the upper floors. These environmental features combine to promote a lively streetscape and social milieu for creative activity. A stroll down these streets discloses glimpses of the workings of a still-robust artisanal district, with the workforce for the most part consisting of middle-aged and older (and mostly, but not exclusively, male) craft workers, with the design and fabrication processes conducted in small workshops (Figures 3.2 and 3.3).

The most salient operational feature of these workshops concerns the production of art, crafts, and design, congruent with the distinctively Florentine cultural traditions. Historically, artistic production in Florence consisted of a high culture stratum as represented by the most accomplished and influential artists, architects, and writers in the *quattrocento*, supported in part by a larger platform of apprentices and skilled artisans engaged in craft specialization, including plaster work, picture framing, engraving and lithography, garments, and leather work, among other product lines.[3] In contemporary Florence, many of these artisans are now themselves principal artists and designers, at the vanguard of cultural production, rather than merely supportive or subcontracting workers. Two of the enterprises I interviewed in 2005, Giancarlo Giachetti metal fabricators ('*lavorazione artistica in ferro*'), on the via Toscanella, and Bijan, of 'Firenze of Papier Mâché', cater in part to the needs of filmmakers in Italy and in the US for crafted set and costume works.

New forces of global-local interaction

While the overall impression in the Oltrarno is one of robustness in its long-established artisanal role, there are, however, signs of potentially transformative

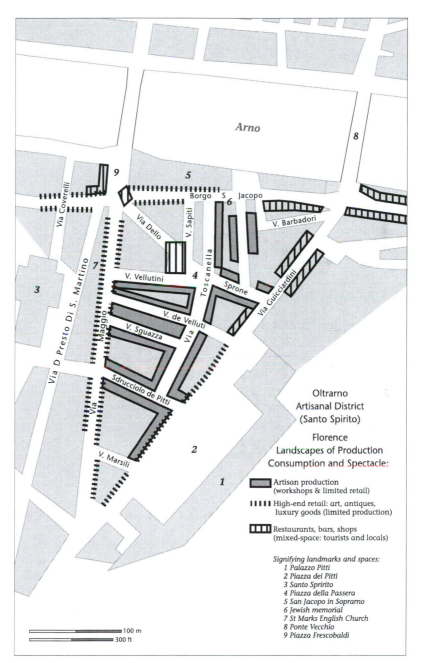

Figure 3.1 Oltrarno Artisanal District, Santo Spirito, Florence.

Figure 3.2 Lithographer, via de' Velluti, Oltrarno Artisanal District: Florence.

processes at work. First, the remarkable expansion of Florence as an iconic centre for cultural tourists has brought with it a tendency for visitors to seek out less-congested sites and attractions. There are now public tours of artisanal workshops available (including on-line registration), and web-sites for the purchase of crafts from consumers 'anywhere in the world'. (This can also be seen in long-established leatherworks to the north of the Arno.) The Florentine workshop has transitioned, at least in part, from an enclosed, almost hermetic place of specialized production, to one of scheduled spectacle and experience.

Second, it is clear that the area's cafés, bars, and coffee shops are being drawn into the ambit of the expanding circuit of tourist consumption outlets, intro-ducing new users, and prospectively, price inflation and displacement for local retail outlets. For the moment at least, these local spaces of consumption are shared by the Oltrarno's resident artisans as well as visitors. Finally, there are now English-language artisanal and craft education programs offered *in situ*, offering

Figure 3.3 Artisan and apprentice, Piazza della Passera, Oltrarno.

a 'Florentine cultural experience' for those visitors with the time and resources to engage in the life of the district at this level. Observation of workshops in Santo Spirito also discloses the presence of young foreigners, including numerous Chinese, working as students or apprentices in some of the workshops, reinforcing the emergent role of the Oltrarno as site of global experience, as well as local production.

Hanoi's Ancient Quarter: from artisanal production to Internet landscape

Hanoi's Ancient Quarter has a history of craft production dating from the division of Hanoi into two sections in 1010. The Ancient Quarter was part of the 'Commoners City' as distinct from the 'Royal Citadel', and included the organization of craftspeople into guilds with specific territories according to guild association, an early form of spatial division of production labour in the capital city. Papin observes that each guild or craft association performed customs, festivals, and other local social and cultural practices (Papin 2001, quoted in Turner 2006), similar in some ways to the organization of the *contrade* in Siena.

Sarah Turner has written about the sometimes tumultuous history of Hanoi's Ancient Quarter, including both continuities and disjuncture, and including its more recent episodes of transformative change (Turner 2006). Upheavals included the destruction of much of the commercial area by Chinese pirates in the early 1880s, and the French conquest of Vietnam in 1883. But through these and other disruptions, the Ancient Quarter sustained a growing cluster of skilled craftspeople and artisans, with specializations including lacquer work, mother-of-pearl inlaying, and embroidery. The craft workers of the Ancient Quarter catered to royalty and to wealthy merchants, suggesting a parallel with the market orientation of artisans in Clerkenwell and Hatton Garden[4] in London.

Some foreign penetration of the market spaces of the Ancient Quarter was evident by the 1930s, including fruit imported from places as distant as San Francisco, and high-end French goods exported from the centre of empire in metropolitan France. Worse was to come, in the form of the Indochinese wars, with the first war (1946–1954) resulting in the destruction of Hanoi's industry and essential services, and the evacuation of the Ancient Quarter. Following the war, Vietnam was partitioned, with North Vietnam emerging as a socialist state, influenced by the ideologies of the larger COMECON[5] 'second world', but also developing according to Vietnamese characteristics. Thus, in 1958, the Hanoi government initiated a comprehensive collectivization process, and 'by 1960, the collectivization of Hanoi's small industries and handicrafts was nearly complete, with 95 per cent of craftspeople having joined production or service collectives' (Turner 2006). With the onset of the Second Indochina War in 1965, and periodic heavy bombing inflicted by the U.S. Air Force, the area was heavily damaged. With an end to the bombing in 1973, the craft industries and businesses were among the first to recover, reflecting their relatively low reliance on the urban infrastructure which had been largely demolished in the air campaign.

The reunification of Vietnam in 1975 led to a period of renewed state (or social) production, but the proclamation of the 'economic renovation' (*doi moi*) program in 1986 initiated a new period of transformation within the Ancient Quarter's production sector. These changes included shifts in highly localized production and trading sectors, in response to consumer demand, rather than continuing traditional activity patterns.[6] Further, the patterns of urban–rural linkages characteristic of the guild era were now largely sundered, as the influence of household registration system was now considerably diminished.

Turner's narrative of transformative experiences in Hanoi's Ancient Quarter concludes with an acknowledgement of continuities as well as ruptures. The ancient production and trade in lacquerware continue, while other artisanal production (such as jewellery and textiles) which had come under co-operative management in the socialist era are now continuing within new management structures reflecting the *doi moi* regime.

Global forces and Internet landscapes in the central city

There are also new actors and activities, reflecting the insertion of market influences and global processes in the reproduction of Hanoi's central city, including
tourism. The Hanoi story-line is taken forward another step in Björn Surborg's
(2006) account of 'the New Economy and the built environment' in the central
city. Here a series of processes have combined to reproduce the structures of the
three principal central districts. The Ancient Quarter described in Turner's work
has now been recolonized by small firms, notably 'mini-hotels' patronized by
backpacker tourists, while head office and producer services cluster in the CBD,
and a New Economy construct of software developers and Internet companies
inhabit 'the culturally vibrant "south of CBD area" ' (ibid.: 249). The tight
kinship and production linkages situated within the regional setting described in
Turner's paper have been supplanted in part by new patterns of international
demand, reproducing in turn the spaces of the central and inner city. This spatial
configuration (Figure 3.4) is suggestive of reconfigured central city patterns
observed in many cities subject to global processes of restructuring. These
include reconstructed production spaces, and new divisions of labour and task
specializations, as well as emergent spaces of consumption. But Hanoi's central
area also incorporates a distinctive landscape shaped by French colonial interests,
Soviet planning, and a peculiarly Vietnamese form of the market economy,
demonstrating once again the complexities of global-local interaction.

Evolution of the industrial district discourse

The industrial district represents a core concept of economic geography, with its
origins in Alfred Marshall's work on the spatial organization of industrial production (Marshall [1890] 1972). The extraordinary richness of Marshall's theoretical
work continues to attract concerted critique and search for contemporary applications (see, for example, Arena and Quéré 2004, reviewed in Caldari 2004). But
for the purposes of this study we will confine our references to his role in initiating
a century-long debate over the development of industries within space, place, and
territory.

Industrial districts: from Marshall to Markusen

For Marshall, 'place' encompasses a hierarchy of nations, regions, cities, and
industrial districts (Marshall 1927), within which (as Bellandi observes) 'the
local level seems to function here as the basic unit' (Bellandi 2004: 245). In the
Marshallian view, localization of industrial production carried with it three sets
of supply-side benefits: first, knowledge spillovers among individuals, including
those working for different companies; second, access to pools of common factors
of production, notably land, labour, capital, energy and transportation; and third,
productivity gains accruing from specialization. Under these shared or communal
conditions, the unit production costs of individual producers will be lower than if

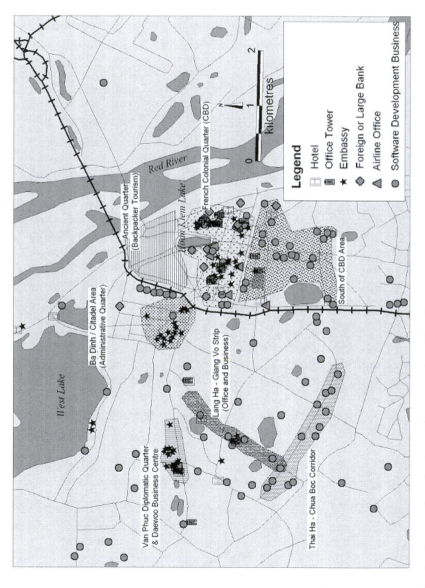

Legend

□ Hotel
⊞ Office Tower
★ Embassy
◆ Foreign or Large Bank
▲ Airline Office
● Software Development Business

West Lake

Red River

Hoan Kiem Lake

Ancient Quarter
(Backpacker Tourism)

French Colonial Quarter (CBD)

South of CBD Area

Ba Dinh / Citadel Area
(Administrative Quarter)

Lang Ha - Giang Vo Strip
(Office and Business)

Van Phuc Diplomatic Quarter
& Daewoo Business Centre

Thai Ha - Chua Boc Corridor

N

0   1   2
kilometres

Figure 3.4 New landscapes of production and consumption, Hanoi.

Source: Surborg (2006).

these producers were obliged to provide these factors on their own, or to import these inputs from external sources.

The progression of scholarly thought on the nature and operation of the industrial district dating from Marshall's seminal writing was eloquently explicated in Bennett Harrison's classic *Regional Studies* article of 1992 ('Industrial Districts: Old Wine in New Bottles?'). In this treatment, Harrison identifies the following benchmarks of conceptual innovation, dating from the immediate postwar period. First, Harrison acknowledges the importance of Scitovsky's idea of 'pecuniary external economies' (1963), i.e. 'the dynamic impacts of one firm's new investments on the possibilities for enhanced profitability of, and therefore expansion by, other firms' (Harrison 1992: 472). These might be represented, for example, by the products of this new investment enabling a lower-priced input to other local firms, constituting an *agglomeration economy*.

Second, Harrison credits Perroux (1955) with the application of Schumpeter's views on the importance of innovation with the concept of the 'growth pole' (*pôle de croissance*) and that of the 'propulsive industry' (*industrie motrice*). Growth poles (sectors, rather than places) grow at faster rates than other industries, and create a 'positive feedback system' mediated through input–output relations, a concept inserted into innumerable local and regional economic development strategies over the past half century.

Third, Harrison acknowledges the contributions of scholars working within the project framework of the New York Metropolitan Region Study, including Chinitz, Hoover, and Vernon, in developing the product cycle theory, which describes a temporal sequence of experimentation, diffusion, and maturity. Harrison's summary of the work of this influential group includes this observation:

> In the early stages, firms tend to be small, work is often organized in a craft mode, specialized skills are especially needed (and valued) and intense face-to-face interactions are essential since an important part of the innovation process involves the rapid exploitation of unexpected (serendipitous) exchanges of ideas.
>
> (Harrison 1992: 473)

Chinitz, Hoover, and Vernon extended this theorem to describe the changing fortunes of industries and regions within extended developmental cycles, but there is in this extract a clear signal for our investigation of new industry formation processes within the contemporary inner city.

Bringing the 'social' into the industrial district discourse

While these early theorists of the evolution of the industrial district included at least an implicit acknowledgement of actors such as managers, artisans, and operatives, the social dimension of inter-firm interaction that lies at the heart of the concept was largely subsumed within the more austerely economic vocabulary of industry growth and firm behaviour in space. Harrison cites Polanyi's (1944) recognition that over the extended industrial district discourse, economic behaviour had become disconnected or 'disembedded' from social relations, as a consequence of the neoclassical economics position that 'in order for decentralized market exchange to work efficiently, the behaviour of buyers and sellers *must* be subject solely to the pursuit of self interest by rational atomistic individuals' (Harrison 1992: 476). The challenge to this position was taken up by 'outsider' entrants to the industrial district discourse, including anthropologists, sociologists, and political scientists, among others, a tradition that has continued to the present.

Harrison acknowledges Granovetter's seminal paper of 1985, which introduced the concept of social embeddedness, an idea 'key to understanding how the theory of industrial districts differs fundamentally from neo-classical agglomeration theory' (ibid.: 476). In the most successful of these industrial districts, there is an individual internalization of group norms, and sense of trust demonstrated through multiple transactions and exchanges, that work to the advantage of all who share these norms, representing 'concrete personal relations and structures (or "networks")' (Granovetter 1985: 490, quoted in Harrison 1992: 477). To illustrate, Harrison cites Lorenz's research on subcontracting relationships in the mechanical engineering sector in the Lyons region (Lorenz 1988; 1989) in which interviews with firm owners and managers disclosed an 'emotive vocabulary' which continually underscored the pervasiveness of values such as *partnership, loyalty, morality,* and *mutual trust.*

The defining quality of embeddedness is widely acknowledged as critical to the functioning of industrial districts at a regional level, with the 'Third Italy' (Becattini 1989; 1990) comprised of central and north-central regions, advanced as a prototypical model in Piore and Sabel's *Second Industrial Divide* (Piore and Sabel 1984). At the city-region level, the 'social economy' of Bologna–Emilia Romagna has been widely acknowledged as a paragon of trust-based economic development within the Third Italy. Below the regional level, embeddedness has also been acknowledged as key to the functioning of community economic development strategies, based on the development of trust and shared experiences of local actors.

Emergence of the 'new industrial district' discourse

Recognition of the embedded social dynamics of regional industrial development in the Third Italy, and the growth of technology-based production zones in the US and elsewhere, stimulated a new conceptual turn in the industrial district discourse. Here the intent was in part to link contemporary features of the industrial district as a distinctive form of territorial production, including the centrality of social network formation and operation, to the literature on post-Fordism and flexible specialization. The basic concept of the new industrial district took the form of a regional platform of innovative small enterprises, linked through embedded networks typified by high levels of co-operation and collaboration, suggesting a lineage from Polanyi and Granovetter.

Rather than assuming a common developmental pathway or trajectory, new industrial district researchers recognized the reality of variegation in structure, function, and operating characteristics, and therefore proposed a lexicon which accommodated a number of variants. Michael Storper and Richard Walker proposed four patterns of location which flow from economic growth: (1) localization, where development occurs at points some distance from existing concentrations; (2) clustering, which takes place when one emergent site develops a base of firms ahead of other such areas; (3) dispersal-growth peripheries, formations which develop centrifugally from a core site; and (4) shifting centres, where new rivals to established centres emerge (Storper and Walker 1989). Neil Coe acknowledges the utility of this concept, but also affirms the need for concepts which have the power to describe and explain the 'internal structural dynamics' of spatial configurations of industry (Coe 2001: 1756).

Here Coe cites Ann Markusen's research (1996) on industrial districts in Brazil, Japan, and the US which yielded a new typology to accommodate innovation in the spatial configuration of production, including: (1) Marshallian districts, comprised of dense networks of exchange and co-operation between local firms entrenched within a largely internalized labour market, as well as an 'Italianate variant' typified by innovation and creativity, and institutional organization, including unions and local government; (2) 'hub-and-spoke' districts, in which industrial structure is largely shaped by a relatively small number of major, vertically integrated firms; (3) the 'satellite platform' structure, characterized by a cluster of branch plant operations, captive to executive decisions made elsewhere; and (4) the 'state-anchored' district, defined by dependence on a major institution, such as a major university, research laboratory, or defence establishment (summarized from Coe 2001: 1756–1757).

Markusen accepts that the regional reality of industrial organization is more complex and 'messier' than this typology might imply, with a strong likelihood that 'real' industrial districts comprise a blend of types, creating 'sticky mixes', as well as the potential for mutation from one form to another (1996). Coe contributes to an elaboration of these possibilities by describing the film sector in Vancouver as a 'satellite-Marshallian' industrial district. In Coe's narrative, Vancouver has developed as a significant site of 'runaway production', hived off from the original Los Angeles site as a means of achieving significant economies, as well as suitable locations for 'shoots'. In Coe's analysis, the Vancouver film production sector comprises elements of both the classic Marshallian industrial district, with its locally embedded firm networks and labour market characteristics, coupled with the 'instability and external dependency' of Markusen's satellite platform model. Despite the instability of the latter, though, Coe affirms that since the 1960s Vancouver has emerged as a sustainable site of film production, based on advantages of proximity (2.5 hours flying time from Los Angeles, within the same time zone), suitable climate, a range of photogenic locations within two hours drive of the city, and local agencies (including unions) prepared to vigorously promote Vancouver as a runaway production site following the vertical disintegration of the Los Angeles studios. Within the Vancouver region, there are major production centres in the inner suburbs, plus creative and technical staff located in the inner city (see Chapter 8), and a local labour pool which can maintain 35 production units – a substantial platform for a satellite film production sector.[7]

Our necessarily concise review of the industrial district literature, drawing on Harrison's influential review and synthesis, offers a number of significant conceptual references for the investigations of industry formation at the urban district level. Certainly, the original idea of the industrial district, the concept of agglomeration and external economies, and the tenets of social embeddedness, can each contribute to our understanding of the dynamics of industrialization processes in the contemporary metropolis. But there are also limits to the extent to which these concepts can explain the workings of new industry districts, on the evidence of recent development experiences. The ideal of relatively stable industrial districts, in which social relations and networks promote a measure of insulation from the ravages of recessions and external competition, has been subverted by recurrent rounds of restructuring and globalization, not only in the small artisanal quarters described above, but also in a number of leading regions.[8]

The industrial district in the city-region: contours of change

For our purposes, processes of contemporary reindustrialization within the metropolitan core take place within the former production sites of the classic 'industrial city'. These typically include former factory, warehousing, and distribution buildings and sites whose provenance can be traced to the nineteenth century, following the introduction of new production technologies, machinery, and processes. These industrial districts bore the brunt of the collapse of Fordist manufacturing and ancillary industries over the 1960s, 1970s, and 1980s, and have also comprised in many cases the sites of new industry formation over the last decade and a half or so, a theme to be explicated in the case studies to follow in this volume.

While advanced cities have suffered major contractions of Fordist manufacturing, many have also experienced growth in specialized creative industries, exhibiting a number of the defining features of the classical industrial district first described by Alfred Marshall, as well as the contemporary refinements proposed by scholars writing about the 'new industrial district'. We will conclude this section with reference to two case studies which illustrate the dynamic qualities of industrial agglomerations in the metropolis, which illustrate both durable and more volatile aspects of spatial organization.

Cluster relations in Leipzig's media industry

Harald Bathelt has written extensively about the behaviour of advanced industrial production systems operating in urban-regional space, including Chinese examples as well as case studies in German cities, including Munich's media industry. His work on Leipzig follows in part the traditions of the new industrial district discourse, emphasizing the centrality of social networks and embeddedness of norms and practice, and also brings in 'relational' aspects of cluster behaviour, a theme which occupies a prominent position within contemporary economic geography scholarship.

In Bathelt's account, Leipzig has functioned as an important trade and service city since the medieval times, when it emerged as a principal location for trade fairs, and then as a book publishing centre before the Second World War. By the 1930s, Leipzig's publishing industry comprised more than 300 firms, with 500 allied companies numbering more than 3,000 employees. These firms were concentrated within a *Graphisches Viertel* (Graphical Quarter), which, as Boggs has observed, functioned in the manner of a nineteenth-century industrial district (Boggs 2001).[9]

Leipzig's media economy was greatly compromised, first, by the destruction of the Second World War, and then by its inclusion in the German Democratic Republic, cut off from traditional markets in the much larger and more prosperous West Germany. Leipzig maintained a prominent position within the GDR, but effectively lost its national and international roles, consistent with the internal production orientation and market distribution characteristic of COMECON states.

Bathelt notes that following German Reunification in 1990, Leipzig's book publishing industry 'was not well positioned for market-driven competition' (Bathelt 2005: 112). As an institutional support initiative, the *Förderverein Medienstadt Leipzig* (Development Association of the Media Industry Leipzig), a public-private partnership, promoted the rejuvenation of the Graphisches Viertel as a book publishing site. But this effort largely failed, as the Graphisches Viertel 'was neither able to grow into a centre of traditional media branches nor did it develop into a significant location of the flourishing electronic and new media sector' (ibid.).

This failure notwithstanding, Leipzig's media sector experienced a measure of growth following Reunification, associated with the location of West German operations aspiring to secure contracts in the region, and with the privatization of the television and film industries of the former GDR. Bathelt cites an estimate generated by Bentele *et al.* (2003) based on a postal survey exercise that Leipzig's media cluster, augmented by start-ups and university spin-offs, employed 23,100 permanent employees and 9,700 freelancers, equalling about 15 per cent of the regional labour force, and comprising 750–1,350 media firms, depending on definition and data source.

These data indicate a degree of success in the (re)formation of Leipzig's media cluster, but Bathelt's judgement is that 'this new media industry does not (yet) have the potential to create self-induced growth, involving enhanced processes of learning and knowledge creation' (2005: 120). Specific deficiencies include 'over-embeddedness', characterized by a focus on nearby customers rather than potential clients elsewhere in Germany, coupled with a lack of effort to secure local strategic alliances, creating 'the worst of all possible configurations of localized networking' (ibid.). To remedy these deficiencies, Bathelt urges a regional policy initiative 'directed towards both the generation of local networks to provide opportunities for interactive learning and the formation of trans-local pipelines to secure longer-term growth potentials' (ibid.: 121), suggesting a model of 'induced' rather than 'spontaneous' media industry development.

Changes in graphic design firm agglomeration in Melbourne

Graphic design is widely acknowledged by scholars as a particularly instructive example of creative industry formation in the metropolis. This exemplary value is associated with its intrinsic design functions and links to artists and art work; with its dense inter-industry linkages, notably to printers and publishers; with its role in re-imaging and rebranding, including services to corporate and institutional clients; and with its sensitive locational requirements and tendencies.

Peter Elliott has written an insightful account of the behaviour of graphic design firms in Melbourne, a leading metropolitan centre of cultural production and performance in Australia. More specifically, Elliott offers an analysis of changes in the location, linkage patterns and growth dynamics of graphic design agglomerations within the metropolis, including a mapping exercise showing patterns of change from 1981 to 2001. In the initial phase, graphic design firms were highly concentrated within a single inner city cluster, the South Melbourne Agglomeration, conforming in important respects to the inner city creative industry locational model. Over the following two decades, this initial agglomeration experienced both growth in the number of firms, and a modification of its spatial configuration, but otherwise exhibited the behaviour of what Elliott describes as a 'persistent' agglomeration in the inner city (Figure 3.5).

Over the two decades following this profile of the first agglomeration of graphic artists, Elliott demonstrates that both the number and location of graphic designer agglomerations in Melbourne have evolved. The changing geography of graphic designers in the metropolis includes new formations in the inner suburbs, as well as in the inner city which nurtured the initial cluster, as shown in Figure 3.6. The inner suburban agglomerations were formed in response to new commercial and industrial clients in this zone of the city, as well as to rent and other cost differentials. That said, the expansion of graphic design firms and clusters in the inner city demonstrates the persistent appeal of this territory for creative enterprise in Melbourne. Elliott also cites a wider context for this most recent phase in the development of graphic design firms in Melbourne, including a secular decline in manufacturing within the inner city. In 1949, Melbourne's inner city production sector encompassed 143,000 manufacturing workers (60 per cent of the metropolitan manufacturing employment), a figure that had decreased to approximately 30,000 (12 per cent of the regional total in this sector) in 2001. Elliott notes that services such as 'management consulting, computing and graphic design are reshaping the economic geography of the inner city, often utilizing the spaces left by manufacturing' (2005: 150), a trend which he observes has a parallel in southern Manhattan.

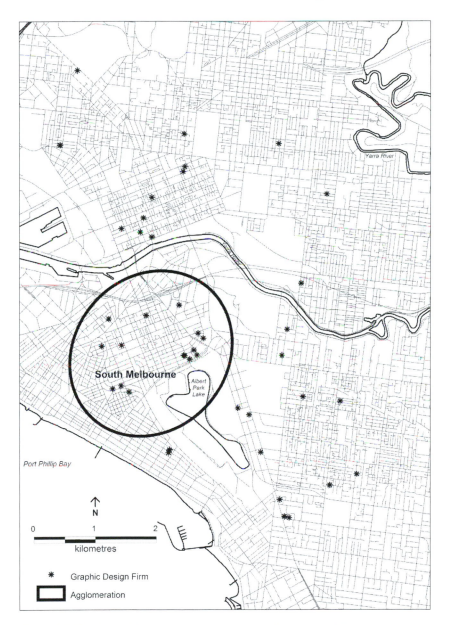

Figure 3.5 The South Melbourne graphic design agglomeration, 1981.

Source: Elliott (2005).

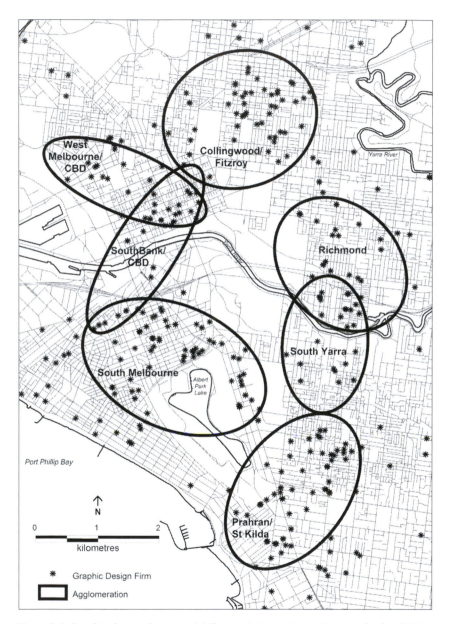

Figure 3.6 Graphic design clusters in Melbourne's inner city and inner suburbs, 2001.
Source: Elliott (2005).

Three narratives of the industrial district in Manhattan

If the compact spaces of the Oltrarno and Hanoi's Ancient Quarter disclose signifiers of globalization and restructuring at a modest scale, and if the evolution of the Leipzig (media) and Melbourne (graphic design) clusters exemplify certain aspects of dynamism in the space-economy of the metropolis, then the multiple production spaces of Manhattan may offer commensurately larger narratives of change. Manhattan's development history includes distinctive story-lines which reflect in different ways New York's world/global city status: two with their origins in the nineteenth century, the garment district and the corporate office complex, and one with a more evanescent late twentieth-century genesis and evolution, the new media firms and dot.coms of 'Silicon Alley'. Each of these is illustrative of larger experiences of urbanism in the metropolis, and has naturally attracted significant scholarly attention, as well as producing potent imageries in the media and public imagination.

Narrative no. 1: the garment district and the 'old industrial economy'

New York is one of the apex global cities, together with London and perhaps Tokyo, underpinned by specialized banking, financial, and head office functions, as well as other advanced intermediate services. But over the nineteenth century (and well into the twentieth) New York experienced significant industrial development in manufacturing, with some parallels to the London industrialization story-line. Like London, New York specialized in light, labour-intensive manufacturing, as opposed to the heavy industries which developed in Chicago and Philadelphia (and in Glasgow, Manchester and Liverpool in Britain).

Again like London (see Chapter 4), New York's chief manufacturing industries included food and beverage production, skilled metalwork and precision trades, and garment manufacturing, typically undertaken within the congeries of workshops and small factories of lower and mid-town Manhattan. Much of this light manufacturing survived into the 1960s in a form recognizable to observers of New York's industrial history, not without contractions and high social costs, as well as periodic downturns and disjuncture, including the Great Depression of the 1930s, and the effects of war on domestic consumption.

But as in London, the 1960s represented a benchmark for two divergent trajectories in Manhattan: the rise of a specialized service economy based on finance and producer services, and the near-collapse of traditional manufacturing. Saskia Sassen has chronicled these consequential (and deeply wrenching) events in her influential portrayal of New York as global city (Sassen [1991] 2001). Since 1960, New York has suffered a loss of over 500,000 manufacturing jobs, cutting total manufacturing employment by more than one-half. Much of this loss occurred early in the restructuring period, but manufacturing employment contracted by 22 per cent in the period 1977–1985, and a further 18 per cent from 1993–1997 (Sassen 1999: 207), demonstrating both the protracted and deep nature of industrial decline in America's largest city.

Continuities and restructuring pressures in the garment district

Norma Rantisi has written an instructive account of New York's women's garment industry, concentrated within a four-block area of Midtown Manhattan bounded to the north by 40th Street, to the south by 34th Street, to the west by Ninth Avenue, and to the east by Fifth Avenue (Figure 3.7).

Despite the massive contractions in New York's manufacturing industry noted

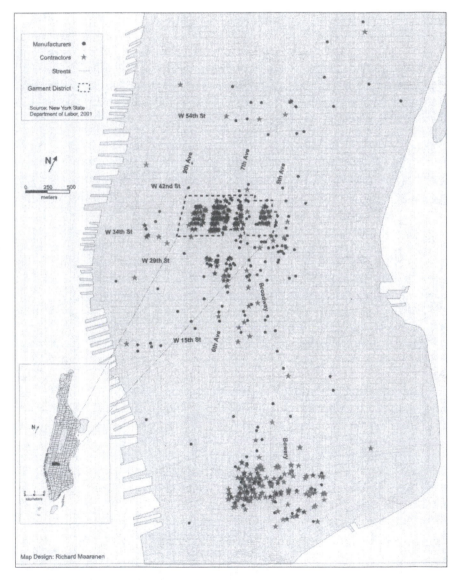

Figure 3.7 Manhattan's garment district, New York.

Source: Rantisi (2002).

above, a substantial portion of Manhattan's garment sector has survived, with an ensemble of approximately 4,000 firms in 2001 (two-thirds of the total number of firms in the area), comprising an integrated network of apparel manufacturers and contractors, as well as retailers, textile suppliers, resident buying offices, forecasting services, trade publications, and complementary business services (Rantisi 2002: 448). Rantisi quotes a senior city planner who described the Manhattan garment district as the densest manufacturing centre in America (ibid.).

Rantisi offers explanations for the relative buoyancy of the women's garment sector of Manhattan, against the general run of manufacturing decline in New York. First, apparel companies derive two principal sets of benefits through co-location in Manhattan's Midtown: 'both *directly*, in the form of a shared production culture (conventions, norms and common expectations), and *indirectly*, through an ability to monitor and track their respective competitors', aided by the key intermediary agencies and institutions in the district cited above (ibid.: 442). These embedded sociocultural norms are illustrative, at a more intensely localized scale, of the 'concrete social relations and structures' characteristic of the classic industrial district in Granovetter's network model, cited above. Second, the garment industry clustered within Manhattan does not represent a homogeneous production sector, in terms of product sectors and market orientation, but rather a highly segmented ensemble of firms, including four categories within the high-end segment (defined, from top to bottom as 'couture', 'designer', 'bridge', and 'better' price points), and a moderate-to-low end (MTL) segment divided into 'moderate' and 'budget' price points. This hierarchical variegation has enabled the Manhattan women's garment industry to cater to increasingly segmented consumer markets, and more particularly to address high-end markets as the lower end producers are undermined by cheaper imported goods. Finally, the saliency of *design* as a competitive advantage, especially within the high-end segment, which positions at least a significant element of the Manhattan industry within the applied design sector, 'underscores the continuous and recursive relationship between product and place associated with the increasing commodification of culture in the new economy' (ibid.: 441).

As a postscript to this account, at one level the women's garment sector in Manhattan seems positioned within the relatively robust *cultural economy* of New York, rather than the *manufacturing economy* of the metropolis firmly ensconced within a trajectory of secular decline over the past four decades. Seventh Avenue alone encompasses major designers including Bill Blass, Liz Claiborne, and Ralph Lauren. That said, there are an increasing number of competitors seeking entry to the cultural economy and trade in cultural products, and this will inexorably place greater competitive pressures on even the most durable and high-integrity industrial districts. The last decade, for example, has seen inroads made by offshore competitors combining low price points with increasing design and fabrication quality into the cultural products sectors and regions in Italy and France, a syndrome experienced earlier in, for example, electronics, textiles and automobiles. Cultural products may enjoy a measure of insulation from this tendency, but it seems unlikely they can enjoy primacy indefinitely though innovation, refinement, or cost-cutting.

Social upgrading and property market pressures

Nor are the competitive pressures faced by Manhattan's fashion industry limited to those of international product markets. The external stresses generated by competitors are accompanied by the tendencies of the local property market, which increasingly favour upscale housing conversions among the industrial land-scapes of mid-town Manhattan. In an article published in the *The New York Times* entitled 'The Shrinking and Fading Garment Center', Joseph Gerber depicts these external and localized pressures as follows:

> The Garment Center, the fabled heart of the city's largest manufacturing industry, continues to shrink, buffeted by cheap labor overseas, rising rents and an exodus of skilled workers and entrepreneurs. The talk in the district is that it is going the way of SoHo or TriBeCa, drawing pioneering apartment-dwellers, artists, actors and white-collar professionals into the cavernous lofts where immigrants sat hunched over pummeling needles to make the nation's clothing.
>
> (Gerber 2004)

The story has many angles, as Gerber tells it, including the historic role of the garment district as a point of entry for many of thousands of immigrants to New York's labour markets and thus an entrée to American urban society. A second is the relentless process of outsourcing which undercuts the complex networks of suppliers and intricate task specializations unique to the garment industry. The overall profile is one of waxing pressures and secular decline. Finally, there is a connection to another of the industrial districts and restructuring experiences of Manhattan, in the form of the New Economy sector to be elucidated below: here Gerber writes of landlords who missed the dot.com boom now advocating the 'condensation' of haute couture firms in a more restricted area, thus freeing up more space for the lucrative loft market, another reminder of the entangled markets and stringent filtering processes of the new inner city.

Narrative no. 2: the corporate complex and the 'global services economy'

The genesis of New York's global-scale specializations (and national primacy) in banking and finance can be traced to the emergence of important regional service centres in nineteenth-century America. During the early post-colonial period a number of rivals prospered, including Boston (with its long-established port functions and trading role, notably with England), Philadelphia, and Baltimore. But the speed and scale of growth in banking and finance shaped the clear ascendancy of New York by mid-century. New York surpassed Boston and Philadelphia in bank-held capital by 1924, and by 1857 was the 'unchallenged financial center of the country' (Abu-Lughod 1999: 38). New York's financial rise was reinforced by the relocation of the New York Merchants Exchange to Wall Street in 1841, presaging the 'professionalization' of stockbrokers as an accredited occupation and the rise of the New York Stock Exchange (NYSE), while in insurance the total capital of institutions insured by New York companies exceeded that of Philadelphia, Baltimore, and Boston combined by 1824 (ibid.).

While the growth and international power of New York's corporate complex have attracted the attention of many scholars, perhaps the most compelling analysis is that provided by Saskia Sassen in her work, *The Global City* (2001). In this second edition she is concerned to address the shifts in global city structure and function from the conditions of her first volume, set in the particular restructuring context of the 1980s. Sassen's point of departure is New York's fiscal crisis of 1975–1976, which saw major job losses as well as the City itself approaching the brink of bankruptcy. In the two decades following this crisis a clear divergence in the fortunes of New York's economy and employment emerged, favouring many (but not all) service categories, accompanied by the steady attrition of manufacturing industries and labour noted above. Sassen notes that in 1950 'manufacturing supplied almost one job in three while services supplied one in seven' in New York (ibid.: 209), underscoring the dimensions of restructuring in the second half of the twentieth century.

Manhattan forms part of New York City, along with the other four boroughs, and is also situated within the larger New York–New Jersey metropolitan region. Manhattan shares a number of characteristics of labour formation with the larger American economy. But for our purposes here the most salient fact is the exceptional concentration of industries and employment within finance, insurance, and real estate (FIRE) and producer services encompassed within Manhattan. If Manhattan's garment district has been positioned as the densest manufacturing district in the United States, we can also acknowledge Manhattan's status as the dominant corporate complex in America, reinforcing the borough's primacy as specialized production site par excellence. Sassen cites as important measures of Manhattan's specialization the fact that 23 per cent of Manhattan's workers in 1996 were employed in FIRE, compared to 17 per cent for the city as a whole (which of course includes Manhattan), and 7 per cent for the country as a whole (ibid.: 207). Just under 11 per cent of Manhattan's workers were employed in business services, and 3.4 per cent in legal services, compared with figures of 8.3 per cent and 2.3 per cent (respectively) in New York as a whole, and 7.1 per cent and 0.9 per cent (respectively) for the US as a whole. Further, New York as a whole increased its share of the New York–New Jersey regional growth in the 1990s, reversing the trends of the 1970s and 1980s, reflecting in large part the power of Manhattan's production economy (ibid.: 209).

The image here is one of Manhattan's sustained strength as a centre of specialized services production, in FIRE, legal, and business services. But to conclude this vignette we must also acknowledge some of the 'stresses and shocks' to which Manhattan's corporate complex is subject. The former include the competition for corporate control, in which New York has experienced attrition in manufacturing head offices especially dating back to the 1960s (see Drennan 1987). Pressures of competition, downsizing, and the intensification of capital have all contributed to lower growth or even decline in some advanced service employment (62 per cent increase in legal services 1977–1985, compared with –2 per cent 1993–1997). But the most spectacular 'shock' of course was the destruction of the Twin Towers of the World Trade Center on 11 September 2001, with the loss of about 25 million square feet of office space,[10] although the earlier WTC attack in 1993 should be acknowledged as a precursor.

Narrative no. 3: Silicon Alley and the New Economy

Our third Manhattan narrative concerns (what proved to be) a more ephemeral experience than the previous two, but one which nonetheless demonstrates New York's saliency as a site of industrial experimentation and innovation. In some important respects the central story-line follows a pattern observed across a diverse range of cities in the volatile New Economy phase. But this reference case also underscores the distinctive interdependencies between the New Economy of the late twentieth century and the workings of the property market, place-marketing, and 'buzz' in the global metropolis.

Manhattan, along with San Francisco's South of Market Area (SOMA) and north-east Mission district, emerged as one of the world's leading New Economy sites during the middle years of the twentieth century's last decade, with 'Silicon Alley' structured as the epicentre of New York's New Economy. Michael Indergaard has written an insightful and compelling monograph (Silicon Alley: The Rise and Fall of a New Media District, 2004) on the astonishing rise and even more precipitous collapse of this new industrial district, situated within more secular processes of industrial restructuring in the metropolis. The narrative is also informed by the operation of New York's property market and players within the development game, the polarizing tendencies of neo-liberal governance and policy values, and of course the trauma of 9/11. As a framing device, Indergaard asserts that the story of Silicon Alley 'requires an account of both place-making and industry-building' (ibid.: 3), underscoring the intersections of economic geography and urban studies incumbent in the evolution of the new industrial district.

At one level, Silicon Alley is fundamentally about people, institutions, and place, rather than a fetishization of the power of technology in the New Economy. The dramatis personae include not just the stars of Manhattan's new media sector, but also representatives of the many ancillary industries and institutions, drawn from the worlds of finance, property development, politics, and marketing. That said, a partial list of the members of the leading cast included Jamie Levy, a pioneer of the Web and social convener for Silicon Alley's new media community; Jason McCabe Calacanis, founder of the *Silicon Alley Reporter*; Josh Harris, who established the Internet studio Pseudo.com; and Craig Kanarick and Jeff Dachis, founders of the webshop Razorfish in 1995. Despite their individual technological savvy and new media acumen, most of the protagonists of the Silicon Alley story 'were liberal arts types rather than programmers – principled slackers, arty punk rockers, and deconstructionists from "good families" (several were grads of Brown's Modern Culture and Media department)' (Indergaard 2004: 1).

Intersections of socioeconomic processes and urban geography

Urban geography is key to the Silicon Alley saga, demonstrating the intersections between 'process' and 'place' acknowledged in Chapter 1. As Indergaard observes, the rise of new media in Lower Manhattan 'was abetted by a larger postindustrial transformation that left the area's physical and human assets unattached or underutilized as of the early 1990s' (Indergaard 2004: 5), conditions similar in many ways to the state of London's City Fringe during the same time. Silicon Alley also featured an 'iconic building' in the form of the Flatiron Building, 'formerly a testament to the pioneering age of skyscrapers . . . recoded as Silicon Alley's preeminent landmark' (ibid.: 4). A feature of the Silicon Alley story as recounted in Indergaard's narrative concerns the overspill effects of innovation in the micro-spaces of the city, and its tendency to contribute to territorial change in the city. The geography of Silicon Alley in its initial phase encompassed an area of Lower Manhattan south from Broadway from the Flatiron district through Greenwich Village and Soho. But as the energy generated from the New Economy firms waxed, '[m]ajor offshoots sprang up in the Financial District and elsewhere in Manhattan; the city government tried to encourage the formation of new satellites in the other boroughs' (ibid.: 4).

Michael Indergaard's engrossing narrative of Manhattan's Silicon Alley suggests that the motive development forces included (in addition to the raw technological features and systems), first, the centrality of social relations in the extension of contacts, collaborations, and deals among an 'ever-widening circle of workers and friends', and, second, a New Economy 'cultural mobilization', a movement which was 'as much about the financialization of the economy as it was about digitalization' (ibid.: 85). Where these forces come together is in the hydridization of relationships, in which 'bosses' can become sponsors, friends, and colleagues, in which lawyers provide legal advice and transform into 'network partners', and in which business service firms and start-ups formed a 'ring of faith' in the expectation of 'stock market riches' (ibid.), perhaps a more opportunistic form of social network associated with earlier versions of the industrial district.

As regards the structure of enterprise in Silicon Alley during its heyday as New Economy exemplar, Indergaard identifies innovative business models, as well as new technologies and occupations, situated within the following typology of practice and agency. According to Indergaard (ibid.: 49), these are:

> *Web services* (e.g. Agency.com, RareMedium, Razorfish): Many web design shops came to call themselves 'agencies' – a model derived from the advertising industry. They typically did contract work for corporations, such as designing splashy websites, developing banner ads, and planning online ad campaigns.

Advertising networks (e.g. 24/7, DoubleClick): These firms assembled and profiled groups of websites so as to offer advertisers a selection of sites by their demographic profiles.

Community networks (e.g. iVillage, StarMedia): on-line communities organized networks of sites according to some theme such as ethnic or women's interests.

E-commerce firms (e.g. Alloy, Barnesandnoble.com, Bluefly): Online retailers sold goods to consumers (e.g. books, CDs, or clothing), and typically tried to get themselves linked to clusters of websites such as online communities or portals where large numbers of users entered the web (e.g. browsers).

The novelty of these practices was accompanied by extraordinary growth in enterprises and employment. Michael Indergaard cites the measures of New Economy growth in the city from a PricewaterhouseCoopers study published in 2000, which showed new media firms increasing from 2,601 in 1997 to 3,831 in 1999, an expansion of 47.3 per cent, and the new media workforce growing from 55,973 to 138,258 over the same period – a startling 147 per cent in two years.

The technology crash and 9/11: the end of the New Economy saga

The central causes of the New Economy crash are well known, including oversupply, unsustainable demand, and grossly inflated nominal company stock levels that bore little relationship to real market value. But in New York there were one or two distinctive features. As Indergaard notes, the AOL and Time Warner merger announced in January 2000, seen initially as the harbinger of new synergies between old-line and New Economy communications, proved instead to be the most dramatic precursor of disaster, a reaffirmation that the old rules of business practice concerning value, markets, and 'fit' might still apply. In parsing the meaning of the merger, Indergaard underscores the following anomaly:

> The New Economy's biggest deal bore a big contradiction: One of its champions was taking over an old media power; yet, its leaders had agreed that its currency was worth 50 per cent less than that of the old media conglomerate. Was AOL conquering new domains or cashing in while it could?
>
> (Indergaard 2004: 134)

What ensued in New York was not so much a 'crash' on the lines of which occurred in some other places, but rather a 'slow erosion' of the New Economy, with Silicon Alley exhibiting a 'resilient quality' lacking in some of the other 'national contenders' (ibid.).[11] But the attack on the Twin Towers of the World Trade Center dealt a terrible blow to the New York economy, including many of the Silicon Alley survivors linked to Wall Street through subcontracting arrangements and partnerships as well as social networks. As we have seen, the world metropolis encompasses multiple districts of specialized production, which at a number of levels occupy discrete urban space, have their provenance in different historical periods, and exhibit contrasts in industrial structure, labour and task specializations, and in product sectors. It is equally clear that these otherwise quite different sites share common developmental attributes, including localized production networks, embedded social norms and values, and constant pursuit of innovation. But as our vignettes disclosed, urban production sites also share in some measure vulnerability to shocks and stresses, including competition, the vagaries of property markets, and disastrous events and episodes.[12]

Conclusion: from 'new industrial district' to 'cultural production quarter'

As we have just seen in the preceding digest of selective cities and sites, it would appear that the idea of a revival of the inner city industrial district, some decades following the collapse of traditional manufacturing, constitutes a tenable thesis, although the experience is replete with major swings of fortune, in contrast to the ideal of the stable industrial district epitomized in Granovetter's writing. As a recent evocation of the conceptual implications of these contemporary inner city districts, Peter Hall's model of the 'Polycentric City' includes among the key sectors 'creative and cultural industries', as well as the long-established core industries of finance and business services, 'power and influence', and 'tourism' (Figure 3.8). As a means of elaborating upon Hall's creative and cultural industries' model, we can also discern increasing differentiation of scale and specific industrial and enterprise structure, as shown in Table 3.1. The proliferation of cities and sites encompassed in this typology testifies to the global dimensions of new industry formation, and to the intricate rearticulation of industries and firms produced by the complex interdependencies described in Chapter 2.

The contemporary inner city new industry site functions in large part as a zone of experimentation, creativity, and innovation in the metropolis, rather than as contemporary evocations of the durable Marshallian industrial district of the last century. Further, this repositioning of the inner city as a territorial innovation system can readily be identified not just among western cities such as New York, London, and Paris, but (as noted in Table 3.1) also increasingly among the leading cities of the Asian growth economies, notably Tokyo, Seoul, Shanghai,[13] and Singapore.[14]

As observed, there are in most cases significant contrasts between the reconstituted industrial district of the early twenty-first century and those of previous eras, notably in the volatility of industry formation and mix of firms, in divisions of production labour, and in the influence of public policies, among other factors. But the emergence of the contemporary industrial district tends to follow certain pervasive or structural patterns, shaped by new rounds of industrial restructuring, global processes, and the 'power of example' which has encouraged increasing levels of inducement and mimicry of well-known or

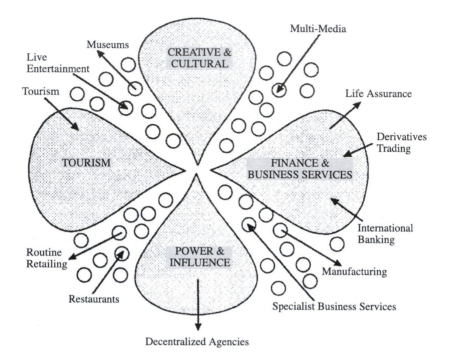

Figure 3.8 Major service clusters in the polycentric global city.
Source: Hall (2006).

iconic sites. Changes in ideology, governance, and spatial planning also act to reshape the urban space-economy more along lines of market-driven societies, as evidenced in Thomas Ott's (2001) work on the reshaping of urban structure in 'post-socialist' German cities, in Fulong Wu and Anthony Yeh's (1999) analysis of urban structure in the 'transitional economy' of Guangzhou, and in the Hanoi example documented separately by Turner and Surborg presented earlier in this chapter.

The discussion presented in this chapter offers a selective rather than comprehensive review of the evolution of the industrial district, including reference to the sometimes tumultuous history of inner city sites. There are also many instances of 'ruptures' in bundles of regional specialization, as the history of media clusters in Leipzig described by Bathelt attests. That said, the reassertion of specialized industrial districts within the core areas of the metropolis suggests some important continuities, as well as signifying departures from older models. The contemporary industrial district of the twenty-first century stills functions as a spatial construct for organizing specialized production and labour, much as it always has. Inner city production firms maintain supply linkages with companies on the periphery of the region, relying upon (in some cases) courier deliveries, as well as the deployment of digital means of sourcing inputs (information, labour)

Table 3.1 New territorial forms of industrial production in the inner city

A. *Typology of spaces*	B. *Representative sites and cities*	
I. *Extensive New Production Districts*		
• concentrated and dispersed patterns of new industries • integrated production ensembles • diverse activities and land use patterns	1. 'City Fringe' (Clerkenwell, Bunhill, Shoreditch)	London
	2. 'Multimedia Gulch' (South of Market Area)	San Francisco
	3. Chinatown Heritage Area	Singapore
II. *Compact New Economy Clusters*		
1. *'Spontaneous' Clusters*		
• essentially market-driven clusters of New Economy firms	1. Victory Square – Gastown	Vancouver
	2. Soho	London
	3. Shibuya	Tokyo
	4. Suzhou Creek	Shanghai
	5. Silicon Alley	New York
2. *'Induced' Clusters*		
• central role for public policy (rezoning, land use, equity role in property/ buildings, heritage policies)	1. Cite Multimedia	Montreal
	2. False Creek Flats	Vancouver
	3. Design Exchange site	Toronto
	4. 22 @ Project	Barcelona
	5. Far East Square	Singapore
3. *'Signifying' New Economy Precincts and Cultural Quarters*		
• typically 0.5–1 kilometre square in extent • concentrations of leading-edge firms • environmental and consumption amenities • cultural markers and 're-imaging' effects • leading role in 'reterritorialization' processes within the inner city	1. Yaletown	Vancouver
	2. Hoxton	London
	3. South Park	San Francisco
	4. Telok Ayer	Singapore
	5. Belltown	Seattle
	6. Ehrenfeld	Cologne
	7. Temple Bar	Dublin
	8. Wicker Park	Chicago
	9. Liberty Village	Toronto
III. *'Incipient' New Industry Districts and Sites*		
• early forms of transition evident ('pioneer' gentrifiers and New Economy firms)	1. Deutz	Cologne
	2. Mid-Main District	Vancouver
	3. Little India	Singapore

Source: adapted from Hutton (2004a).

internationally, but they also enjoy the benefits of proximity to complementary enterprises within the compact inner city district. There is still a persistent 'local' in the production systems of specialized firms in the city, as well as reliance upon more spatially extensive linkage systems (including outsourcing, international labour recruitment, and global information networks) required to maintain a competitive edge.

4 Restructuring narratives in the global metropolis

From postindustrial to 'new industrial' in London

Introduction: transformative change in the global city

London's unique saliency within the history of urbanization is attributable in large measure both to its early development as a world city in the age of empire, and to its contemporary status as one of a handful of legitimate global cities.[1] London can also be positioned at the global vanguard of transformative urban change in the modern era. A vast and influential research literature underscores London's significance as locus for a series of restructuring events, marked by wrenching social costs, urban policy crises, and deep theoretical implications (Fainstein and Harloe 2000).

London's record of transformative change and policy innovation is complex and multi-faceted, associated with interdependencies of scale, specialization and primacy, the magnitude of industrial restructuring and recessionary oscillations, and the complex governance relationships between London and the central government in Westminster. Benchmark events and processes in the modern era of London's development include the following: (1) a context of metropolitan expansion and industrial development in the interwar period, inspiring strategic plans for the containment of the broader London region, including postwar social and physical reconstruction programs which have been widely emulated internationally (Hall *et al.* 1973); (2) the decline and eventual collapse of traditional manufacturing and ancillary industries from the 1960s to the early 1990s, impelled both by market forces and macroeconomic policy factors (Thornley 1992); (3) the early onset of gentrification in London's inner city, first reported by Ruth Glass in 1963, associated with the effects of a rent gap in central city housing markets, and more exigently with the changing fortunes of social classes in the metropolis (Hamnett 1991); (4) a truncated experiment in metropolitan government, initiated by the establishment of the Greater London Council (GLC) in 1965, and terminated by the abolition of the GLC by the Conservative Government of Margaret Thatcher in 1986 (Hall 1998); (5) the expansion of banking, finance, and producer services which underpinned London's ascendancy as a first-order global city in the 1980s, accelerated by the Thatcher Government's monetarist policies and the deregulation of financial markets in 1986 (Sassen [1991] 2001); and (6) growth in international immigration from diverse overseas societies, shaping London's development as a global-transnational metropolis, a movement which has included local conflicts and tensions (Jacobs 1996), but which has also contributed powerfully to London's cultural energy, global connectivity, and new narratives of contemporary urbanism (Butler 2003).[2]

Each phase of London's transformation, shaped by a fluid mélange of market, social, and policy forces, has produced substantial dislocation and displacement, only partially mitigated by policy and local planning efforts and by welfare supports. In particular, the working-class communities of East London (and in Bermondsey and other districts south of the Thames) were eviscerated by the collapse of the inner city industrial sector over the second half of the twentieth century, and have been subject to the dual transformative influences of gentrification and the settlement of new immigrant groups. But the postwar period has also demonstrated London's capacity to secure new economic vocations following major restructuring events; to accommodate a new urban social morphology characterized by increasing diversity and complexity; and to cope with major shifts both in the nature of political control at Westminster and in the structures and systems of regional and local governance.

London's government structure includes an elected Mayor and Greater London Assembly, as well as 32 London Boroughs (each with a Leader, Council, and staff), and the City of London Corporation (see Figure 4.1 for a map showing the London boroughs). The political forces shaping London's development have been manifestly influenced by the ideologies of national governments, including, over the past decade, changing policy values of Labour politicians at both central and local government levels, who combine commitments to sustaining London's global status with progressive practices in housing, social policy, and transportation. Indeed, the reproduction of twenty-first-century London, including the Millennium Dome (Thornley 2000), the 2012 Olympics, and other megaprojects, constitutes a prime exhibition piece of the transition of the Parliamentary Labour Party (PLP) from a commitment to socialist ideology, to the more artfully blended policy practices of 'New Labour' under the aegis of Tony Blair and Gordon Brown, and their sometimes uneasy cohabitation with Ken Livingstone's faction within the Greater London Authority (GLA), just as the development of the global financial centre at Canary Wharf represents a totemic project of the Conservative Government of Margaret Thatcher and Michael Heseltine (Hall 1998).[3]

Prior to the Thatcher regime commencing in 1979, both the national government and the GLC exercised a strongly regulatory approach to London's development, with the capital's relatively high growth interpreted both as a fundamental regional policy problem at the national level, as well as the source of increasing negative externalities (congestion, property and rent inflation, displacement) within the London region itself. Vestiges of this posture remain, but the dominant political discourses at Westminster and the Greater London Authority now emphasize policies supportive of London's key economic roles at the national, European, and global scales, and, increasingly, the use of pricing and other market-based forms of allocation rather than the more stringent development controls of the pre-Thatcher era.[4]

Figure 4.1 Greater London and the London Boroughs.

Source: Clayton (1964).

New economic spaces in twenty-first-century London

The evolution of London's metropolitan space-economy represents a critical dimension of its industrial restructuring processes and more comprehensive socioeconomic, cultural, and physical transformation. The broad spatial contours of London's restructuring record since the 1960s conform to the basic postindustrial model of inner city industrial decline and the concomitant expansion of the central city corporate office complex (Figure 4.2), although the scale of change in each of these key terrains in the London case far eclipses the experience of all but a few cities. The socioeconomic dimensions of London's inner city industrial collapse, and the spectacular rise of the City's banking and financial sector, have been articulated in scholarly, literary, and popular forms, including film and television.[5]

The spatial organization of London's economy has of course always been more empirically complex and variegated than this bipartite model implies (Simmie 1985). We can reference here the establishment of manufacturing within the inner north-west of London in the 1930s, and the industrial development in Outer London following the Second World War (Hall 1962b), as well as the myriad retail, personal services, and public sector activities concentrated in the metropolitan core but strongly represented throughout Greater London. The

Figure 4.2 'Heavy metal': Lloyds Bank, City of London.

economy of the large metropolis is typified after all not only by intensive specialization but also by its unique diversity, supported by local consumption and final
demand at all points across the income spectrum, and by very substantial tourist
and visitor markets. The successful revitalization (or adaptive re-use) of Covent
Garden, Spitalfields, and the Borough Market, as well as the renewal of numerous
smaller and more localized markets, gives testimony to this effect, as does the long-
running attraction of Oxford Street and other West End retail precincts both for
visitors and for Londoners. The consumption sector assuredly comprises a large
share of the contemporary metropolitan economy, a key sector of constituent
industries and labour in its own right (Glaeser *et al.* 2001). Consumption industries
also perform essential roles in the expansion of amenity-seeking activities (such as
New Economy firms and labour; Mugerauer 2000), in the expression of new urban
lifestyles and identities (Ley 2003), and in the reproduction (or recovery) of the
inner city as site of urban spectacle and theatricality (Hutton 2006).

The reconfiguration of London's space-economy at the strategic level includes the emergence of new industrial clusters and ensembles, moulded by global as well as domestic (national and local-regional) processes. The City of London, though spatially constrained, continues to experience growth in specialized financial and business services, enabled by strategies encouraging taller buildings and higher densities. London's second established commercial centre, the City of Westminster (and parts of the adjacent London Borough of Camden) encompasses important concentrations of creative industries, including mature industries and labour cohorts (for example, film and video production, graphic design, and advertising in Soho; 'boutique' hedge fund traders in Mayfair), as well as New Economy activities such as computer graphics and imaging, software design and Internet services, and specializations associated with established sectors (for example, special effects enterprises linked to the film industry). The continuing expansion of the Canary Wharf financial district, situated in Docklands, provides London with a second major international business cluster (Figure 4.3), while the development of industries adjacent to Heathrow (including business hotels and offices as well as freight forwarders, air courier services, and specialized, high-value manufacturing) constitutes an additional regional growth pole within Greater London. Other strategic development sites in London include the major cultural enclave of Bankside in Southwark, and the long-delayed King's Cross mega-project, which will transform an obsolescent and vestigial landscape of London's N1 district into a mixed-use, medium- and high-density residential community

Figure 4.3 'Manhattan on the Thames': Canary Wharf, London Docklands.

and business centre, including high-speed train services linking London with Europe. As a final example, the large-scale redevelopment of Stratford in the East End for the 2012 Olympics represents the latest episode in the production of global space in the metropolis.

The historic bipolar spatiality of London's economy, concentrated in the cities of London and Westminster, has thus been supplanted since the 1980s by an increasingly polynucleated structure, although the density of firms and wealth-creating capacity (sales, profits, incomes, and revenues) of the City of London is without equal anywhere in the UK or Europe. These development sites exemplify the relayering of capital in London, and contribute to new spatial divisions of labour in the metropolis.

Industrial restructuring and the 'new inner city'

The evolving space-economy of Greater London includes not only the strategic-scale projects and development sites noted above, but also a new chapter in the long-running saga of industrial urbanism in the metropolis: the reassertion of industrial production in London's inner city. A sequence of restructuring pro-cesses and episodes, following the late twentieth-century collapse of traditional manufacturing, has been inscribed within the postindustrial landscapes of the metropolis. Further, industrial innovation and restructuring in London's inner city include increasingly intense and multifaceted relations between enterprises and the larger social, cultural, spatial, and physical transformations of the global metropolis.

The experience of industrialization within London's inner city contributes to the increasing organizational and spatial complexity of production in Greater London. As observed in an annual report of the GLA's Economics Division, the space-economy of London's central area has recently spread from the City proper 'southwards across the River into riverside Southwark and Lambeth and east-wards into Tower Hamlets' (GLA 2006: 23). An evaluation of industrial pro-cesses and outcomes within the new economic spaces of inner London evokes the rich empirical dimensions of industrial redevelopment in the metropolis. Further, industries in London's inner city operate not in a context of pure economic space comprised of firms, production systems, and markets, but instead interact in highly complex ways with consumption industries, housing markets, and public agencies, as well as with non-governmental organizations (NGOs) and com-munity-based organizations (CBOs). The location patterns and growth of nas-cent industries of London's inner city are also shaped in part by the distinctive spatiality, landscapes, and built environment of these districts.[6] These interactions implicate ascendant industries in the production of the 'new inner city', contribut-ing to the reconfiguration of the postindustrial urban core with respect to indus-trial regimes and the urban space-economy, product sectors, divisions of labour, and social class.

Chapter structure: processes of multiscalar change

Following this introduction, a context-setting discussion of shifts in London's positioning within national, European and global settings will be presented, including illustrations of the interdependencies between multiscalar processes of industrial restructuring which influence the reshaping of space in the metropolis. Next, the chapter offers an essay on the changing nature of industrialization experiences in London's inner city, including developmental continuities and discontinuities, the relational geographies of specialized production in inner London (see Chapter 2), and the evolution of the inner city industrial district (Chapter 3). Key reference points include the important scholarly treatments of East London's distinctive industrial districts, notably industry and product sector specializations at the localized level, elucidated by Peter Hall, J.E. Martin, and Allen Scott, among others. The narrative includes expert opinion concerning the apparent robustness of this important industrial economy, expressed just on the cusp of its comprehensive collapse, as well as a rehearsal of factors implicated in the deindustrialization of London's inner city. The chapter concludes with a new story-line concerning the reassertion of production in London's inner city, including its uniquely rich mix of production regimes, industries, and labour.

Dynamics of London's repositioning within global, European, and national contexts

London's status as a world city is underpinned by its primacy in specialized economic functions at the national, European, and global scales. However, there have been significant fluctuations in the course of London's positioning within systems of production, trade, and labour formation at different spatial scales. Further, sequences of industrial restructuring influence not only shifts in global urban hierarchies (Sassen [1991] 2001), but also the reconfiguration of space within the metropolis.

Evolving global city discourses

London has occupied a prominent position in the evolving scholarship on world cities (Hall 1966) and global cities and city-regions (Friedmann and Wolff 1982). Research orientations and discourses within the genre encompass of course certain constants, including notions of primacy, hegemony, and associated normative implications, but have also mutated in response to contestation, critiques, and new viewpoints (Fainstein *et al.* 1992).

In important ways London is the archetypical global city, defined by its early provenance as a world city in the era of colonialism and empire, and its contemporary global projection. New York shares with London the peak position within the global urban hierarchy, but is embedded within the world's largest national economy, and also has significant American competitors (notably

Chicago and Los Angeles) in all but the most specialized financial functions.[7] As Janet Abu-Lughod has observed:

> In comparison with the other global cities . . . New York operates under some important disadvantages. Situated within a nation far vaster than England or Japan, it must compete with other major American cities for primacy in 'postindustrial' producers' services, while internationally it must compete with other world cities for financial and business-services supremacy.
>
> (1999: 289)

Tokyo, conventionally included within the triumvirate of first-order global cities, is tightly integrated within the world's second ranking industrial economy, and its international financial and corporate functions are largely directed to regional (e.g. Pacific Asian) markets. As is well known, Tokyo has also suffered a major blow to its development momentum arising from the liquidity problems of its domiciled financial sector and other corporations dating from the late 1980s, and a related collapse of the commercial property market, although there have been signs of a recovery in the early years of the twenty-first century. Tokyo is still the primary city-region of the Asia-Pacific, but is facing competition from Shanghai, Seoul, Hong Kong, and Los Angeles, and is arguably not quite at the same peak level of global city status as London and New York.[8]

London's global (or world) city position rests both on a unique historical legacy of empire and hyper-specialization in critical financial functions, concentrated within the privileged confines of the square mile of the City of London, including foreign exchange, equities, derivatives, hedge funds, and brokerage. According to a recent City of London report, London's international competitive advantages include:

> Its highly skilled and flexible labour force; its ideally located time zone; its native tongue being English, the language of business; the relatively low levels of corporate and personal taxation in the UK; the availability of low cost communications technology; and the effective but not overly onerous regulatory environment.
>
> (City of London 2004: 1)

These competitive advantages are, of course, offset to an extent by more problematic factors, including the ferociously expensive London housing market, and the unreliable (and on occasion dangerous) London public transportation system.

To be sure, many of the merchant banks, insurance firms, and other specialized financial services which cluster within the City and Docklands are now (wholly or jointly) owned by offshore interests, as is an increasing proportion of office space in the City. But this foreign ownership has not to date compromised London's pre-eminence, and in some respects even serves to underscore its global status, as it demonstrates that for some specialized financial services a London address is essential to sustaining a competitive market positioning and elite corporate identity.[9]

In support of this thesis, P. J. Taylor's research addressing the functions and rankings of cities as nodal points within economic, cultural, political, and social networks indicates that London and New York are 'clearly above all others' as principal nodal cities within multiple networks, emerging 'as the most important "all round" global contributors' (Taylor 2004: 5). According to Taylor, London and New York are 'well-rounded' global cities, while Los Angeles, Paris, and San Francisco rank somewhat lower and are imbued with 'cultural bias' (presumably a deeper domestic cultural embeddedness), and Amsterdam, Boston, Chicago, Madrid, Milan, Moscow, and Toronto are classified in the Taylorian taxonomy as 'incipient' global cities. At the same time, Tokyo is positioned as a 'global niche city', along with the much smaller (but more fully internationalized) Asia-Pacific cities of Hong Kong and Singapore (ibid.). This analysis may understate the underlying strength of Tokyo, but serves to emphasize the multifaceted nature of London's claim to the highest levels of global city positioning within evolving discourses of primacy and projection.[10]

International immigration is widely acknowledged as a motive force for the development of 'globalizing cities' (after Taylor), and is powerfully linked to the concept of the 'transnational city' or city-region proposed by Michael Peter Smith (2001). Clearly, *transnational corporations* represent an essential feature of London's global positioning, but *transnational populations* and their constituent stocks of social, cultural, economic, and intellectual capital also constitute defining attributes of the twenty-first-century world city. Historically, London has benefited from the talents, entrepreneurship, and connections of immigrants dating from its earliest period as centre of empire and global trade. These contributions include, to illustrate, Dutch and German immigrant financiers in the seventeenth century who helped shape the City of London's specialized mercantile and banking roles; Huguenot and Jewish refugees in the eighteenth century, cohorts which included highly skilled artists, artisans, and craft workers; and waves of Irish immigrants in the nineteenth and twentieth centuries, most of whom initially worked as labourers in the lowest paid occupations, but many eventually attaining more rewarding work in the expanding economy of the metropolis.

The current period has been nominated as the 'age of international immigration' as a consequence of the unprecedented scale and diversity of migrant flows, and London has been one of the principal beneficiaries. To be sure, international immigration is a significant growth factor for many British and European cities, but London's experience is distinctive, with the metropolis taking in about one-third of the British total (with the South-East Region as a whole accounting for approximately one-half of the UK total). Aside from the scale issue, London's immigrants comprise (relative to most European cities) an exceptionally diverse ethno-cultural mix, enhancing London's 'productive diversity' and its role as centre of inter-cultural production in the arts, design professions, and business. London is also thoroughly multi-cultural in the constitution of its class structure and social morphology, relative to the more segregated American experience.

London in the European context

London is the fastest-growing large metropolis in Europe, following a long period of decline in the 1960s, 1970s, and 1980s, although in some important measures of development (e.g. incomes, GDP per capita) the British capital is only a mid-table European performer (Morgan 2006). Like most major European cities, London is overwhelmingly (and increasingly) a 'service city', following the collapse of traditional industrial production in the late twentieth century. London is clearly further along the pathway of advanced tertiarization than most of its continental counterparts, with less than one-tenth of its employment in manufacturing (Hamnett 2003), in contrast to other European cities, such as Stuttgart, Munich, Turin, Milan, and Barcelona, which sustain a significant (though shrinking) manufacturing base.

London's economic relationships with European cities are in some respects mutually beneficial. The City of London report cited above suggested that the City's contribution to European Union GDP was €31 billion, and '[w]ithout London, the EU would lose 18 per cent of its City-type business to competitors elsewhere in the world and a further 12 per cent would be lost altogether as higher costs make some transactions unviable' (City of London 2004: 2) The specialized banking, financial, and business services clustered in London also contribute to the productivity of European enterprises.[11]

The scale and velocity of physical redevelopment represent salient points of contrast between London and major European cities. In general, European cities have been growing much more slowly than those in North America and Asia, so demographic pressures for development are relatively low. Further, many of the continent's principal cities favour strongly preservationist policies, especially in the urban core, while carefully managing growth and development in the interests of maintaining the heritage quality of the built environment: here we can cite examples such as Rome, Florence, Vienna, Salzburg, and Munich. In other cases, notably Paris and Amsterdam, corporate offices have been accommodated in sites located some distance from the historic city core.[12]

London, on the other hand, has elected to encourage high levels of development, as a means of supporting its global financial and business roles, as demonstrated in the expansion of Canary Wharf and in new office towers for the City and other central districts, and also as a means of promoting social goals, observed in residential development. Examples of high-impact development projects in London abound, notably the 2012 Olympic sites, but there is also a pattern of redevelopment at the more localized scale which demonstrates the comprehensive nature of urban regeneration within the metropolis relative to most continental cities. In his history of Europe in the postwar period, Tony Judt offers the following commentary on London's development pathway over the 1990s:

> Despite keeping its distance from the Euro zone, the British capital was now the unchallenged financial capital of the continent and had taken on a

glitzy, high-tech energy that made other European cities seem dowdy and middle-aged. Crowded with young professionals and much more open to the ebb and flow of cosmopolitan cultures and languages than other European capitals, London at the end of the twentieth century appeared to have recovered its Swinging Sixties sheen – opportunistically embodied in the Blairites re-branding of their country as 'Cool Britannia'.

(2005: 755)

London's growth and development are acknowledged as defining features of the New Labour regime, key to the reconstruction of Britain's political values, as well as to shaping a more dynamic national imagery. Among major European cities, perhaps only Berlin (and possibly Barcelona) are comparable to London with regard to levels of redevelopment and reconstruction, and to openness to experimentation in architecture and urban design.[13]

London in its national setting

London is the largest British city-region, and its national primacy constitutes a key dimension of its growth and change. For much of the postwar period, however, London experienced year-on-year declines in population and employment, and more particularly (since the 1960s) a secular contraction of manufacturing capacity and employment. As Saskia Sassen observed in her seminal study of the global city phenomenon, London suffered a contraction of 800,000 manufacturing jobs in the quarter century following 1960s, 'in a city that was once an important center for light manufacturing' (Sassen [1991] 2001: 209–210). London's share of UK employment fell steadily from 17.4 per cent in 1971 to 14 per cent in 1992, punctuated by a recessionary period 1990–1992 which cost some 100,000 jobs in the City. At least one study indicated a secular decline in London's agglomeration economies (Crampton and Evans 1992). But London has since increased its share of UK employment to about 15 per cent by 2004, and as noted in a report published by the Corporation of the City of London, 'the turnaround since ... [1992] ... has been remarkable, and has transformed London's place in the UK economy' (Oxford Economic Forecasting 2006: 8).

While market factors are commonly referenced as crucial to London's economic revival, there is also a political context which forms an essential part of the new urban-regional development conditions in the UK. The broad momentum of regional policy in the postwar period, with the Barlow Report as the benchmark policy statement, aimed at suppressing London's development, initially to reduce Britain's vulnerability in time of war (in terms of over-concentration of industry, and proximity to continental enemies), and then to allow other regions a larger share of national development. This restrictive policy was followed in spirit at least by central governments until the deregulation initiatives and rolling back of regional and local policy powers by the Thatcher Government, tacitly endorsed by successor administrations, which, combined with the devolution of governance, 'gave London a freedom and a political voice on the national stage, which have unshackled the London economy in national and European terms' (K. Morgan, personal communication, 2007).

The *State of the English Cities* report, released in 2006 by the Office of the Deputy Prime Minister, noted that the population of England has been growing at a faster rate than at any time since the 1970s. Between 1991 and 1997 the resurgence in growth in the south and the east, coupled with declines in growth rates for the north and the west, underpinned processes of inter-regional divergence in England, while a continuation of these trends produced in the early years of the twenty-first century 'the widest regional differential of the 22-year period' (Office of the Deputy Prime Minister 2006: 35). These regional measures of growth are significant to an understanding of London's role in the national context, as there are multiple connections between London and the southern and eastern regions of England, with respect to labour markets and commuting patterns, growth spillovers, knowledge transfers, and inter-industry and inter-firm input–output linkages. London's growth, economic specialization, global projection, and transnational urbanism are the cardinal factors in the ascendancy of the south and east of England since the early 1990s.

The *State of the English Cities* report notes that

> while the proportion of England's population living in London has grown at an increasing rate, elsewhere the picture is overall population deconcentration, with higher growth rates as one moves down the urban hierarchy from large to small cities and to towns and rural areas.
>
> (Office of the Deputy Prime Minister 2006: 27)

Thus, London's recent strong growth record, underpinned by growth in financial, business and cultural services, runs counter to the declines experienced in most other large English urban areas, although Manchester and Glasgow have each experienced a selective revival, based on cultural development, business services, and higher education. A report by Oxford Economics for the City of London disclosed that London has (relative to the UK as a whole) a greater specialization in financial and business services, part of the 'power sector' among cities at the peak of the global urban hierarchy, and less dependency on manufacturing, a sector in secular decline since the 1960s (Table 4.1).

London performs important national roles, most notably that of Britain's capital city, as well as head office, transportation and distribution functions. As one measure of London's relationship with the rest of the UK, London maintains a positive 'balance of trade' with other regions, with the value of intra-UK 'exports' (in 2004) equalling £125.3 billion (with financial and business services accounting for about one-half the total, and manufacturing and wholesale and retail trade following), while 'importing' goods and services to the value of £110.4 billion from other UK regions (Oxford Economic Forecasting 2006: 14, 20).

But London's urban scale, degree of economic specialization, magnitude of primacy, and recent growth rate set it apart from normal considerations of national urban system relationships. In addition to its concentrations of high-

Table 4.1 Change in London's jobs by sector, 1971–2001

	1971	1981	1991	2001
Manufacturing	22.5	16.2	9.3	6.6
Other production (including construction)	7.6	7.0	6.3	5.1
Distribution and hotels	19.7	20.7	20.5	21.0
Transport and communications	10.9	10.1	8.6	8.0
Financial and business services	15.9	19.1	27.2	33.1
Non-market and personal services	23.1	26.6	27.8	26.2
UK Shares				
Manufacturing	30.5	23.6	17.4	13.7
Other production (including construction)	12.9	12.0	10.7	8.6
Distribution and hotels	19.4	21.4	22.5	23.2
Transport and communications	6.9	6.4	5.9	6.2
Financial and business services	9.0	11.3	15.6	19.3
Non-market and personal services	20.3	24.4	27.1	28.5

Note: (% of total London jobs).

Source: Oxford Economic Forecasting/City of London Corporation (2005).

Table 4.2 Recent changes in London's jobs by sector

	2001	2005
Manufacturing	6.6	5.4
Other production (including construction)	5.1	5.5
Distribution and hotels	21.0	21.2
Transport and communications	8.0	7.5
Financial and business services	33.1	32.1
Non-market and personal services	26.2	28.3

Note: (% of all jobs in London).

Source: Oxford Economic Forecasting/City of London Corporation (2005).

value intermediary banking, corporate head offices, and multinational enterprises, the London–South-East region encompasses a disproportionate share of Britain's growth industries, and scores at the high end of the rankings of national wages and earnings, industrial productivity, new business formation, and innovation. Employment in manufacturing continues to decline (Table 4.2), underscoring London's specialization in services. The broader London region also enjoys far better global 'connectivity' than other British city-regions, measured by the quality of international air services, telecommunications systems, and diasporic networks. And, while Britain as a whole can boast of a distinctive international identity, the London 'brand' has a resonance of its own, presenting a unique metropolitan imagery: a mélange of power signifiers incorporating resonances of empire, royalty, and aristocracy; the influence of financial, business, political, and cultural elites; and the narrative value of iconic landscape features. Arguably, London has in some important ways outgrown its national setting.[14]

Multiscalar processes and the reproduction of space in the metropolis

It seems clear enough that London's overall development trajectory is recurrently shaped by a mixture of processes and flows, which in turn underpin new rounds of industrial restructuring and divisions of labour. These multiscalar processes also shape in complex ways the reproduction of economic space within the metropolis. The spatial imprints of London's global city functions can be observed in the robust clustering of specialized finance and business services in the City, and in the formation of new advanced services agglomerations in Canary Wharf (Figure 4.4), while the construction of sites for the 'hallmark event' of the 2012 Olympics represents a new episode of globalization.

Second, the effects of these global processes and events are powerfully reinforced by London's increasing primacy at the national level. Over twenty years ago, at the end of London's postwar decline and at the advent of its ascendancy, Doreen Massey advanced a thesis of an emerging 'spatial division of labour' at the national level which would clearly favour the growth industries and specialized services of London and its satellites in the south of England, portending an era of increasing inter-regional divergence in a small unitary state, with profound normative consequences (Massey 1984). A recent paper by Kevin Morgan endorses the basic tenets of Massey's thesis, identifies the costs of increasing divergence and disparities, and calls for a new regional policy approach which addresses 'territorial justice and the north-south divide' (Morgan 2006: 29).

Third, a new set of processes of change in the 'post-postindustrial' period – the technological-deepening legacies of the 'New Economy', the creative industries and labour of the 'cultural economy of the city', the enterprises and institutions of the 'knowledge-based economy', and the socioeconomic consequences of immigration and transnationalism – can also be discerned within the economic spaces of the metropolis. These processes represent in many cases sublations of current and past trajectories, and thus present complex and in some respects volatile urban outcomes, in contrast to the more clearly delineated features of postindustrialism. Further, the imprints of these forces on the spaces of the city reflect intensely localized conditions (including policy factors) as well as more pervasive tendencies. In sum, London has over the past decade clearly strengthened its international competitive position in specialized banking, financial, and producer services, while at the national level capturing disproportionate shares of ascendant sectors, industries, and labour.[15]

The evolution of London's industrial geography

In keeping with the metropolitan character of London's economy, the peripheral areas of the city-region have attracted most of the expansion in manufacturing in recent years, and now encompass about three-fifths of industrial employment

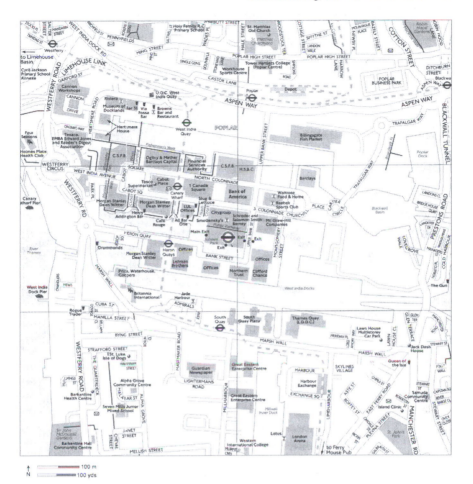

Figure 4.4 The corporate geography of a global financial centre: Canary Wharf, London Docklands.

in Greater London. But the recent industrialization experience of inner London (broadly conforming to the old County of London area established in 1888, and incorporated in the Greater London created in 1965 [Figure 4.1]) presents a uniquely rich, evocative, and instructive entrée to the study of restructuring and its impacts on the city. Broadly, the extraordinary value of the London experience is embedded within: the particular structure, labour profile, and geographies of London's nineteenth- and twentieth-century industry formation; the intricate connections between production industries and the social morphology of the metropolis, from the earliest days to the present; the catastrophic scale of industrial restructuring and collapse in the quarter century after 1960; and the diversity of reindustrialization processes over the past two decades.

The structure of inner London's industrial districts

So sweeping was London's industrial collapse in the late twentieth century, and so pervasive the industrial and social reconstruction of London's inner city since the 1980s, that it might be tempting to proceed directly to a consideration of contemporary development experiences. But a concise rehearsal of the tumultuous development of industry in inner London is of value, both in acknowledgement of Saskia Sassen's injunctions concerning the importance of the 'deep economic history of the metropolis', implying developmental continuities as well as ruptures; and also in recognition of the more specific and locally-contingent paths of development in the metropolitan core. In support of Sassen's exhortation, sequences of new industry formation in parts of inner London have followed certain historical patterns, in terms of location and siting, the extensive stock of heritage industrial buildings which provide preferred environments for many forms of creative and knowledge-intensive production, the renewal of traditions of skilled, artisanal (and neo-artisanal) occupations and labour cohorts, and the operative features of localized production networks and supply chains. A historically informed approach to urban industrialization serves to underscore the co-presence of differentiated production regimes, 'old' and 'new' economy industries, and 'high-value' and contingent labour, among even the most advanced urban economies positioned at the upper echelons of the global city hierarchy.

The configuration of London's traditional industrial economy

Well-known studies in the genre of London's industrial geography include M.J. Wise's work on London's industrial specializations within the broader context of British manufacturing (see, for example, Wise 1956), and O.H.K. Spate's account of the early (to 1850) industrial development of London (Spate 1938), while the following century of industrial development (1861–1961) was captured in Peter Hall's research (Hall 1962a). In addition to this historical scholarship, Hall contributed studies of specific industries and product sectors, notably a perspective on the development of clothing trades in London (Hall 1960), and his classic study of the East London footwear industry (Hall 1962b). Allen Scott cites this latter study as an example of the changing fortunes of light industries, in which the shift from essentially integrated, craft-based forms of production to fragmented tasks was signalled 'not so much by changes in technological hardware as it was by insistent divisions of labor and the development of specialized labor-intensive units of production' (Scott 1988: 68).

Scott acknowledges the tight bonding between traditional industry, skilled labour, and the social morphology of the inner city. To illustrate, the shoe and footwear industry in England, Europe, and North America was: 'drawn to major cities where it tended to occupy distinctive industrial quarters which were, in turn, the geographic foci of dense residential neighborhoods from which the main source of labor was drawn' (ibid.: 69).

The congeries of small-scale industries, industrial workers, and working-class neighbourhoods provided the formative mechanism for the particular social networks of family and kinship in East London, articulated in the seminal ethnographic work of Michael Young and Peter Willmott in Bethnal Green (1957), a precursor to the present-day social reconstruction of the metropolitan core.

Unlike the leading industrial cities of northern Britain (including Glasgow, Manchester, Liverpool, Derby, Sheffield, Leeds, and Nottingham), which featured heavy industries as centrepieces of the urban economy, London's nineteenth-century manufacturing sector specialized in light industries and consumer goods production.[16] The early nineteenth century also saw the rapid expansion of the East London docks, consistent with London's waxing status as principal world centre of trade (Fox 1992). The development of the extensive system of docks along the Thames generated a very large stock of riverside and inner city warehouses, mostly obsolescent since the 1960s, preserved for adaptive re-use on a large scale, and now an integral element of the highly inflationary London property market.

J. E. Martin's essay, 'The Industrial Geography of Greater London' (1964), written just on the cusp of London's late twentieth-century industrial collapse, offers an insightful perspective on inner London's traditional industrial economy. Martin's essay opens with a declaration that 'Greater London is today the first manufacturing region of Great Britain' (ibid.: 111). He was concerned to correct an impression that London's economy was principally about higher-order services:

> It is commonly believed that manufacturing is in some sense less fundamental to London's economic life than other activities such as administration, commerce and finance. Both types of activity, manufacturing on the one hand, services and 'tertiary' occupations on the other, provide the urban economic base. Both provide a range of 'exportable' commodities and services entering the arteries of interregional and international trade.
>
> (ibid.: 111)

Martin's essay included maps showing the distribution of larger factories (more than 100 workers each) in the old County of London area, in 1898 (Figure 4.5) and 1955 (Figure 4.6). The distributions show a significant dispersion of factories toward the periphery, especially in the north-west zone, reflecting in part the interwar development of manufacturing and industry in Park Royal and Willesden, demonstrating the growing industrial strength of suburban locations, but also disclosing the persistence of fairly tight clustering within the core of the County of London over this extended historical period.

Of particular relevance to the themes and spatial emphasis of this volume is Martin's detailed profile of the distinctive sites and systems of specialized production within Inner North-East London, comprising the City of London and ten of the old metropolitan boroughs, bounded on the east by the River Lea, and on the west by Parliament Hill and Regent's Park, as shown in Figure 4.7, roughly

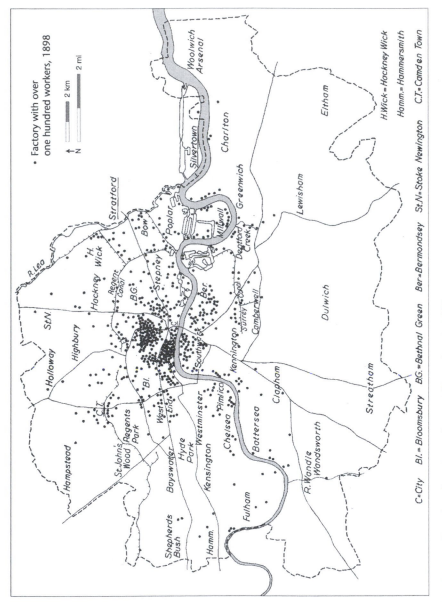

Figure 4.5 The County of London: factories with over 100 workers in 1898.

Source: Royal Commission on London Traffic, Vol. V (1906), from Martin (1964).

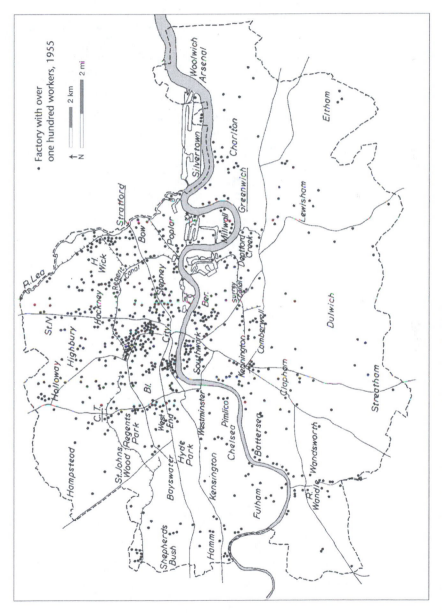

Figure 4.6 The County of London: factories with over 100 workers in 1955.

Source: Martin (1964).

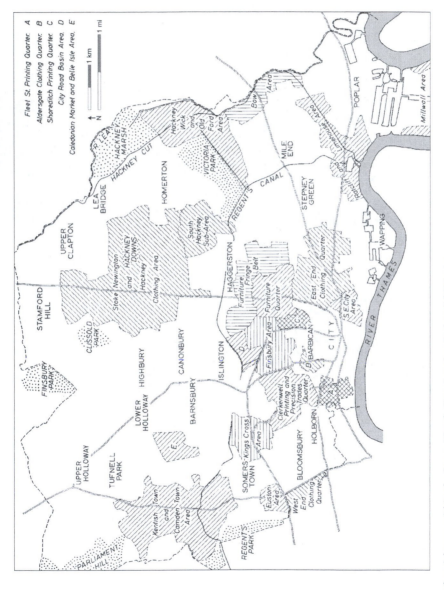

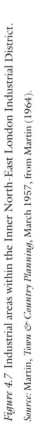

Figure 4.7 Industrial areas within the Inner North-East London Industrial District.

Source: Martin, *Town & Country Planning*, March 1957, from Martin (1964).

congruent with the 'City Fringe' districts of twenty-first-century London. The key sites of production in Inner North-East London at mid-century included the following:

The East End clothing clusters, encompassing one square mile centred on Whitechapel, and containing in 1955 about 900 establishments and 17,000 workers. These clusters comprised 'special character' areas defined in part by the nature of premises, varying from 'workshops, occupying a room in adapted houses or located in outbuildings located at the rear, to modern blocks erected in the "thirties", with the latter used by an individual firm or as flatted workrooms' (Martin 1964: 129). Product sectors and sites included ladies' tailoring in Whitechapel, Bethnal Green, and Spitalfields, involving principally skilled male tailoring labour, and including a network of ancillary enterprises conducting buttonholing, pleating, and button and buckle covering; the Hackney clothing area, which experienced initial growth in the 1920s, and involved mainly female labour in increasingly larger factories; and the small Aldersgate clothing quarter which included wholesaling as well as high-grade production of samples, ties, and millinery, with manufacture often conducted in top floor workrooms and factories in Finsbury.

Printing activity concentrated within a crescent of firms adjacent to the City, with marked features of localization of linkages, including newspaper production in the Fleet Street quarter, and commercial printing in Clerkenwell and Shoreditch, including typesetters; but by the 1960s subject to decentralization to the 'Home Counties or provinces' (ibid.: 134).

The furniture quarter of Shoreditch and Bethnal Green, a defining industry of East London since the early nineteenth century (including timber shops and cabinet makers dating from 1832), a sector in decline from its heyday in the 1880s and 1890s, but comprising in Martin's survey some 400 establishments and just under 4,000 workers, with premises situated typically in back-street locations, engaged in short-run, customized production. The East End furniture industry was characterized by clusters of intricate labour subdivisions (e.g. upholsterers, polishers, glaziers), supported by ancillary (and proximate) suppliers of plywood, veneer, tools, and polish.

The Clerkenwell precision trades quarter, focusing on the use of glass and non-ferrous or plated metal, and including the 'renowned jewellery trade in the Hatton Garden area of the quarter' (ibid.: 134). Key precision industries and trades in Clerkenwell included makers of surgical and precision instruments, machine tools, printing machinery, and manufacturing opticians, but much of the production was in the form of 'parts and accessories rather than complete products and is specialized in the extreme' (ibid.: 134).

Engineering, with distributions to the north of Clerkenwell in Camden Town and Holloway, as well as constructional engineering in Poplar and Millwall.

Food, drink, and tobacco manufacture, situated in Limehouse and in the inner industrial crescent from Finsbury to the London Docks.

Martin's narrative included an acknowledgement of rising costs, and the growing attractions of larger spaces and sites beyond the County of London area, but overall presented a picture of industrial vitality in mid-century East London.

As a further demonstration of the durable qualities of localized production in London, Martin's essay exhibited patterns of location for East End tailors in 1888 and 1955 respectively (Figures 4.8 and 4.9), within a mile eastwards of Liverpool Street Station, concentrated in Spitalfields and Bethnal Green. These tailoring concerns were typically small, even household-based operations, and therefore offer a somewhat more nuanced spatial perspective of production than the distributions of medium-sized and larger factories shown in the preceding figures. As Martin observed, 'the limits of the main concentration then [in 1888] were almost identical with those of today' (ibid.: 129), presenting a quite striking image of spatial integrity, in the face of depressions, two world wars, and social and political upheavals. Overall, the state of industry in mid-century London presented few signs of imminent collapse, or of the wrenching social dislocations and theoretical disjunctures associated with the deep restructuring of the metropolitan economy which soon followed.[17]

The causes of London's industrial collapse seem fairly clear, and recitations of metropolitan de-industrialization typically include a familiar list of factors: increasing international competition, inadequate investment, managerial deficiencies, unsatisfactory labour relations, rising land costs, and the growing attractions of smaller, less congested cities and towns (Gripaios 1977; Dennis 1978). The role of the Thatcher Government in accelerating the running down of traditional manufacturing and industry in London and Britain as a whole also forms part of the narrative of causality.[18] In a well-known analysis of London's industrial collapse, Fothergill and Gudgin (1982) assigned some part of the decline to site constraints within inner London. Nicholson *et al.* (1981) suggested that inner London's traditional industrial incubator role had lapsed by the early 1980s, apparently signalling the demise of industrialization as an agency of development within the territories and landscapes of inner London.

The New Economy of inner London: phoenix or chimera?

London's period of industrial contraction had essentially run its course by the early 1990s, although (as we shall see) new rounds of industrial restructuring were to follow. The physical legacy of postindustrial London, in the form of extensive precincts of Victorian warehouses and factories, was subsequently revalorized in new phases of social upgrading. The new social groups have included recent immigrant cohorts,[19] and the rise of what Tim Butler and Loretta Lees have described as 'super-gentrifiers' (2006), individuals characterized by increasingly higher wealth thresholds, as opposed to the more mundane affluence of the new middle class emblematic of the postindustrial restructuring of the

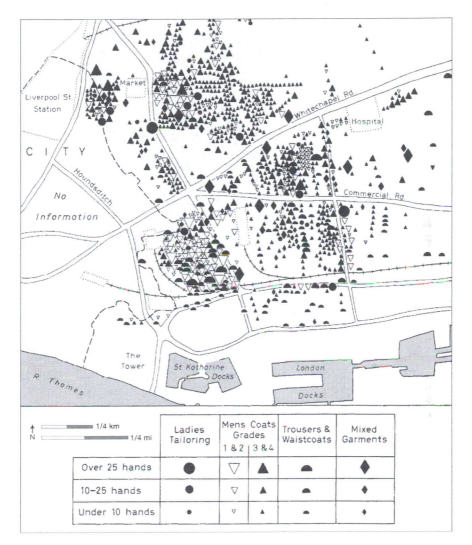

Figure 4.8 East End tailoring, 1888.

Source: Booth Collection, Group B, vols 108 and 109, British Library of Political and Economic Science, in Martin (1964).

1980s.[20] These new social groups generally supported the reconstruction of the inner city as a domain of livability, conviviality, and spectacle, and were sympathetic to the launch of a renewed cultural development trajectory for central and inner London, exemplified by investments in consumption and amenity, the proliferation of artists studios (and more modest habitats), the emergence of 'cultural quarters' in certain districts of the inner city, and high-profile institutions such as the Tate Gallery of Modern Art in Bankside. Inner London will likely never recover the population thresholds and neighbourhood densities (nor replicate the

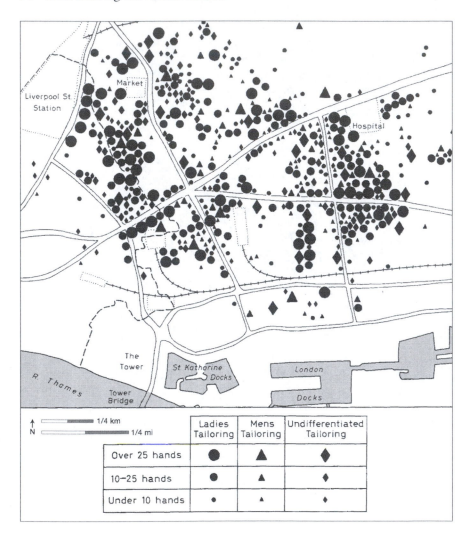

Figure 4.9 East End tailoring, 1955.

Source: Martin (1964).

squalor and deprivation) of its late nineteenth- and early twentieth-century peak levels,[21] but has been reconstituted as a site of exceptionally lively, diverse and transnational communities.

'Precarious reindustrialization' within the City Fringe

While inner London's status as site of rejuvenated social space, spectacle, and consumption can readily be acknowledged, interpreting its contemporary positioning as a locus of industrial production may present a more problematic task.

London's early twenty-first-century inner city space-economy presents a complex and variegated admixture of production regimes, product sectors, firms, and labour, characterized by elements of competition as well as complementarity. In some respects the industrial geography of the inner city bears some similarity to the patterns of distinctive, area-based industries and product sectors of the nine-teenth- to the mid-twentieth-century London light industrial regime depicted in Martin's essay cited above (Figure 4.7), although it is clear that there is more volatility in the mix of industries, and signs of competition with new, upscale housing abound. There is also the emergence of Canary Wharf as a major global financial space in Tower Hamlets, although this complex was essentially imposed upon inner London by central government and certain corporate actors, rather than generated organically from within.

We can rehearse the range of *generative processes* and conditions of industrialization within the City Fringe, including the interaction of exogenous and domestic factors, as follows.

The persistence of a residual presence of traditional industries

In inner London these include high-value custom product lines, as well as major sectors (notably printing and publishing) experiencing ongoing restructuring pressures associated with new production technologies, and advanced telecom-munications systems which enable outsourcing, although these activities are now largely residuals of the inner city economy.

Power of the commercial office sector

Continued growth in the service economy of central London has generated over-spill commercial development within certain 'City Fringe' districts, notably Holborn, Clerkenwell, and Shoreditch, subject to the supply and demand oscilla-tions of the Central London office market, and the appeal of competitor sites in Docklands and elsewhere in the larger London–South East Region.

Imprints of the technologically-driven 'New Economy'

The 'New Economy' is evidenced within inner London in the growth of tele-coms, multimedia, and Internet service providers, as well as a more general technological deepening of the production economy of the metropolis as a whole. The impact of the New Economy within Inner London can be seen in the presence of telecommunications companies located in, for example, Shoreditch (along Great Eastern Road), and the growth of small subcontractor firms in the technology sector catering to larger enterprises in the City and elsewhere.

The rise of the cultural economy of the city

This economic trajectory constitutes a principal motive force in the reproduction of inner London's economy, including the establishment of Europe's largest artist community in East London (notably in Tower Hamlets and Hackney), the formation (spontaneous and induced) of 'cultural quarters' at the neighbourhood level (for example, around Cowcross Street in southern Clerkenwell, on the City Fringe), and the emergence of parts of inner London as major staging areas for London's cultural and creative industries, as seen in Bankside in northern Southwark and other sites. London's emerging geography of cultural production also includes the formation of 'art zones', each characterized in part by individual identity and profile, and by the presence of studios and/or galleries of leading artists, as well as lines of spatial articulation (Figure 4.10).

Specialization of production and labour formation at the localized scale

Specialization was a defining feature of the light industrial economy in nineteenth-century London, and has been replicated to a degree in the contemporary era, following the processes of innovation which drive changes in the internal structure of the metropolis articulated by Allen Scott (1988), and more specifically new divisions of production labour in the twenty-first-century metropolitan core.

The evolving role of the state in economic development

Within London, policies of the central government and its ministries, which construe the success of London and its economy as essential to the well-being of the

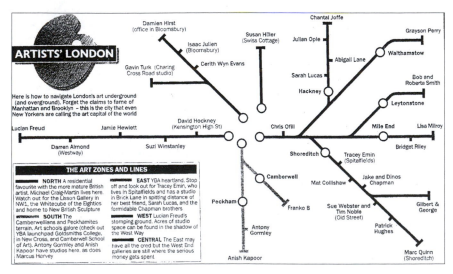

Figure 4.10 Spatial representation of 'Artists' London': artists and art zones.
Source: © *The Times* (London).

larger nation, support the high-powered financial and business services of the central and inner city, as well as the capital's cultural economy. The Mayor's economic development strategy, investment program, and cultural policies, although metropolitan in scope, contain significant measures for inner London boroughs and districts. Local (borough) policies include regeneration policies and programs, heritage programs, land use policy, and education, among others. In the aggregate, these development policies represent a sharp contrast to the stringent controls of the pre-Thatcher era.

The behaviour of the London property market

Although prone to broad swings and cycles, the London property market tends to promote 'highest and best' uses (in terms of rates of return on investment) of land in London's expanding core districts and communities, inserting a destabilizing influence on small and marginal land uses, increasingly squeezing lower-return activities, and in recent years favouring high-end housing within an increasing number of City Fringe districts.[22]

The role of non-governmental organizations (NGOs) and community-based organizations (CBOs)

NGOs and CBOs represent important instruments of local development and regeneration. These include, notably, agencies committed to the sustainability of artisanal production and craft labour within the City Fringe, via training, network development, marketing and procurement programs, and other means (see Chapter 5 for examples).

Summary

These forces shape an evolving industrial geography within the inner city which comprises, at the broadest level, a residual presence of Fordist industries, an assortment of small-scale, craft, and artisanal industries, a substantial clustering of flexible, post-Fordist intermediate services, and a generally dynamic set of creative industries. A profile of inner London's production regimes and representative industries illustrates the extraordinary specialization and diversity of economic activities in the metropolis (Table 4.3), reflecting London's dual status as both 'world city' in Peter Hall's interpretation (with industrial scale and degree of specialization as important measures), and also as first-order 'global city', with strong representation in the 'power sectors' such as banking, finance, and other intermediate services.

While the industrial representation within the global metropolis is uniquely diverse, however, there is of course considerable unevenness in the significance, power projection, and growth performance among these industry groups. In the London case the basic platform of the metropolitan economy comprises the post-Fordist industry groups included in Table 4.3, listed under category III, with the

Table 4.3 Production regimes and representative industries for London's inner city

I *Pre-Fordist industries*	II *Fordist industries*	III *Post-Fordist industries*	
		A *Intermediate service industries*	B *Cultural / New economy industries*
(1) Skilled artisans, artists, apprentices	(1) Operatives: skilled, semi-skilled labour, managers and supervisors	(1) Segmented labour: executives, managers, professionals, technical, sales and clerical	(1) Specialized neo-artisanal labour, design professionals, scientific and IT staff, artists, sales, managers
(2) Workshops, shops, residential space	(2) Factories and plants	(2) Office buildings	(2) Studios, workshops, live-works, work-lives, offices
(3) Artists Bespoke tailors Bookbinders Jewellers Milliners Model-builders Musical instrument makers Perfume and scent makers Precision instrument makers Silver plate engravers	(3) Food and beverage production • bakeries • breweries • food-processors Garment production (long-run, mass market) • factories and plants • sweatshops Printing and publishing • mass-market integrated Fordist production	(3) Corporate control: head and regional offices Intermediate banking and finance • merchant banking • fund managers • stock exchange • insurance Producer services • legal firms • accountants • marketing • management consultants Property • development companies • property managers • estate agents • research and market intelligence	(3) New media/multi-media Internet services and web-design Computer graphics and imaging Software design Digital arts Graphic design and arts Digital publishing and printing Film production and post production Video game production Music studios Galleries: curatorial services Specialized food and beverages • organic food • specialty bakeries • specialty coffee roasting • ethnic foods and beverages

Note:
(1) labour.
(2) infrastructure.
(3) representative industries.

specialized intermediate services constituting the dominant group, asserting itself not only in the City but also in Canary Wharf, Westminster, and 'inner edge city' formations such as Paddington Basin and Hammersmith. But the cultural and New Economy industries acknowledged in column B add to London's portfolio of growth sectors, reinforcing its national primacy.

Many of these industries have exhibited significant resilience in the face of broader restructuring forces and industry-specific competition, but overall there is considerable pressure on the City Fringe's firms. A report by the City Fringe Regeneration Audit Group stated that in 1993, at the end of the long era of London's industrial decline, the eight wards of a relatively restrictive definition of the Fringe (including wards situated within Camden, Islington, Hackney, and Tower Hamlets) contained 123,700 jobs – just over one half of the total number of jobs in the City of London. Within this City Fringe employment base, business and professional service firms represented by far the largest proportion, at just under 40 per cent, followed by financial services (just under 15 per cent), health, social services, education and public administration (about 12 per cent), and publishing, media and cultural production (approximately 7 per cent) (City Fringe Regeneration Audit Group 1997: 13). Manufacturing, once the lead sector in this heart of London's light industrial economy, accounted for less than 5 per cent of the total employment within the City Fringe in 1993, underscoring the depths of the restructuring experience of the inner city.

Restructuring processes in contemporary London are vividly demonstrated by the contrasting fortunes of printing and publishing. A recent publication of the City of London Corporation observes that printing and publishing, conventionally viewed as an integrated production sector, are increasingly operating as separate industrial ensembles. Publishing in the City Fringe experienced growth both in employment and businesses in the period 1998–2002 (and indeed grew faster than the national average), with a 2004 employment base of 27,097 workers, about 40 per cent of the regional employment in publishing (City of London Corporation 2004). The publishing sector, which includes three distinct components, magazine and journal publishing, book publishing, and book retailing, has thrived in a competitive environment, with firms orienting their business toward 'the specialist vertical markets for which they cater, rather than operating as part of something called the "publishing sector" ' (ibid.: 6). Sales and marketing are conducted within international channels, increasingly through the Internet and by digital means, effectively transcending the limitations of the regional market (Driver and Gillespie 1993). In contrast, printing (which numbers just under 8,000 workers within the City Fringe) follows more closely the traditional, highly localized inner London industrial clustering patterns, reliant upon regional markets, and increasingly pressured by forces which include declining demand for 'hard copy' (associated with the ascendancy of digital media), over-capacity, congestion, high wages and land costs, and increasing competition from overseas producers (ibid.: 8) – a mix of problems ominously similar to those underpinning the larger industrial decline of late twentieth-century London.

The swings of the business services and financial service industries provide another illustration of volatility in the industrial structures and systems of the City Fringe. The report of the City Fringe Regeneration Audit Group cited above noted that together these important specialized service industries comprised over one-half of the total employment within the City Fringe wards of Camden, Islington, Hackney, and Tower Hamlets in 1993. These intermediate services represented in large measure 'overspill' activities from the adjacent City of London, taking advantage of significantly cheaper rents and proximity to Central London, in the context of constrained commercial office supply conditions in the City proper. But over the following decade office development in the City has generated new opportunities for commercial tenants, while the expansion of Docklands provides another option for firms seeking to locate in one of London's dominant corporate-financial complexes, diminishing the appeal of generally less prestigious areas (Hamnett 2006). There are also opportunities for office development within emerging 'inner edge cities' that lie outside the City Fringe, such as the Paddington Basin, as well as within established office centres such as Hammersmith. The declining attractions of the Fringe for corporate activity may serve to increase the likelihood of residential conversion, clearly a pervasive phenomenon in the City Fringe as a whole.

The cultural economy of the City Fringe

If finance, business services, printing, and publishing represent established industries of the City Fringe, albeit subject to competitive and technological pressures, then cultural and creative industries are widely acknowledged as comprising the ascendant trajectory. The City Fringe (incorporating Bankside and Bermondsey in Southwark) represents a principal domain for London's cultural economy, including centrepieces such as the Tate Gallery of Modern Art (TGMA), the Globe Theatre, and the Design Museum in Shad Thames; sites of spectacle and consumption, such as Spitalfields, Hay's Galleria, and the Borough Market; signifying clusters of creative industries and cultural production, notably those ensconced within Shoreditch and Clerkenwell; and a proliferation of cultural businesses, agencies, institutions, NGOs, and CBOs that occupy smaller sites within London's inner city.

Reflecting the volatility of recent restructuring processes in London and in advanced urban economies more generally, though, the development pathway of the cultural economy of the City Fringe presents recurrent episodes of industrial 'churning' and re-assortment of firms, rather than an orderly progression of enterprise establishment, growth, and maturation. The latter years of the 1990s saw the emergence of a vibrant cultural economy of creative industries within the City Fringe, led by the ascendancy of multimedia enterprises, driven by synergies of technological innovation and design skills, and supported by a rich institutional platform of public programs and associative agencies. A report prepared by the Arts Business Limited identified the comparative advantages of the City Fringe for 'successful cultural clusters' as including: (1) the high concentrations of resident artists; (2) the growth of cafés, bars, pubs, and restaurants which contribute to the 'metropolitan buzz of the Fringe's cultural clusters'; (3) cheap accommodation in the form of an extensive local supply of studios, warehouses, and factories; (4) investments in amenity and infrastructure; (5) the broader resonance of existing clusters, as well as ascendant sites such as Hoxton and Spitalfields, 'accepted newcomers to the picture with increasingly high reputations'; and (6) the industry-leading position of a number of cultural firms and organizations situated within the City Fringe (the Arts Business Limited 1997: 45). As the report observed, the City Fringe 'can lay claim to an unprecedented concentration of innovative training provision at all levels' (ibid.: 47), supporting skill enhancements for creative workers, enhancing the competitive advantage of the City Fringe, and imparting a degree of resilience to the industries of the Fringe in the face of recurrent restructuring pressures.

A more recent report (2005) published by the City Fringe Partnership underscores the City Fringe's continuing robustness as a site of creative industrial production, offering an estimate of 4,100 firms and approximately 44,000 employees as an empirical measure of saliency (TBR Economics Report 2005: 1). Within the City Fringe creative sector, according to this report, the largest industry groups by employees include (in order) publishing, fashion, architecture, advertising, games and software, and furniture and interior design, with a sharper drop to the next larger industry groupings, which include designer fashion (a more upscale market classification than 'fashion'), jewellery, radio and television, and music and visual arts (Table 4.4).

As a further demonstration of developmental significance, the TBR Economics Report acknowledged the City Fringe's high location quotients for creative firms and employment in the London context (1.5 and 1.4, respectively), and high

Table 4.4 Estimated employment in the City Fringe creative sector

Segment	Firms	Estimated total employment
Publishing	310	7,938
Fashion	881	6,737
Architecture	372	5,912
Advertising	557	5,436
Games and software	330	4,865
Furniture and interior design	363	4,611
Designer fashion	285	2,846
Jewellery	488	2,367
Radio and television	46	2,196
Music and visual arts	402	2,005
Crafts	397	1,122
Film	125	1,010
Art and antiques	74	476
All creative	4,105	43,686

Source: TBR Economics, for the City Fringe Partnership (2005).

Gross Value Added (GVA) productivity levels of City Fringe creative firms relative to the London average. At the same time, new enterprises comprise a higher proportion of the base of firms in the City Fringe's creative sector over the period 2001–2004, suggesting 'shorter life-spans of businesses' (ibid.) and implicitly high turn-over rates, consistent with a thesis of recurrent churning of industries and businesses.

This observation concerning high rates of business births and deaths underscores the role of the contemporary inner city industrial district as a territorial zone of experimentation and innovation – as a bellwether of change in the modern urban economy subject to global influences and recurrent restructuring processes – rather than as a theatre of deeply entrenched firms and defining production systems. There is a resiliency of creative and design-based production within the City Fringe, but the specific mix of product sectors, industries, and labour is subject to recurrent restructuring and reformation, following episodes of technological innovation, global and domestic competition, and shifts in consumer market preferences. As in other New Economy territories, too, such as South Park–SOMA in San Francisco, the dot.com crash of 2000 eroded the City Fringe's base of technology-intensive firms, although the greater industrial diversity (and richer institutional supports) of the London sites have provided in some cases a degree of insulation from the effects of this recessionary experience, and a platform for regenerative development.

Conclusion: from 'postindustrial' to 'new industrial'?

As the preceding text indicated, London's inner city can be characterized by a uniquely rich and diverse structure of production regimes (pre-Fordist, Fordist, and post-Fordist), industries, and labour, exhibiting important developmental continuities as well as disjuncture. That said, over a compressed four-decade period, London's inner city experienced a sequence of deep swings in its defining development trajectory, from an apparently robust, world-scale light manufacturing economy at mid-century, followed by a protracted and calamitous period of industrial decline played out over three decades, out of which a partial reassertion of industrial production has emerged since the early 1990s, largely encompassed within the inner city's traditional industrial districts and communities. In contrast to London's light manufacturing economy sustained in large part over a century and a quarter, though, this 'New Economy of the inner city' is markedly less stable, subject to recurrent, abbreviated restructuring processes, punctuated by recessions and downturns, and increasingly vulnerable to the vagaries of the London property market, within which the financial-commercial sector and upscale residential development are paramount shapers of demand.

The London property market is crucial to the fortunes of industry and employment within London's inner city, but other factors are germane. These involve the complex framework of development policies, including those pertaining to the production industries, designed to influence the configuration of the metropolitan economy, as well as local regeneration, cultural, and land use policies. It is also the case that while structural forces shape broad new trajectories for the inner city and the metropolis as a whole, there are quite distinctive experiences of industrial change occurring at the more localized level. What follows in Chapter 5, then, is a presentation of illustrative sketches or vignettes of new industry formation situated in three instructive districts of London's inner city, with a view to drawing out the saliency of local contingency in shaping micro-scale processes of industrial innovation and restructuring.

5 London's inner city in the New Economy

Introduction: industrialization and socioeconomic change in inner London

The discussion in Chapter 4 depicted the broad strokes of industrial change within the City Fringe zone as a whole, emphasizing the centrality of cultural production and creative industries to the regeneration of the area, but acknowledging as well the velocity of market pressures on the Fringe's constituent firms. The influences of London's global city status and national primacy, entailing scalar issues and deep specialization, were acknowledged as cardinal factors in the shaping of the inner city's production economy, as was the specific experience of London's history of industrialization and restructuring.

At this broader metropolitan level there is a remarkable spatial congruence between the patterns of manufacturing and fabrication characteristic of Inner London's heyday as a light industrial district as elucidated by Hall, Martin, and others, and the contemporary spaces of reindustrialization within the broadly defined City Fringe (Figure 5.1) Part of the narrative value of the revival of specialized production in areas such as Shoreditch, Clerkenwell, and Bermondsey lies in this juxtaposition of new industries within the postindustrial landscapes of the old East London light industry and warehousing districts, with at least some carryover of area-based specialization, and with adaptive re-use of much of the old industrial building stock. There are also interdependencies which generally shape the systems of production and the rearticulation of economic space within the Fringe in the early years of the twenty-first century, as there were in the late nineteenth and early twentieth centuries. Imprints of the abrupt restructuring cycles of the past decade and a half can be discerned widely throughout London's inner city, in the form of IT firms and new media, creative industries, and cultural quarters, each of which corresponds to new dynamics of innovation and restructuring in the metropolis.

A socioeconomic profile of inner London

While innovation and restructuring within inner London now take place within some of the industrial spaces of the nineteenth and early twentieth centuries, and

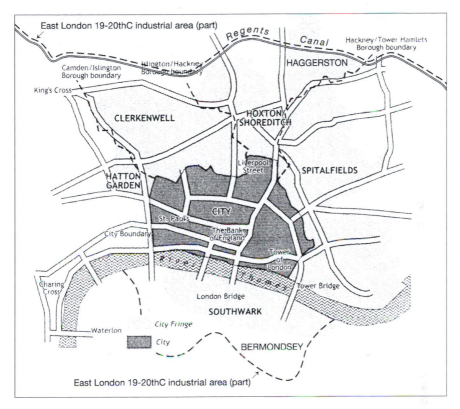

Figure 5.1 The 'City Fringe', selected local areas, and London's traditional inner city industrial districts.

Source: City of London and Martin (1964).

even follow some of the production practices of the past, the socioeconomic structure of the zone presents a number of redefining contrasts. The global city experience tends to insert greater polarities of wealth and poverty, as elucidated in the well-known treatment of London and New York by Fainstein and Harloe (1992). Deep poverty still exists, and some of the inner London boroughs, such as Hackney and Tower Hamlets, number among the most socioeconomically deprived areas in Britain. In particular, deprivation levels within some visible minority groups, and among single-parent families, constitute particularly stubborn social policy issues within the metropolis, reflecting entry barriers to the better-paying jobs within the labour market, the increasingly exclusive nature of the London housing market, and persistent prejudice, as well as other factors.

Table 5.1 shows the distribution of employee jobs within inner London, compared to comparable data for London as a whole and for Great Britain. As a measure of change ensuing from the wrenching restructuring processes of the late twentieth century, manufacturing jobs within inner London, the former heart of the metropolitan production economy, now account for proportionately fewer

Table 5.1 Employment by industry, for Inner London, Greater London and Great Britain

	Inner London (employee jobs)	Inner London (%)	London (%)	Great Britain (%)
Total employee jobs	2,381,400	–	–	–
Full-time	1,826,000	76.7	74.0	67.9
Part-time	555,400	23.3	26.0	32.1
Employee jobs by industry				
Manufacturing	100,300	4.2	5.0	11.1
Construction	44,000	1.8	3.0	4.6
Services	2,231,200	93.7	91.7	82.9
Distribution, hotels & restaurants	452,400	19.0	21.7	24.1
Transport & communications	151,700	6.4	7.7	6.0
Finance, IT, other business activities	940,900	39.5	32.7	20.7
Public admin, education & health	508,000	21.3	23.0	26.9
Other services	178,200	7.5	6.6	5.2
Tourism-related†	218,300	9.2	8.4	8.1

Notes: † Tourism consists of industries that are also part of the services industry.
   a: % is a proportion of total employee jobs.
   b: Employee jobs excludes self-employed, government-supported trainees and HM Forces.

Source: annual business inquiry employee analysis / www.nomisweb.co.uk.

jobs (4.2 per cent) than for London as a whole (5 per cent), and less than half the proportion of manufacturing jobs in Great Britain as a whole (11.1 per cent). The proportion of inner London jobs in service industries (93.7 per cent) slightly exceeds the level for London, and is a full 10 points higher than for Britain as a whole. The gap in the key category of 'finance, IT and other business activities' is especially wide, with almost 40 per cent of inner London employee jobs situated within this grouping, as against 32.7 per cent for London as a whole, and only 20.7 per cent at the national level.

London's pronounced shift from manufacturing to advanced service industries is complemented by a distinct professionalization trend, a central motif of Chris Hamnett's portrayal of 'London in the Global Arena' (2003). Inner London shares in this trajectory of socioeconomic change, although again not all residents reap the benefits of this restructuring tendency. Table 5.2 shows employment by occupation, drawn from a survey period encompassing April 2005 and March 2006. At the upper end of the occupational hierarchy, employment levels within inner London as a whole tend to slightly exceed those for London as a whole, with a larger margin when compared against national levels. Employment among those occupations which comprise much of the new middle class, including managers and senior officials, professionals, and associate professionals and technical workers, exceeds one-half of the total within inner London, as against about two-fifths of employment in Britain as a whole. On the other hand, employment among administrative and secretarial occupations, and personal services and sales

Table 5.2 Employment by occupation (Apr. 2005–Mar. 2006), for Inner London, Greater
London and Great Britain

	Inner London (numbers)	Inner London (%)	Greater London (%)	Great Britain (%)
Soc 2000 major group 1–3	718,600	56.3	52.3	41.9
1 Managers and senior officials	235,900	18.5	17.6	14.9
2 Professional occupations	218,800	17.1	16.1	12.7
3 Associate professional & technical	263,900	20.7	18.6	14.3
Soc 2000 major group 4–5	217,700	17.1	21.1	23.4
4 Administrative & secretarial	137,500	10.8	13.3	12.5
5 Skilled trades occupations	80,200	6.3	7.8	10.9
Soc 2000 major group 6–7	160,800	12.6	13.2	15.6
6 Personal service occupations	90,600	7.1	7.0	7.9
7 Sales and customer service occs	70,200	5.5	6.2	7.7
Soc 2000 major group 8–9	167,500	13.1	12.8	18.9
8 Process plant & machine operatives	52,100	4.1	4.7	7.5
9 Elementary occupations	115,400	9.0	8.1	11.4

Notes: a: Numbers and percentage are for those of 16+.
   b: Percentage is a proportion of all persons in employment.

Source: Annual population survey / www.nomisweb.co.uk.

and customer services, represent a somewhat smaller proportion of inner London
employment by occupation than in London as a whole and the nation, underscoring the overall specialization of labour within inner boroughs of the metropolis.
And, as might be inferred from the previous table of industrial employment, the
proportion of inner London workers employed in goods producing occupations (process, plant, and machine operatives), 4.1 per cent, is slightly lower
than for London as a whole, but significantly below that for Britain as a whole
(7.5 per cent).

While unemployment and socioeconomic deprivation within areas of inner
London rank among some of the most serious within Britain, there is also greater
buoyancy of incomes for many of the professional, managerial, and entrepreneurial workers. For those in employment, average earnings within inner London
compare favourably with those for London and for Britain as a whole. For male
full-time workers domiciled in inner London, weekly earnings of £603.7 exceed
those for London as a whole (£579), and comfortably surpass those for the country as a whole (£490.5). The margin favouring full-time female workers is even
greater: £529.6 for female workers resident within inner London, as against
£498.7 for London as a whole, and £387.6 for Britain. These data present only
part of the socioeconomic reality of inner London, however, as against these
favourable figures one has to account for the income pressures on the many
part-time workers, the effects of short-term and structural unemployment, and
of course the quite ferocious price points of the London private sector housing
market.

Interrogating new industry formation at the local level in London

As in the case of London's initial nineteenth-century industrialization experience, we can readily identify important aspects of localized development within the districts of the Fringe, with correspondingly differentiated storylines, signifiers of change, and insights for theories of contemporary urbanism (Foord *et al.* 2005). The diverse production landscapes of inner London, encompassing a mix of industrial regimes, firms, and labour (Table 4.3, p. 96), suggest a contemporary evocation of Scott's earlier model (1988) of the internal specialization of the metropolis, with increasingly fine-grained spaces of activity. These inner city economic spaces include the intimate juxtaposition of production industries amid a rich field of consumption uses, public institutions, and housing, marked by interdependency as well as competition. What follows is a sequence of illustrative vignettes of precarious (or at least volatile) industrialization within London's City Fringe, reflecting dynamics of innovation and competition in the advanced urban economy, as well as the relentless pressures of the property market and larger forces of transformation in the global metropolis.

The case studies presented in this chapter are as follows: (1) Hoxton, in Shoreditch (London Borough of Hackney), widely acknowledged as the iconic New Economy site in London in the 1990s, and a model of accelerated transition and succession processes in the 'new inner city'; (2) Bermondsey Street, in Southwark, which demonstrates the influence of heritage-built environment and conservation policies upon new industry formation in the inner city; and (3) Clerkenwell, in the southern part of Islington (with small extensions into Camden and the City), which replicates in exemplary fashion the complexity of industrial organization in the metropolitan core, and exhibits signifying attributes of industrial restructuring in the inner city. Each of these three sites forcefully exhibits the tensions of coincident reindustrialization and the social reconstruction of the twenty-first-century metropolis, and the competition for space in the revalorized inner city. In each case study the narrative will include a sketch of the character of the area, emphasizing the influence of spatiality, built form, and industrial legacy features on the reshaping of development, the mix of industries and representative firms and institutions, and the particular swings of fortune observed over the past decade. The concluding section will extract salient observations from these vignettes of industrial innovation and restructuring at the district level in the global metropolis.

Hoxton: the rise and fall of a cultural district

Hoxton, situated in the London Borough of Hackney, represents perhaps the most iconic New Economy site in London. Hoxton's exalted status as a signifying industrial site is derived in large part from the media hype and buzz of the New Economy era. But there is also a more authentic saliency of Hoxton's experience associated with its late twentieth-century provenance as a centre of creative production connected to local artists and cultural institutions; with its intended role as an engine of regeneration in postindustrial Shoreditch; and, finally, with the cautionary implications of its more recent redevelopment.

Character and setting of the Hoxton–Shoreditch site

Hoxton is situated in the London Borough of Hackney, in the old metropolitan borough of Shoreditch. The heart of the New Economy site which emerged in the 1990s comprises (from north to south) the Hoxton Square precinct, a triangular enclave bounded by Old Street, Curtain Road, and Great Eastern Street, and the Leonard Street area (Figure 5.2). Shoreditch was an integral district of inner London's industrial economy of the nineteenth century, as depicted in J.E. Martin's treatment of London's East End presented in Chapter 4. Shoreditch specialized to a significant degree in furniture and wood products, incorporating a dense network of workshops, suppliers, and warehouses. Shoreditch was also a significant site of East London's tailoring and garment production prior to the collapse of the metropolitan manufacturing economy commencing in the 1960s. These traditional industries now constitute at best a residual presence in Shoreditch (Figure 5.3), displaced (or eradicated) by a confluence of exogenous structural forces and internal pressures for change, but producing a tangible legacy in the form of an industrial built environment susceptible to adaptive reuse and conversion purposes.

Processes of transition and succession

The recent development of Hoxton and the adjacent environs of the Shoreditch district takes the form of multiple phases of restructuring and dislocation, which vividly demonstrate the volatile nature of the contemporary inner city and its insistent experiences of transition and succession. First, rapid industrial decline represented a precondition for the recolonization of the Hoxton precinct and other neighbourhoods of Hackney by artists, following the classic model of inner city gentrification articulated by Hamnett, Butler, Lees, Bourne, Ley, and others.[1] A growing population of artists was attracted by the combination of cheap rents, gritty inner city ambience, and suitable studio space potential in the area's Victorian residential and industrial built environment. In time, this sociocultural movement produced clusters of artists, in the form of co-ops and studios, the commercialization of artistic production, and the institutionalization of the artistic presence in galleries and salons, notably on or proximate to Hoxton Square, the centre of Shoreditch's arts community.

Second, the ascendancy of a formal design sector, acknowledged as an integral feature of the cultural economy of the city, can be perceived as a logical extension of the artistic trajectory of the community, but also represented a powerful agent of dislocation, given the commensurately larger resources of professional design firms. This applied design sector introduced a harsher competitive edge to the struggle for space in the Hoxton area, encouraged the growth of high-end amenities acknowledged as accoutrements of high-value production in the urban core,

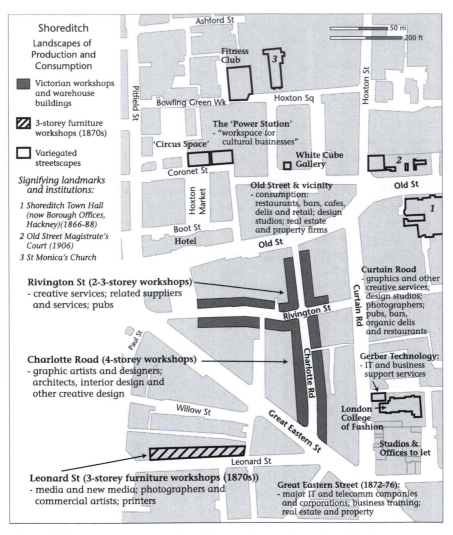

Figure 5.2 Hoxton and the Shoreditch Triangle, London Borough of Hackney.

Source: Adapted from Hutton (2006).

and provided a more affluent clientele for the area's burgeoning property market players. Competition between artists and design professionals was exacerbated by the rise of the technologically-driven 'New Economy' in the 1990s, which introduced a new set of actors (including Internet firms, and the so-called dot.coms) competing for space.

Third, in a new phase which takes us to the present, the relentless incursion of gentrification, facilitated by developers and estate agencies ever sensitive to the higher return possibilities of conversion and adaptive re-use, promotes the physical and social relayering of the area for upscale residential purposes and for

Figure 5.3 Garment production (coats and jackets), Shoreditch, 2006.

complementary amenities, compromising in turn the viability of the industries, firms, and labour acknowledged as central to the regeneration mission. Here the evanescent qualities of contemporary urban development are a defining feature of the new inner city.

Profiles of twenty-first-century change in Hoxton–Shoreditch

A program of field exercises conducted over the period 2000–2006, including multiple site visits and survey work, mapping, and interviews, discloses both the broad contours of transformation and more specific markers of change in Hoxton and its environs. For the purposes of this research project, a study area centred on Hoxton Square and the 'Shoreditch Triangle' was deployed as a means of generating profiles of development in this iconic district of London's New Economy. Enterprises in this area maintained linkages with clients and suppliers beyond the City Fringe, but also operated in an environment characterized by highly localized production networks.

Following initial scoping visits in 2000 and 2001, a detailed survey exercise in 2003 generated a profile of the diverse structures of enterprise in the district. The survey work disclosed significant heterogeneity of land use and activities in this district, but also three fairly distinctive precincts with characteristic specializations and identities. First, the Hoxton Square precinct accommodated concentrations of artists and designers, including prominent gallery space (notably the White Cube), but also an assortment of amenities including bars, restaurants, cafés, and delis, as well as a large fitness studio common to new industry sites across a broad range of cities (Figure 5.2). Emblematic enterprises and institutions within this precinct include 'Circus Space', formerly the St Leonard's power generating station, and now a school for prospective circus performers which vividly evokes the theatricality of the inner city;[2] the White Cube gallery on the south side of Hoxton Square, a principal exhibition space for modern art in Hoxton's heyday as site of artistic production (Figure 5.4); and the 'Power Station' in Hoxton Market, a site providing 'workspace for cultural businesses' (Figure 5.5). Around Hoxton Square was arrayed a promiscuous assortment of activities, with bars, cafés, and restaurants as the principal uses, but also including St Monica's Basilica, a fitness centre, a number of untidy-looking textile and garment operations and outlets, and housing.

Within this precinct, Hoxton Square itself presents evocative imageries of Lefébvre's (1974) 'mixed space', with a heterogeneous and animated population characterized by diversity in terms of gender, race, and ethnicity. The park space in the square was well populated during the site visits (admittedly on mainly sunny, summer days). There was a generational bias toward younger individuals and groups (twenties and thirties), many of whom worked locally as artists, designers, or ancillary technical staff, as disclosed by a series of conversations over the study period. The interactive social space of Hoxton Square thus complements the business and consumption spaces of the adjacent streets, providing opportunity for knowledge sharing (including job market opportunities and other forms of market intelligence) as well as recreation.

Figure 5.4 The White Cube Gallery, Hoxton Square.

Figure 5.5 The Power Station, Hoxton ('space for cultural industries').

The 'Shoreditch Triangle' situated just to the south of the Hoxton artists' precinct offers a salient example of the industry-shaping power of space and the built environment, and the centrality of 'concrete' built form in the development of creative industries, following (respectively) the injunctions of Ed Soja[3] and Ilse Helbrecht.[4] The enterprise structure and streetscapes of the major arterials on the Triangle's perimeter reflect the proximity of the site to the City of London (a ten-minute walk from Hoxton Square to Liverpool Street Station to the south). Great Eastern Street in particular is dominated by larger (six–eight storeys) renovated or modern buildings, occupied by a mix of business services including IT firms and telecomms, business training institutes, estate agencies, and property development companies, together with complementary amenities such as upscale restaurants, bars, and pubs. Old Street presents a somewhat less explicit corporate imagery, with delis and cafés and an assortment of studios and small retail and personal services, but also includes high-rise commercial offices toward the Old Street Underground Station.

The arterial roads encompassing the Shoreditch Triangle present a generic CBD fringe/City Fringe look, feel, and function, dominated by overspill business services removed from the steeper rent gradients of the City. But the internal spaces of the Triangle (along Charlotte Road and Rivington Street) suggest a quite different set of animating tendencies. These more intimate spaces include generally smaller, older buildings, former warehouses and workshops, and during the Hoxton area's ascendancy accommodated a rich mix of activities, including graphic artists, artist studios, architects, interior designers, new media and retail operations oriented toward the arts (e.g. a music store on the south-east corner of Rivington Street and Charlotte Road), pubs, and edgy cafés and coffee shops appealing to the younger artists and designers working in the area (Figure 5.6). (There is also a major arts-oriented institution on Charlotte Road, the Prince's Foundation, which provides training and other support to young artists in Shoreditch and elsewhere in East London especially.[5]) This enterprise profile suggested a comfortable fit with the overarching trajectory of Hoxton favouring artistic production. During multiple observations in 2000, 2002, and 2003, too, the spaces of the Shoreditch Triangle exhibited a high social density, including steady flows of pedestrian traffic and congregations of *bon vivants* around the area's abundant consumption sites, presenting a lively and convivial streetscape, and suggesting in turn a revival of this old quarter of the East London industrial economy.

A third (and smaller) precinct within the Hoxton–Shoreditch Triangle district comprised of a cluster of firms along Leonard Street, occupying three-storey furniture warehouses dating from the 1870s (Cherry and Pevsner 1998: 525).

Figure 5.6 Convivial consumption in the creative neighbourhood, Curtain Road, Shoreditch.

This site is separated from the principal concentrations of firms to the north by the major arterial of Great Eastern Street. Observations generated from the program of fieldwork in 2002 and 2003 disclosed that this compact site included a measure of functional diversity, but specialized in primary design firms (such as architects, interior design, and graphic artists and designers) as well as technical services such as commercial photographers, suggesting a modest role for this precinct within the creative industry networks of inner London.

Epilogue: markers of change, 2003–2006

The most recent period in the developmental history of Hoxton and its environs has seen a new sequence of changes, which in important respects dilutes the district's industrial focus and identity. Even in 2003, an architect I interviewed (Oxford Architects, Bateman Street) recounted that he had offered an apology of sorts to an artist he had encountered in a local café, acknowledging the inevitable dislocation tendencies of the incursion of professional enterprises in an area of artistic enterprise and production, symptomatic of larger processes of change within inner London.

Andy Pratt has chronicled with insight the evolution of the area over the past decade, starting with Hoxton's emergence in the 1990s as 'world-class cultural innovation space' (Pratt 2007). But this period has included episodes of 'industrial upgrading' that have displaced local workers as new media firms have edged out less affluent artists, and the recent conversion of industrial premises for residential development, including upscale lofts. Pratt concludes from his detailed survey work that the relayering of development in Hoxton has been '[f]rom a cultural production perspective . . . a disaster', a failure of an overly narrow regeneration agenda, resulting in (as Pratt affirms) the critical loss of a major cultural economy site for north-east London.

Further evidence of the changing development trajectory of the Hoxton area was generated from survey work in 2006. Within the Shoreditch Triangle, many of the creative industry firms had closed or relocated,[6] several of the more edgy cafes had been replaced by up-scale restaurants, and a substantial program of building renovation was in place, making way for a new cohort of residential tenants. The pattern was repeated along Leonard Street, to the south, where several sites were under reconstruction, and the presence of production sector firms appreciably diminished. There was noticeably less pedestrian traffic and 'people presence' in the area on the multiple visits to the area in the summer of 2006 vis-à-vis the previous years, reflecting, inferentially, a decline in the numbers of artists and design firms.

It may be worth stating that the fortunes of industrial enterprises in the Hoxton-Shoreditch Triangle area have been reshaped by a combination of localized processes and broader tendencies. First, the Hoxton experience of the past decade vividly demonstrates the volatility of industrialization cycles in the contemporary inner city. The rise of new industries, and their displacement effects on pre-existing activities, are inscribed upon the landscapes of the metropolitan core.

Second, it seems clear that new industries and firms are increasingly vulnerable, not only to dislocation as a consequence of industrial upgrading of the kind Andy Pratt describes, but also a more comprehensive experience of residential conversion, as Tim Butler and Chris Hamnett have chronicled for much of inner London.[7] The precise experience of conversion and succession in land use differs from place to place, reflecting localized factors such as borough policies and planning permission practices, new foci of development industry attention, and shifts in the location of where (and what) is considered 'cool'.

Third, the legibility of Hoxton as the epicentre of Shoreditch's New Economy has been appreciably blurred, not only because of the internal shifts described above, but also because of the spread effect of new industry formation in Shoreditch and the City Fringe as a whole. A recent publication by the Shoreditch Business Network (part of the larger Business Junction consortium) provides a directory of businesses for 2006–2007, including a predictable listing of mainstream business services and (especially) estate agencies and property firms, as well as the array of bars and restaurants which comprise the convivial consumption orientation of the 'new inner city'. The directory also lists firms classified within the rubric of new media and creative industries. Included is a selective representation of software development firms, emblematic of the New Economy, and 15 design consultants, with such evocative company names and diverse locations as 'Exploding Monkey' (Buttesland Street N1), 'Crumpled Dog Design' (Phipp Street EC2), 'phunQube' (Shoreditch High Street E1), 'The Creative Corporation' (Leconfield Road N5), and 'B3 Creative' (Brick Lane E1). Clustering presents one scenario of location and development for new industries in the City Fringe. But as new industries emerge, and as changes in relations between firms embedded in production networks occur due to technological innovation or shifts in the organization of labour, a relayering of activity can produce a more diffuse pattern of firms, resulting in the commensurate erosion of localized connections, identity, and community engagement: a defining contrast to the durable and compact industry ensembles described by Martin, Hall and Scott in Chapter 4.

Bermondsey Street: the conservation ethos and the creative economy

The development of Bermondsey Street in the London Borough of Southwark as a site of London's creative economy takes place in a somewhat more remote quarter of the City Fringe than the more widely known districts north of the Thames, on the northern crescent of the City of London. Even within Southwark, historically beyond the centres of power and influence in the cities of London and Westminster, Bermondsey Street retains something of a backwater ambience, situated just south of the new sites of cultural spectacle and experience in Bankside between London Bridge Station and Tower Bridge.

The emergence of Bermondsey Street as a creative economy site also vividly exhibits both the pervasiveness and specificity of new industry formation in the inner city. The formation of new industry is linked to the distinctive spatiality and heritage built environment of this section of Bermondsey, and typifies the uneasy co-existence between industry and residential development. The emergence of a creative industry enclave within and proximate to Bermondsey Street can also be acknowledged in part as a corollary outcome of a 'cultural turn' in the Borough of Southwark's development pathway, expressed in the production of new global spaces of spectacle in Bankside, facilitated by progressive urban design policies, and the promotion of a stronger aesthetic in shaping the Borough's built environment. This aesthetic is shaped in turn by conservation programs (in conjunction with other agencies) designed to commemorate signifying heritage resources and values.

Character and setting of the site

Bermondsey Street, situated just south of the Thames in the old Metropolitan Borough of Bermondsey (incorporated within the new London Borough of Southwark in 1965), functioned as the old High Street connecting the riverside with the parish church of St Mary Magdalene. Bermondsey developed as an important warehousing district for London's burgeoning international seaborne trade in the nineteenth century, specializing in the storage, processing and distribution of spices, leather goods, and food products. In particular, a large proportion of perishable foodstuffs imported for London's markets and stores were landed in Southwark and Bermondsey, notably at Butler's and Hay's wharfs. Bermondsey also encompassed fabricating and processing industries, notably in leather goods, constituting one of the few significant manufacturing quarters south of the Thames, although not on the scale of the dense clusters of specialized production ensconced within East London north of the Thames.

Consistent with the tight spatial linkages between industry, employment, and housing within much of inner London since the early nineteenth century, the growth of warehousing and ancillary occupations generated working-class residential communities in Bermondsey, Southwark, and Camberwell. Present-day Bermondsey contains one of the last white working-class communities in London, a vestige of the great swathes of industrial communities and neighbourhoods which had characterized the social morphology of East London for most of the nineteenth and twentieth centuries.

Bermondsey Street contains a substantial stock of high-integrity industrial buildings, which are eminently suitable for adaptive re-use and conversion purposes, as well as residential precincts. Writing toward the end of the twentieth century, Bridget Cherry and Nikolaus Pevsner observed that Bermondsey Street 'still has a recognizably village character, even though the older houses are interrupted by C19 warehouses and factories and by C20 lorry forecourts in front of even bigger buildings' (Cherry and Pevsner 1983: 608). But two decades or so on from this observation, the village character of Bermondsey Street is manifestly under pressure from the social and industrial transformations common to the City Fringe districts.

Space, structure and processes of change

Bermondsey Street represents the heart of the Bermondsey Street Conservation Area, an axial zone of streetscapes, spaces, and buildings whose preservation is supported by Southwark Borough Council and English Heritage, as well as other agencies. The recent renovation and redevelopment of Bermondsey Street can be situated within a broader policy context and orientation for the area which increasingly emphasizes a suite of urban design, physical regeneration, and cultural economy policies and programs. These programs seek in part at least to celebrate features of Bermondsey's industrial past, but they also contribute to a comprehensive reorientation of the district's economy, social morphology, and imagery.

A key reference point for the new policy orientation concerns Southwark Council's aspirations to a new trajectory based increasingly on the cultural economy and creative industries, supported by heritage preservation policies and sensitive urban design programs, and evidenced in the popular success of the Tate Gallery of Modern Art in Bankside, as well as more lurid tourist attractions along Tooley Street. The new policy approach also seeks a larger role for Southwark in the high-powered corporate economy of Central London. These aspirations come together in Southwark Council's support for the proposed London Bridge Tower, a spectacular 50-storey glass 'shard' designed by Renzo Piano, proposed for the current PriceWaterhouse site near London Bridge station (Figure 5.7). The project's feasibility is in some doubt, but the Council support for Piano's Tower is at least indicative of a level of ambition new to this hitherto unfashionable part of London. In support of this contention, a manager of Southwark's planning and design department advised me in an interview that part of her mandate was to enhance the design culture of development in Southwark, a mission involving frequent discussions with architects and developers engaged in design work in Bermondsey and the borough at large (interview, J. Greer, June 2002).

If the Tate Gallery, the Globe Theatre, and the proposed London Bridge office tower speak to the enlarged ambitions of Southwark and to the broader reproduction of inner London as site of global commerce, spectacle, and culture, then the evolution of Bermondsey Street offers a more localized and nuanced expression of contemporary development trends. Bermondsey Street's spatiality comprises a northern section characterized by a mix of building types, land uses and enterprises, and a more architecturally legible southern part increasingly given over to residential conversions and upscale consumption. Structural elements of Bermondsey Street are shown in Figure 5.8 including spatiality, built form, and major activity zones.

The northern section of Bermondsey Street has accommodated a familiar array of contemporary industries and firms, including multimedia and new media, architectural firms, estate agencies, and restaurants, as well as loft conversions for housing purposes (Figure 5.9). There are also wholesale food and beverage enterprises on the eastern side of Bermondsey Street, a carryover from Bermondsey's heyday as a foodstuffs warehousing and distribution district, and industries in

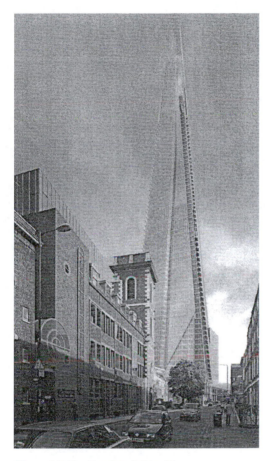

Figure 5.7 The London Bridge Tower.

Source: Hays Davidson and John McLean, Renzo Piano.

tune with the contemporary blend of production, exchange, distribution, and consumption functions characteristic of the metropolitan core.

The southern section of Bermondsey Street encompasses some vestigial uses, including a number of antique shops, corner grocers and retailers, and pubs, as well as some design firms and estate agencies (Figure 5.9). The extreme southern end of the street includes the Ticino Bakery Limited, a presence within the industrial landscapes of Bermondsey for a century, and still enjoying a brisk trade (Figure 5.10). But the dominant trajectory is one of comprehensive residential redevelopment, part and parcel with the larger processes of gentrification and building conversions now well entrenched in London's inner city south of the Thames, as it has been for a longer period in the East End districts of the metropolis. Conversations with workers in about a dozen of Bermondsey Street's smaller firms conducted in 2003, however, suggested that the price points of the new

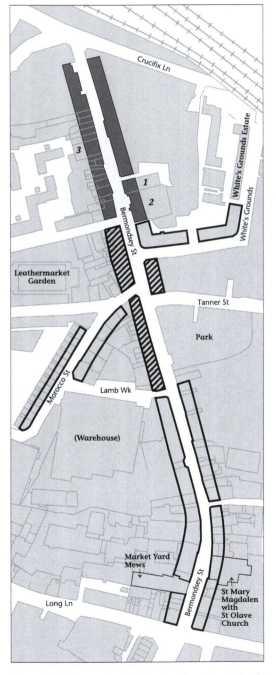

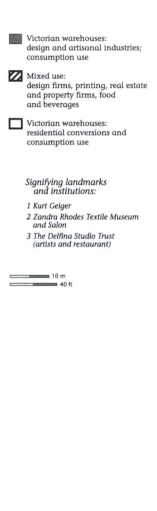

Figure 5.8 Bermondsey Street, London Borough of Southwark: structural elements.

Source: Author's survey 2004.

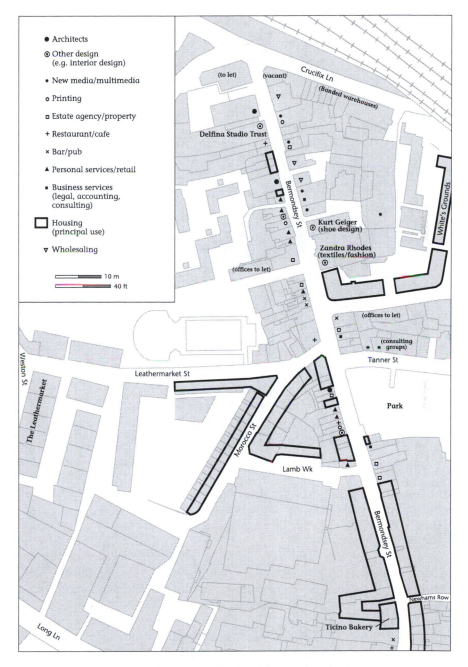

Figure 5.9 Bermondsey Street: location of selected firms and institutions.

Source: Author's field survey 2006.

Figure 5.10 The Ticino Bakery, Bermondsey Street, Southwark.

housing and residential conversions placed these units beyond the reach of the mostly younger staff, underscoring the increasing separation of place of work and residence for all but the most privileged professionals and entrepreneurs in the reconstructed inner city.[8]

While smaller firms comprise the bulk of enterprises, Bermondsey Street encompasses a number of larger companies, institutions, and projects which contribute to the area's saliency as a site of creative production. These include a textile museum and salon exhibiting the work of one of London's leading fashion designers (Zandra Rhodes), and a leading British shoe designer (Kurt Geiger); as well as a prominent institution promoting the work of young British and international artists (the Delfina Trust), and a representative operation of the trend toward 'managed workspace' situated in older warehouse and workshop buildings (The Leathermarket). Each of these operations also demonstrates the distinctive nature of enterprise formation in the reconstructed inner city, shaped by synergies between place, the built environment, and industrial innovation;

interdependencies characteristic of the intricate relational geographies of special-ized production in the twenty-first century metropolis.

First, the co-presence of Kurt Geiger and Zandra Rhodes in the heart of Bermondsey Street generates a significant design cachet and stylistic resonance for this district. Kurt Geiger is well established as an innovator in shoe design both in British and international markets, with recent sales of £110 million in 2005 (*Metrobusiness*, 26 June 2006: 54), and occupies a distinctive former ware-house structure immediately adjacent to Zandra Rhodes in the central precinct of Bermondsey Street. Together, these highly successful designers present an emphatic endorsement of the reproduction of the Bermondsey Street Conserva-tion Area as a site of creative industries and cultural production over the past decade.

The market penetration and success of Kurt Geiger provide an expression of the industrial reconstruction of the inner city for new enterprise, but in some ways the Zandra Rhodes textile museum and salon represent a more evocative storyline in its physical redevelopment and the reproduction of its site. The museum and salon, which jointly occupy a visually high-impact orange and pink converted warehouse, opened in May 2003, and emerged as the signifying presence within Bermondsey Street's heritage streetscape and creative sector. The building is multi-use, with the section fronting on Bermondsey Street accom-modating the museum and salon, and with the back of the structure comprising a complex of residential units. The overall redevelopment project also presents a dualistic profile, with a partnership between Ricardo Legoretta,[9] a leading inter-national architect, and locally-based architect Alan Camp. Legoretta, with a practice based in Mexico City, but a prominent member of what Kris Olds has described as the 'global intelligence corps' of architects and designers engaged in the reconstruction of world cities such as London, Berlin, and Shanghai, was lead architect for the external reconfiguration and the rebranding of the building for Zandra Rhodes, a major international fashion designer; while Alan Camp under-took the complex negotiations with Southwark Borough planners, contributing 'local knowledge' of development conditions and regulations to the project. Camp also developed the idea of providing residential units for the project which would cross-subsidize the construction of the museum (personal interview, A. Camp, June 2002).

The Zandra Rhodes museum and salon project thus exemplifies the (admit-tedly selective) possibility of reconciling multiple uses in the built environment of the inner city, as well as demonstrating the workings of global-local partnerships in site redevelopment within the postindustrial landscapes of London. The Zandra Rhodes project also suggests the potential of establishing in a modest way Bermondsey Street as a niche site within northern Southwark's ensemble of spectacle, experiential culture, and creative industry exhibition, situated as it is a ten-minute walk from the global cultural space of Bankside.

Just up the street from the Geiger and Rhodes' sites, near Crucifix Lane, is the Delfina Trust, situated within a former warehouse (Figure 5.11). The Delfina Trust has established itself as an important institution for the incubation of

Figure 5.11 The Delfina Trust, Bermondsey Street.

promising young artists. The Trust, which is supported by endowments and con-tributions from diverse donors, offers one-year studio residencies to British and international artists, selected annually from a juried competition. The selected artists enjoy the benefits of working in attractive space in a district of established significance for creative production, as well as the larger possibilities of artistic and professional development in a leading global centre of the arts and culture.

The artists studios occupy the higher levels of the Delfina Trust building, with a gourmet restaurant (also operated by the Trust) situated on the ground floor, a source of income for the Trust's stewardship of the arts, and a contribution to the conviviality of the Bermondsey Street conservation area. The Delfina Trust in Bermondsey demonstrates the importance of complementary institutional sup-port for the arts in the cultural metropolis, and more specifically the advantage of the inner city's characteristic institutional thickness relative to the more austere environs of many more peripheral areas of the metropolis. Further, the dual function of the Delfina Trust as a site of creative production (in the form of the studios) and consumption (the restaurant) vividly demonstrates the synergies between the arts, design, knowledge production, and amenity in the reconstructed inner city.

These enterprises can be viewed as the most salient features of Bermondsey Street's evolution as a site of cultural production, encompassing a regional and international scope, but locally embedded within the district. There are also projects within the district that exhibit more pervasive tendencies, notably the Leathermarket, situated on Morocco Street, a one-minute walk west from Bermondsey Street (Figure 5.12). The Leathermarket building and site suggest a strongly localized reference, both in title and in the renovated building which provides a hosting environment for creative firms and other businesses. But the Leathermarket is also one of the 'managed workspaces' within London, owned and operated by the Workspace Group Plc: in essence, a 'chain', of sorts, although imbued with local operating characteristics, historical references, and tenant base. The Workforce Group business model also endeavours to combine

Figure 5.12 The Leathermarket, Bermondsey.

a local regeneration agenda with a viable revenue flow in its operation of The Leathermarket.

The Leathermarket (see Figure 5.9 for location) has been operating for about a decade, and has established itself as an incubator of creative and professional companies, with a clear emphasis on the SME sector widely seen as a critical instrument of local economic development. Leanne Keltie has recently (2006) prepared a report which evaluates some aspects of the Leathermarket's developmental role and history, as part of her profile of diverse workspaces within Greater London. Keltie's panel of interviews, combined with site visits, disclosed developmental principles and operating features of the Leathermarket. With regard to the former, the foundational principles included appropriate building type and internal space considerations, flexible leases and affordable rents for start-up companies, and on-site services provided by the Workspace Group Plc. Here Keltie's informants affirm that the distinctive feature of the 'managed workspace' is an emphasis on 'stewardship' rather than merely 'management'; 'hands-on' and nurturing, in other words, not just enforcing compliance with the site's regulations.

The changing mix of industries and firms within the Leathermarket, however, might require a new definition of regeneration in the reconstructed inner city. If an earlier concept of community economic development relied on the durable presence of industries and enterprise to 'anchor' or stabilize the employment base of an urban district, a contemporary interpretation of the regeneration mission assumes instead the rapid turnover of firms, a return to the notion of the inner city as industrial incubator. Keltie's study of managed workspaces in London (2006) acknowledged the concept of rapid firm turnover or 'churn' as an operating feature of the Workspace Group Plc's model, with one of her respondents noting that if the firm base is stable, 'it is not contributing to the economy. You need growth, dynamism and change. You need churn – it is really important' (quoted in Keltie 2006: 53). The overall annual turnover rate among enterprises within the Workspace Group's London properties approached 25 per cent, ensuring availability of space for start-up firms. A clear implication of these observations is that in important respects the industrial role of the contemporary inner city is one of 'territorial-based innovation' (Morgan 2004), a zone of intense experimentation, available for enterprises associated with the latest phase of innovation and restructuring, but also an area conducive to recurrent transition and succession experiences at the level of the firm.

Epilogue: markers of change, 2003–2006

The trajectory of Bermondsey Street's development is increasingly shaped by two sets of forces, incorporating both complementary and oppositional tendencies: first, the remaking of northern Bermondsey as a site of cultural spectacle and creative industries, and, second, the relentless spread of residential conversions within the postindustrial landscapes of inner London. We can insert between these trajectories the expansion of consumption and amenity, in the form of bars, cafés, restaurants, and open spaces. These amenities are broadly complementary both with new residential development and with amenity-seeking industries and firms. But over the longer term, the development of these amenities is likely to be shaped more by the expanding local residential populations, rather than by the comparatively volatile employment population of the district.

Some hints as to the changing balance of activities and land use in the Bermondsey Street area, and the increasing presence of new cohorts, were disclosed from return site visits in June and October of 2006. The cornerstones of the area's cultural economy and creative industry vocation, Zandra Rhodes, Kurt Geiger, the Delfina Trust, and the Leathermarket, appeared largely intact; although (as noted above) the internal mix of activity in the latter is purposely fluid. But, as in other postindustrial districts of the City Fringe, the residential development trajectory is clearly in the ascendancy. This is most clearly evident in the southern portion of Bermondsey Street, where the processes of loft conversions noted in the earlier site visits were now well advanced (see example, Figure 5.13).

Figure 5.13 Victorian warehouse residential conversion, Bermondsey Street.

This residential conversion syndrome may be unsettling some industries: a sign posted to the wall of the Ticino Bakery affirmed that the family ownership intended to continue the century-old tradition of the manufacturing of baked goods in Bermondsey, emphasized by another sign that the property was 'not for sale', perhaps a declaration of defiance in the midst of adjacent new residential communities who might object to the truck traffic and occasional congestion associated with the daily operation of the bakery.[10] The presence of two new upscale restaurant-bars in this precinct also underscores the momentum of development increasingly favouring housing and residential populations. In this southern part of the area, too, a small enclave of design firms has been established in the mews behind the street front, deferring a more prominent location on Bermondsey Street itself to the luxury loft conversions and condominiums (see Figure 5.9).

There is also visual evidence of the changing class configuration in the area. In the White Estates extending eastwards from Bermondsey Street, an older council flat, its class consciousness proclaimed in its extravagant draperies of the Cross of St George during the World Cup campaign of 2006, a practice emblematic of Bermondsey's white working class, was situated across from a newly-renovated estate occupied (apparently) by members of the new middle class. This latter group exhibited its status by the display of Volvos and BMWs parked neatly in the estate forecourts, and by a more restrained (if indeed not altogether sceptical) affiliation with the England squad and its grandiose quadrennial aspirations.

Clerkenwell: artisanal production and the London property machine

Hoxton and Bermondsey Street exemplify relatively compact new industry sites within inner London, but Clerkenwell represents a larger district within the City Fringe, comprising multiple precincts and clusters of specialized production. Clerkenwell also encompasses a substantial and rapidly expanding array of residential neighbourhoods, a burgeoning consumption sector, a distinctive infrastructure of public and private institutions, some of medieval origin, and, even by London's standards, a particularly rich and at times tumultuous history. Clerkenwell (part of the old Finsbury metropolitan borough, incorporated into the new London Borough of Islington in 1965) encompasses the full range of the diverse industrial production regimes extant within the City Fringe – artisanal, Fordist, post-Fordist and 'neo-artisanal'; comprehensively exhibits the extraordinary complexity of industrial organization and specialization in the twenty-first-century global city; and demonstrates in many ways the interdependencies and tensions between production, consumption, and housing acknowledged as defining attributes of the 'new inner city'.

This specialization has included foremost Clerkenwell's development as an important domain for the growth of cultural industries, artisanal labour, and the arts in London. Graeme Evans has carefully documented the evolution of Clerkenwell's specialized industries and crafts over the last century and a half, observing that the 1861 census disclosed the presence of 877 clock and watch-makers, 725 goldsmiths, 720 printers, 314 bookbinders, 164 engravers, 97 musical instrument makers, and 20 surgical instrument makers – all male; while the same survey identified 1,477 female milliners and dressmakers, 267 book-binders, and 33 embroiderers (Olson 1982, quoted in Evans 2004: 85). As Evans notes, a century later, Clerkenwell still encompassed over 900 arts and crafts-based firms and artist-designers, with just under 50 per cent in the print-design and jewellery-metal craft trades (Evans 2004), constituting a platform for its recent growth as one of London's most important cultural quarters. So Clerken-well's (re)development is in large part illustrative of restructuring episodes, including the upheavals of the first half of the twentieth century and London's comprehensive industrial collapse in the second half of the last century, while in other respects suggesting its resiliency as a zone of experimentation, creativity, and innovation within London's City Fringe.

Character and setting of Clerkenwell

In acknowledging that many districts in London have been proposed as repre-senting a microcosm of the capital, Richard Tames suggests that the claim of Clerkenwell and the Finsbury wards to this emblematic status 'is better than most':

> It has been a religious precinct, an aristocratic *quartier*, a rather raffish resort and, simultaneously, the birthplace of Methodism. It has been the most industrialised borough in the metropolis – and one of the most unhealthy, overcrowded and impoverished. It has been blitzed and built over. It is now very much reviving.
>
> (1999: 6)

Clerkenwell, as Peter Ackroyd (2000) has vividly chronicled, still bears witness to its radical identity in its encompassing of the Marx Memorial Library on Clerkenwell Green, the head offices of Asylum and Amnesty International, and, the Red Star shop on Bowling Green Lane specializing in Soviet-era art and agit-prop artifacts, and, in its prominent Farringdon Road site on the western margins of Clerkenwell, *The Guardian* newspaper.

Clerkenwell's development also reflects the changing social morphology of the district and its impact on industrial enterprise and labour formation, with the nineteenth-century Italian immigration and subsequent growth in the local clock- and watch-making industry constituting an example. The evolution of Clerkenwell's industrial structure and identity has been manifestly shaped by social factors and the distinctive residential communities and neighbourhoods of the area, as well as by its industrial vocation. As Mike Franks, a leading figure in Clerkenwell's regeneration movement, has observed, Clerkenwell's identity has been shaped by an uneasy synthesis between, on one hand, the order and logic embedded within its defining precision crafts tradition, and, on the other, the district's more turbulent social history, creating a distinctive set of tensions and contradictions that have occasionally produced communitarian disruption and anarchy.

Space, structure, and twenty-first-century processes of change

At the broader structural level, the northern half of Clerkenwell, between Pentonville Road and Bowling Green Lane, is dominated by residential uses and retail activity (including the Exmouth Market), encompassing a significant legacy of high-quality public housing in the Finsbury ward, while the southern areas encompass diverse systems of industrial production and commercial activity, consistent with Clerkenwell's prominence within the City Fringe, although residential conversions, vigorously promoted by property market firms, are making inroads in this industrial-commercial zone. Jo Foord's recent work on the evolution of space and land use in Clerkenwell reveals this essentially bifurcated spatial structure of the district as shown by detailed ground-level survey, but also a far greater complexity of activity when the 1st, 2nd, and 3rd floor mixes are included.

The contemporary industrialization experiences of Clerkenwell bear the familiar imprints of successive phases of industrial innovation, restructuring, and dislocation. This experience has included a period of Clerkenwell's development as an important secondary office market following the 'big bang' of 1986, related to its adjacency to the rapidly expanding City of London corporate office complex. This commercial role was underpinned by the redevelopment of obsolescent industrial sites for commercial purposes, enabled by changes in regulation permitting the conversion of industrial buildings and workshops for offices (Hamnett and Whitelegg, in press).

But economic development in Clerkenwell has also been shaped by the district's unique internal spatiality and built environment. In particular, the complex and intricate layouts of medieval streets in the southern half of Clerkenwell, adjacent to Smithfield Central Market and the City of London, have helped shape the formation of multiple clusters of specialized production. The broad distribution of an illustrative set of industrial clusters, precincts, and sites is shown in Figure 5.14. In the area of southern Clerkenwell immediately adjacent to the City of London, we find concentrations of banking, financial, commercial, and professional design firms, mimicking to an extent the high-powered corporate activity profile of the City proper, but accommodated for the most part in more modest and historically resonant buildings. This distribution includes a branch operation of Merrill Lynch and a number of consultancies domiciled within a distinctive 'horizontal skyscraper' on the Farringdon Road, a postmodern horror of the 1980s close to Farringdon Underground Station, which affords convenient access to Liverpool Street Station (two stops to the south) and to King's Cross–St Pancras (one stop to the north) on the Metropolitan Line.

Across Farringdon Road large former warehouses are under conversion, both for office and residential uses. This southern area of Clerkenwell adjacent to the City also includes substantial concentrations of office firms, including architects,

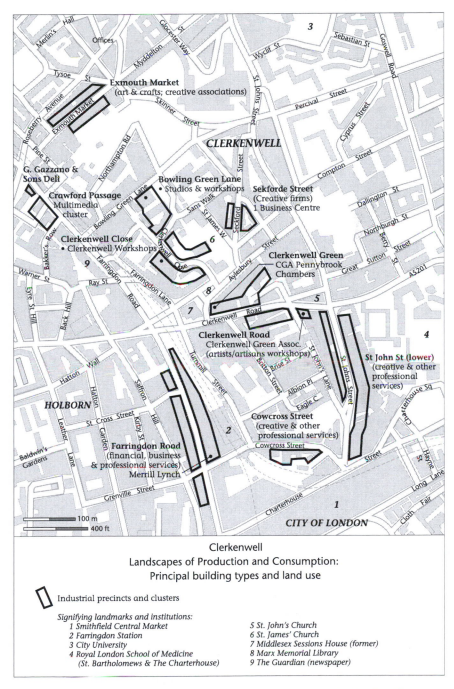

Figure 5.14 Specialized production zones and sites in Clerkenwell, London Borough of Islington.

Source: Adapted from Hutton (2006).

interior designers, and other creative professionals, along the lower reaches of St John's Street, and also within Cowcross Street, the latter ensconced between Farringdon Station and the Smithfield Central Market, although there is evidence of the incursion of consumption activities in this latter precinct symptomatic of what Evans describes as the burgeoning 'café culture' of the district. Further north on the Farringdon Road a precinct of new media firms occupies an enclave on Ray Street, a prosaic enough but authentic demonstration of the more durable features of the New Economy within the City Fringe, while Gazzano's Deli just a few steps to the north has been a Clerkenwell fixture for over a century (Figure 5.15).

These financial, commercial, and professional office operations represent an important feature of Clerkenwell's space-economy. But just to the north, between Clerkenwell Road and Bowling Green Lane, a distinctive set of industrial precincts offers insights into Clerkenwell's role as site of precision trades and craft skills, past and present. These sites of artisanal production include, notably the Clerkenwell Green Association (CGA), an agency established in 1971 to sustain

Figure 5.15 Gazzano's Italian Deli, Farringdon Road, Clerkenwell.

traditional craft skills and production within the district (Evans 2004), with operations in the Pennybank Chambers in St Johns Lane, and also just to the west on the edge of Clerkenwell Green (Figure 5.16).

The former unit includes both the CGA's administrative offices and a dense cluster of workshops, while the second showcases leading artisans, including, currently, jewellery designers and a fresco artist. These operations continue to perform roles in the larger processes of regeneration within Clerkenwell and the broader City Fringe, but also typify in salient ways the more problematic features of precarious industrialization in the 'new inner city'. The insistent pressures of upgrading, both industrial and residential, the relentless dislocations of the property market, and the encroachment of new consumption activity and gentrifiers, all contribute to the pressures on production activity in this historic quarter of industrial London.

Clerkenwell Green, the site of numerous communal riots and disruptions, still represents the heart of the district, a space which includes elite architects, administrative offices of musical ensembles, the Marx Memorial Library, galleries, and of course numerous cafés, pubs, bars, and restaurants. There is no longer any 'green' in Clerkenwell Green, but it still functions as a space of social interaction and conviviality, including performance and exhibition space for events such as the London Architectural Biennale and musical performance. It can also accommodate performances and demonstrations of innovation, exemplified recently by the installation of the 'London Oasis' sustainable development exhibit in the summer of 2006, a facility which invited visitors to experience new concepts of energy conservation.

Figure 5.16 The Clerkenwell Green Association, Clerkenwell Green.

The Clerkenwell Workshops: vicissitudes of artisanal stewardship

The heart of present-day Clerkenwell contains a range of artisanal sites, firms, and institutions, the most resonant vestiges of the district's vocation as epicentre of London's skilled craft labour and production economy. But foremost among the sites of artisanal production has been the iconic Clerkenwell Workshops, situated in Clerkenwell Close (see Figure 5.14 for location), a site which evokes defining episodes of industrial restructuring and more localized agencies of change in the district over the last century.

The Clerkenwell Workshops (nos. 27–30 Clerkenwell Close) occupy four- and five-storey warehouses, constructed in 1895–1897 by the Works Department of the London School Board as its central storehouse. On the brick frontage of the Workshops, 'Furniture', 'Needlework', and 'Stationery' (Figure 5.17) are inscribed in cartouches over the entranceways. As Bridget Cherry and Nikolaus Pevsner observe in their *Buildings of North London* volume, the London School

Figure 5.17 The Clerkenwell Workshops, Clerkenwell Close.

Board Warehouse was converted to workshops in 1975, and were among 'the first to challenge the post-war policy of replacement of Clerkenwell industry by housing' (Cherry and Pevsner 1998: 625), reminding us of the longer history of contestation between industrial and residential development in the district. The three decades that have elapsed since the establishment of the Clerkenwell Workshops have been fraught with episodes of instability and pressures for change, not least being the collapse of inner London's traditional industrial economy over the course of this period. The Workshops' reconfigurations, changing occupancy profile, and periodic rebranding experiences shape narratives of industrial continuity and disruptions within the larger economic landscapes of Clerkenwell.

For a quarter of a century the Clerkenwell Workshops accommodated a mix of artisans, artists, and craft workers, in about 150 workshops and studios, but by the early years of the twenty-first century this traditional vocation appeared to be in terminal decline. The (then) manager of the Workshops advised me in an interview in 2002 that the new owners, Workspace Group Plc (which owned a number of similar sites within the City Fringe) had developed a new business model which combined aspirations of a more 'contemporary' client base with local regeneration. In preparation for the renovation and rebranding of the Workshops, the manager was instructed to remove the external signage which conveyed a decidedly anachronistic identity (Figure 5.18): a symbol redolent of its crafts-based past, but one hardly suitable for the rising professional enterprises in the arts, media, and design sectors now clearly in the ascendant, now seen as prime candidates for the reconfigured Workshops.

Over time the internal building configuration and mix of tenants have evolved, matching the changing trajectories of industrialization within inner London's production economy. The original contingent in the mid-1970s included skilled craftsmen (involving metalwork, printing, weaving, and instrument-making, a direct carryover from Clerkenwell's heyday as centre of London's precision instrument industries and crafts), but this cohort has largely given way to a contemporary tenant base which conforms to the dominant orientations of the twenty-first-century global metropolis, including applied design, cultural production, media and communications, and property management. The relaunch of the Clerkenwell Workshops in June of 2006 included an art exhibition designed to emphasize the new aesthetic orientation of the site, as well as a reconstituted logo (Figure 5.19) conveying a more contemporary business identity than the former calligraphic signage.

The internal reconfiguration of the Workshops for the 2006 re-launch took the form of a consolidation of the small workshops to produce more generously proportioned and glossy studio space. The new studios presented a recognizable heritage look and feel, but were now more conducive to the needs and preferences of the aesthetic professionals who were to supplant the cohorts of artisans and craftsmen. A visit to the Workshops in October of 2006 disclosed a fairly rapid take-up of workspace, with new occupants including media firms, architects, graphic designers, television and broadcasting, and property consultants, as well as the casual but upscale Clerkenwell Kitchens located on the south building front

Figure 5.18 Clerkenwell Workshops: 'artisanal signage', circa 2000.

established to provide a venue for consumption *in situ*, all of which reinforce the dominant trajectory of twenty-first-century development in this archetypical inner London industrial district.

Britannia Row 2: ultra-high amenity for the elite creative professional

If the Clerkenwell Workshops provide an evocative profile of struggle in the effort to maintain a substantial artisanal presence in Clerkenwell, both as business strategy and as an instrument of community regeneration, then Britannia Row 2 presents an exemplar of catering to the most affluent of the creative class in London's new inner city.

Britannia Row 2 is situated within a Grade II listed building at 1 Sekforde Street, an industrial structure that previously accommodated a community service group for senior citizens following its days as a workshop. The building was

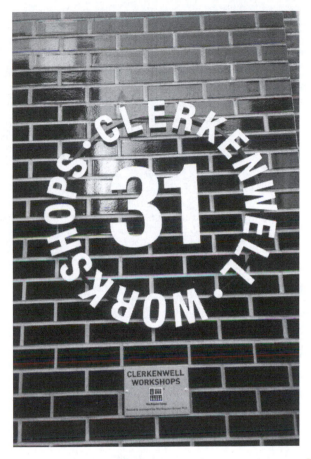

Figure 5.19 Clerkenwell Workshops: 'elite creatives' signage, 2006.

purchased by Nick Mason and Rick Wright of Pink Floyd, an event marking the emergent status of Clerkenwell as celebrity *habitus*, and was comprehensively reconfigured internally in 2002 as an ultra-high amenity workspace for a target clientele of elite creative professionals, including musicians and designers.[11] The array of amenities available to Britannia Row 2's resident professionals included an organic latté bar, a number of 'chill-out' rooms for creatives seeking sanctuary from the quotidian stresses of cultural production, and a massage room for those upper-tier workers requiring a soothing of the muscles as well as the spirit, all ensconced within the enviable and privileged ambience of a secluded corner of Clerkenwell.

In this initial period circa 2002, Britannia Row 2 provided a stark contrast, both to the austere and cramped conditions of its initial workshop vocation, as well as to the vestigial cohorts of traditional artisanal workers of the Clerkenwell Workshops, a two-minute walk to the west. A fresh visit to Sekforde Street in June of

2006 revealed the renaming of the site as '1 Business Centre', suggesting a perhaps slightly more prosaic imagery than the original Britannia Row 2 designation. That said, the building was evidently almost full, with only a few lettings available according to the receptionist, and with a full representation of creative firms and professional staff, including a number of firms at the vanguard of elite and popular cultural expression. To illustrate one of the prime occupants of 1 Business Centre is Bose·Collins, a public relations, branding and design firm catering to the most successful popular culture and musical performers, with a portfolio including artwork for *Sergeant Pepper's Lonely Hearts Club Band*, as well as a more recent work record incorporating the rebranding (and relaunching) of prominent restaurants and bars in Chelsea and elsewhere in London.

Despite a doubling of rents for space in the 'new' Clerkenwell Workshops in the Workspace Plc era, there is still a cost gradient between the Workshops and 1 Business Centre. But the apparent convergence of style and occupancy profile between these exemplary sites nonetheless provides compelling evidence of the strength of Clerkenwell's experience of industrial upgrading, professionalization, and revalorization of property and buildings in this historic quarter of London's City Fringe. Traditional craft production in the established artisanal spaces and sites of the district must now be designated a vestigial feature of the economic landscape, increasingly confined to the Clerkenwell Green Association's workshops and to more scattered sites, effectively supplanted by the upscale design and cultural production enterprises which now dominate Clerkenwell's space-economy.

Epilogue: markers of change, 2003–2006

Clerkenwell's unique value as a site of industrial innovation and evolution as one of London's most important cultural quarters is defined in part by its contemporary representation in each of the generative processes of change outlined in Chapter 4, including, to illustrate: (1) the persistence of traditional artisanal industries, notably supported by the Clerkenwell Green Association, although as the Clerkenwell Workshops experience demonstrates, this presence is now clearly vestigial; (2) the power of the financial-commercial sector, evidenced in a strip of offices in the southern sector of Clerkenwell adjacent to the City of London; (3) imprints of the New Economy, seen in the media cluster on Ray Street just off the Farringdon Road; (4) the saliency of the 'cultural economy of the city', both as an historical and contemporary facet of Clerkenwell's industrial development; and (5) continuing specialization of production and labour formation, a constant in the district's development trajectory for a century and a half.

But from this checklist of processes, we can extract a more refined number of signifying forces and factors that are likely to be more decisive in transforming Clerkenwell's development pathway and identity over the next decade and beyond. First, the rise of commercial office activity has effectively squeezed other uses, with Evans reporting a 40 per cent contraction of cultural sector firms and artists and designers in Clerkenwell between the late 1980s and the turn of the century (Evans 2004: 86). In this scenario the emphasis shifts from *regeneration*, connoting an engagement of corporate investment behaviours with community needs and interests, to *redevelopment*, driven more decisively by narrow financial return imperatives associated with building economics, social demand, and the behaviour of the London property market. Evans also notes a spatial shift of cultural production firms, notably to the eastern margins of the City Fringe and its cheaper premises, and a consequent dilution of the former spatial clustering profile of the original site, a process somewhat analogous to what we observed in the preceding Hoxton–Shoreditch vignette.

Second, the process of residential development continues apace, including conversions and adaptive re-use, driven by a confluence of supply and demand factors pervasive within the City Fringe districts of London, but incorporating some features particular to Clerkenwell. As Chris Hamnett observes, the experience of residential conversions dates from 1991, when Harry Handelsman and his Manhattan Loft Corporation of New York purchased an old print building located at 1–10 Summers Street, immediately to the north of Clerkenwell Road, just west of the Farringdon Road, for a successful commercial loft conversion. Hamnett observes that Handelsman's project was pioneering in two ways: first, in achieving high values, with a selling price ten times the initial acquisition cost, and, second, in the promotion of the new lofts, 'professionally marketed in design and fashion magazines', and with the result a placement of loft living styles 'centre stage' in Clerkenwell (Hamnett 2006).

Third, the mix of forces and trends described above have transformed Clerkenwell's social morphology, with a marked tendency toward upgrading and professionalization (and therefore new housing preferences, lifestyles, behaviours, and identities). Keltie's report (2006) on managed workspace and regeneration in Clerkenwell and other City Fringe districts includes an acknowledgement of the remarkable shift in the occupational structure of the Clerkenwell Ward, from a clear predominance of manual workers in 1971 (57 per cent of the Clerkenwell employment in skilled manual, semi-skilled, and unskilled), to an even more pronounced professionalization structure, with 62 per cent of employment comprised of managers, professionals, and associate professionals in 2001 (ibid.: 82). As regards housing tenure, just over one-half (54 per cent) of households in the Clerkenwell Ward live in rented Council housing in 2001, down from 63 per cent in 1971, while the proportion of owner occupiers increased from a mere 1 per cent in 1971 to 29 per cent in 2001 (ibid.).

Conclusion: implications of the London experience

Viewed through the lens of industrial urbanism, calibrated at a localized scale, what emerges from aspects of the London experience in Chapters 4 and 5 is a fascinating story-line of (selective) reindustrialization and its vicissitudes within the districts of the City Fringe. This distinctive zone encompasses districts which had sustained a world-scale light manufacturing and artisanal production economy and its social cohorts and community structures for over a century, but which had suffered a comprehensive collapse in the late twentieth century, incurring employment contractions in the hundreds of thousands.

The reassertion of production in the inner city, constructed metaphorically on the ashes of East London's traditional industrial economy, provides on its own merits ample material for a stimulating essay on contemporary aspects of urbanism in the metropolis. But the intersections of localized new industry formation with other city-shaping processes – globalization in its economic, social, cultural, physical, and spatial dimensions, the acceleration of industrial restructuring episodes since the early 1990s, and the social reconstruction of the metropolitan core – greatly add to the significance of the London case study. Once again London presents an exemplar of transformative change, with exigent theoretical and normative implications: a unique urban place to be sure, but one which presents lessons which resonate widely within the urban studies realm.

New industry formation and the regeneration agenda

The analysis of the London experience of new industry formation within the City Fringe, including the overview discussion and the selected case studies, points both to the possibilities and limits of reindustrialization as an agency of regeneration in the 'new inner city'. The dominant factors in the reshaping of space in the metropolitan core in London include the power of markets and capital in maintaining the primacy of global financial and corporate control functions, the nascent global spaces and megaprojects represented by Canary Wharf and the 2012 Olympic sites, and, increasingly, the pervasive influence of new cohorts of gentrifiers in the widespread conversion of buildings and space for upscale housing. By comparison, new industry formations within the inner city experience volatility in composition, associated with recurrent phases of innovation and restructuring, with the overriding power of financial and business services, and with the relentless encroachment of upscale housing and amenities, driven by social demand and the London property machine, ever-sensitive to changes in the returns to capital in the real estate market. Our case studies demonstrate that the most recent narratives of industrial urbanism in the global city take the form of a decidedly 'precarious' (or volatile) industrialization experience, rather than the establishment of more entrenched industrial ensembles characteristic of London's apogee as major centre of light industry and engineering, as described by Peter Hall, J.E. Martin and others.

In policy terms, too, there are still aspects of conflict between industrial regeneration as a community development tool and more traditional borough-level development control procedures, particularly where there is a poor fit between the skill demands of new industry occupations and the characteristics of the local labour market. Generally, the boroughs have been 'followers' rather than 'leaders' in this policy sphere, and in some respects NGOs have been more influential through targeted programs (as we saw in the case studies) and a more intimate affiliation with enterprises, entrepreneurs, and workers. The Mayor and GLA also have a more visible profile in creative and knowledge-based industry development than individual boroughs at a strategic level, and central government actors (including the Prime Minister) have been quick to take a measure of credit for London's emergence as a leading cultural metropolis.

The New Economy of the inner city and London's space-economy

The scale and intensity of new industry formation in the City Fringe, including the clustering of firms and labour as well as more dispersed patterns of activity, contribute to the contemporary reconfiguration of London's space-economy. The modest role new industries play in the shaping of London's economy has multiple dimensions. In the aggregate, the new ensembles of specialized industries in the Fringe districts, though subject to processes of transition and succession, represent a significant reassertion of industrial production in the metropolitan core. These new industrial ensembles indicate a partial and selective recovery of industrial production in the inner city, offsetting to a limited extent the massive contractions of firms and labour incurred in the quarter-century or more of industrial restructuring and collapse in the late twentieth century. Second, the new industries of London's City Fringe are connected to more strategic processes of reshaping the core, including interdependencies between new industries, housing, and amenity, although these relations are typified by conflict as well as by complementarity. Creative and knowledge-intensive firms in the City Fringe act as suppliers or subcontractors to corporations in the City and elsewhere in the metropolis. They often function in large part as integral elements of production chains, as Andy Pratt has observed, rather than as isolated and self-contained 'clusters'. Third, the generation of a distinctive employment profile and occupational structure associated with new industry formation comprises a modest element of the emerging spatial divisions of labour within Greater London. Here, the typically hybridized labour of the inner city's New Economy, characterized by a synthesis of creative functions and increasingly advanced technological capacity, and a relatively 'flat' organizational profile, contrasts with the more hierarchical and segmented labour structure of the CBD's corporate office economy. Overall, the new industries of the City Fringe add to the diversity of a metropolitan core economy.

The chapter included a sketch of the positioning of the New Economy of the inner city within the larger contours of London's space-economy. What emerges is a quite distinctive, global city, spatial economic profile, encompassing aspects of dispersion and variegation, as well as multinucleation. Greater London's early twenty-first-century space-economy incorporates the persistence of a 'strong centre' structure as represented by the City of London, as well as the polycentrism represented by the expansion of the Canary Wharf global financial complex, and the more dispersed City of Westminster economy. This polycentric structure comprises creative, cultural, and knowledge-intensive industries as well as central and local government functions, tourism, and retail activity, as depicted in Peter Hall's model presented in Chapter 3 (Figure 3.8). There are also 'inner edge' city complexes such as Hammersmith and the Paddington Basin, as well as more isolated or one-off 'edgeless city' formations such as the BNP-Paribas complex in Marylebone. In this context, the *multiple* inner city formations of new industries and labour, which include our case studies of Clerkenwell, Shoreditch, and Bermondsey Street, reflect the influence of global city scale and specialization, as well as more localized contingencies of development. Higher-order global cities tend to have multiple inner city new industry formations, relative to smaller cities. (For example, in Tokyo there are multiple sites, including Ropponggi, Shinjuku, and Shibuya; Singapore has Telok Ayer and Little India, and so on.)

The New Economy of the inner city and global city discourses

The emergence of a new economy of cultural industries and creative firms in the City Fringe also contributes to new narratives of the global city discourse. The London case demonstrates that creative and cultural industries now comprise part of the repertoire of global city functions and specialization, augmenting the established measures of primacy which include banking and finance, head offices and multinational corporations, and intermediate services. In the London case, the developmental scale and implications of the new industries of the inner city contribute to London's growing primacy within Britain, and also reinforce London's exalted positioning within the European and global urban hierarchies. The growth rates of these cultural industries are uneven and may even be subject to downturns, and are (as we have seen in the case studies) by no means immune to experiences of displacement and succession, but these activities form a part of London's bundle of global economy sectors.

The New Economy and the new middle class

From an urban studies perspective, the emergence of new industry sites and precincts within the City Fringe implies the reformation of constituent social groups in the metropolis, with implications for identity, behaviour, consumption patterns, and housing preferences. There is the suggestion that the qualitative and quantitative dimensions of this cohort, if not exactly constituting a semi-autonomous 'creative class' as Richard Florida proposes, do at least represent a distinctive subset of the new middle class of the global or transnational city. Earlier formulations of the new middle class postulated a strong correlation between professional and managerial workers and specific urban housing market attributes, in terms of location, tenure, and proximate amenity mix, as well as lifestyle, behavioural, and identity signifiers. There is an implied (if anecdotal) affinity between 'new industry' workers and innovative urban housing forms, such as loft conversions and live-work studios.

But in the London case the inflationary nature of the inner city housing market promotes an increasingly exclusive occupancy defined by occupation and income, so this connection between new industry workers and inner city residential neighbourhoods may in fact be quite tenuous. There may be a divergence between the experiences of workplace and place-of-residence relations of the office-based managers and professionals encompassed within the labour and housing markets of the central city, and those of the mostly younger and less well-paid creative worker, a question worth pursuing empirically in London and other cities.

Back to the future? New industries and the revival of London 'places'

Finally, the emergence of these new industries in certain inner city districts may be interpreted as a revival of the identities of former London boroughs now subsumed in the larger local government units of the capital, ensuing from the 1963 London Government Act and the 1965 reorganization which introduced the GLC and new boroughs. These include Hoxton, Shoreditch, Bermondsey, and Clerkenwell, once vital and highly distinctive industrial and social places within the burgeoning London metropolis of the nineteenth and early twentieth centuries, and now recalled to life as centres of experimentation, innovation, and cultural production in London's 'new inner city'. To be sure, the industrial, occupational, class, and ethnic reconstructions of these 'London places' present contrasts to the traditional industrial districts and working-class labour and community structure. But the apparent affinity between new industries and specific places within London's inner city, although subject to pervasive restructuring and displacement, might imply in some small way the persistence of the 'historical local' in the transformations of the global metropolis.

6 Inscriptions of restructuring in the developmental state

Telok Ayer, Singapore

Introduction: mega-projects and micro-spaces in Singapore's development narrative

The postwar record of regional development is replete with policy experimentation, but Singapore's record of innovation and commitment is widely seen as exemplary. From Singapore's tumultuous inception as a sovereign city-state in 1965, the subsequent four decades of (almost unbroken) growth and development have been shaped in large part by state economic policies and programs. These have included statements that articulate progressive economic visions designed to shape the restructuring of sectors, industries, and employment, supported by fiscal and macroeconomic policy adjustments, assertive industrial policies and spatial economic planning, and investments in transportation and other infrastructure. In addition to this panoply of 'hard' policies emphasizing physical infrastructure and capital works, Singapore's development model increasingly features 'soft policies' and programs. These include investments in cultural development, tourism, international partnerships and exchanges (including those with ASEAN partners), marketing and information campaigns, and commitment to higher education as a means of enhancing Singapore's regional and international competitive advantage.

Outcomes of Singapore's development policies and programs are conventionally evaluated in terms of Gross Domestic Product (GDP), export trade volumes, productivity, and employment formation, and for the most part this record is impressive, reflecting the city-state's primacy in south-east Asia according to most indices of development (Perry *et al.* 1997). Singapore's increasingly postindustrial development trajectory serves to further punctuate the city-state's exceptionalist status within the region, and has inspired new research which investigates implications of specialization in advanced service industries for the emergence of a 'new middle class' of professionals, managers, and entrepreneurs, and for notions of citizenship and identity, a task which has included comparative study of occupational shifts and corollary social class reformation between Singapore and advanced 'western' societies (Baum 1999; Ho 2005).

Analyses of Singapore's economic performance and progression have tended to emphasize inter-regional and international trade flows, in light of Singapore's long history of export-led development, from its provenance in 1819 as a British trading port. That said, *internal* measures of development are also significant. These include (in addition to labour force, settlement, and population attributes) physical inscriptions of development on the landscape, which offer tangible evidence of the city-state's rapid development since 1965.

At a macro-level, the physical imprints of Singapore's development history are 'writ large' upon the landscape, in the form of strategic transport installations, industrial infrastructure, and other features of the built environment. Here we can readily enough reference the large drivers of development, including the Port of Singapore, among the world's largest and most efficient; the network of express-ways constructed to facilitate movement of goods throughout the island; the emergence of Changi as a global-scale airport; the expansion of industrial parks on the periphery of the city-state; and the formation of a modernist high-rise corporate office complex in the Central Business District (CBD), encompassing Singapore's multinational corporations, banking and financial institutions, and other intermediate services. To these we can add the Mass Rapid Transit (MRT) system, higher educational institutions (notably the National University of Singapore [NUS], and the Nanyang Technological University), and the Housing Development Board (HDB) estates, which represent critical markers of Singapore's social development.

These mega-scale elements of Singapore's development landscape provide impressive evidence of the city-state's rapid progression from colonial settlement to global metropolis. But more localized and finely textured landscape features can also generate compelling developmental storylines. These include, notably, accounts of district- and community-level change which reflect both intensely local and broader, more strategic development effects, such as the loss of the original Kampung settlements as a feature of Singapore's Malay past, Orchard Road's reconstruction as site of experience and spectacle, and, evocatively, the 'expatriatization' of Holland Village (Yeoh and Kong 1995).

In this chapter I will be developing a perspective on the saliency of Telok Ayer, one of four designated sub-districts of Singapore's official Chinatown heritage district, situated proximate to the Central Business District, as a site of new indus-try formation. Following this introduction, a concise outline of Singapore's dis-tinctive development pathway will be presented, emphasizing the role of the state and its complex interactions with domestic, regional, and global actors in shaping a sequence of benchmark restructuring episodes since the mid-1960s. Broadly, this periodization of Singapore's development encompasses three major eras: first, 1965–1985, in which the emphasis was placed on policies supporting regional entrepôt functions and the development of a (largely branch plant) export-oriented manufacturing sector; second, an assertive state program promoting specialized service industries and a shift from regional to global mar-kets following the brief (but sharp) recession of 1985; and, finally, a new phase of policy experimentation, which includes a suite of programs in support of the knowledge economy and creative industries. Next, the chapter offers a sketch of the Chinatown district, incorporating a recitation of its historical provenance, its recasting as an official heritage area situated on the western margins of the CBD, and its contemporary realization as a zone of cultural production and consumption as well as spectacle.

The substantive section of this chapter will elucidate Telok Ayer's unique role as zone of cultural production and creative industry formation, with special reference to episodes of experimentation since the late 1990s. These include the emergence of Telok Ayer as a social space of conviviality and spectacle at the end of the twentieth century, amid the general decline of Chinatown as a place of clan and kinship association, following the migration of the resident population to suburban HDB communities and estates. This vocation was disrupted by a brief metamorphosis of Telok Ayer as a site of the technology-driven New Economy 1999–2000, expressed as landscape manifestations of the global dot.com phenomenon. Following the subsequent tech crash, a small echo of the cataclysmic collapse of more prominent global sites of the New Economy, Telok Ayer was by 2003 reconstituted as a prime locus of creative industries and cultural production, incorporating an expansion of consumption amenities as well as new production enterprises, while the latest survey (December 2006) disclosed intensification of industry representation and a 'lifestyle' orientation. A subplot of this story-line throughout the narrative concerns the comparative fortunes of Far East Square (part of the larger China Square project), immediately adjacent to Telok Ayer – an 'induced' new industry site which contrasts with the largely 'spontaneous' nature of Telok Ayer's development. Chapter 6 concludes with an exposition on the localized significance of Telok Ayer's experiences of restructuring, as well as larger implications for Singapore's aspirations as a global centre of cultural production.

Singapore as exemplar of the developmental state

While Singapore's scale, regional position, governance culture, and political economy all underscore its exceptionality, its manifest successes in promoting a succession of new vocations over the past forty years have inspired emulation (as well as a fair degree of envy and resentment), both within Pacific Asia and elsewhere. As is well known, the provenance of Singapore's modern development can be traced to the desperate measures imposed by Lee Kwan Yew and the People's Action Party (PAP) in the midst of civic turmoil, economic crisis, and external threats following the city-state's expulsion from the post-colonial federation of Malaysia.

In the early years of independence, Lee and the PAP government pursued with remarkable commitment three cardinal public policy goals, including, first, the establishment of a military capacity sufficient to deter would-be aggressors; second, an accelerated industrialization program; and third, the development of public housing as a cardinal element of social and economic policy. These defining policy values were in turn supported by key institutions, respectively the Singapore armed forces, the Economic Development Board (EDB), and the Housing Development Board (HDB). Singapore's development over the past forty years can be attributed in large part to the efficacy of these state institutions, as well as to the leadership of Lee and his successors, and to the energies, skills, and productivity of the Singapore labour force. The external image of Singapore's government is one of a near-monopoly of state power in the form of the Prime Minister and PAP, and indeed there are limitations to the Singapore model of democracy judged against western tenets of governance. That said, PAP has at times responded to changing public sentiments and attitudes, demonstrating a capacity to periodically shift the course of policy to suit evolving civic aspirations.

Restructuring and globalization episodes in the 1980s and 1990s

As in other high-growth Asian states, notably Japan, South Korea, and Taiwan, Singapore's development story-line presents a spectacular realization of the industrialization paradigm. A combination of massive infrastructure investments over the 1960s and 1970s, combined with state management of fiscal and monetary policy, and manipulation of labour wages in the interests of attracting foreign investment, underpinned the expansion of increasingly higher value-added manufacturing, principally for export trade. In support of the latter, Singapore built upon and extended its regional entrepôt capacity, incorporating the development of the Port Authority of Singapore as one of Asia's world-leading container ports (along with Yokohama, Pusan, Hong Kong, and Kaohsiung), and the expansion of Changi as a global airport ('Airtropolis') for international passenger traffic and air freight.

This industrial production and export trade trajectory served to underpin Singapore's development as one of the exemplars of the Asian economic miracle of the second half of the twentieth century, conventionally included as one of the Asian 'tigers' and 'dragons' along with South Korea, Taiwan, and Hong Kong, or (as another bestial trope) a high flyer following the 'lead goose' of Japan in Akamatsu's 'flying geese' analogy. But the limitations of this vocation were exposed by a brief but sharp recession in 1984–1985, an experience which led to the formation of a new policy model in the form of the 'New Directions' statement articulated by industry minister (now Prime Minister) Lee Hsien-Loong in 1986. The New Directions program embodied assertive new policies to support Singapore's banking, finance, intermediate services, higher education, and international gateway functions. The scope of Singapore's economic development program was decisively reoriented from the regional to the international stage, in an effort to transcend the limits of traditional markets, and to sharpen the city-state's competitive advantage in the attraction of new growth sectors in an era of market globalization and industrial restructuring.[1]

Singapore's ascendancy as a global city since the mid-1980s has been as remarkable as its earlier development as an industrial production and regional entrepôt, perhaps more so given the exigency of competitive pressures in the global arena relative to the more modest regional competition the city-state faced in the industrialization period. Singapore is the smallest global city in Pacific Asia, but its international projection in banking, financial, and corporate control functions greatly exceeds its population ranking in world city terms. As observed in an influential account of the city-state's development history:

Singapore is now a world city whose fate is dependent on events in New York, London, Tokyo and connected nodes in the international economy. Its trading, investment and information links to distant countries are far more important than those to its immediate regional neighbours.

(Perry *et al.* 1997: 1)

Singapore's disengagement (in relative terms) from its regional hinterland in favour of increased linkages and connectivity with international circuits of capital, trade, and culture has been emulated by other coastal city-regions in the Asia-Pacific (Hutton 2004c).

New development policy discourses for the twenty-first century

Singapore succeeded in realizing its strategic objectives for the principal phases of its post-independence economic development, including, first, the establishment of a high-productivity, export-oriented manufacturing sector; and, second, the shaping of a sophisticated, specialized services economy over the 1980s and 1990s, including intermediate banking and financial activities. At the same time, the HDB has been instrumental in Singapore attaining the highest levels of housing quality and home ownership among advanced Asian societies, supporting the city-state's social and economic policy ideals.

But successive rounds of restructuring on the global scale act to recurrently reshape industries and labour among advanced societies, so Singapore's economy is perpetually a 'work in progress' rather than an end-state construct, stimulating recurrent policy visioning exercises and programmatic experimentation. Exigencies of competition are reflected in policy discourses which seek to maintain (and if possible enhance) Singapore's positioning near the leading edge of global economic development, as well as keeping two or three steps ahead of regional competitors. As changing cost factors favour the relocation of production and labour among mature sectors and industries to less expensive jurisdictions, there is a pressing need to replace this 'lost' (or obsolescent) activity with new growth industries, when the limits of productivity gains in these long-established industries are reached. In response, the EDB and other state agencies seek to build Singapore's competitive advantage in order to attract investment associated with New Economy industries and labour.

Since the early 1990s, new episodes of industrial restructuring have reshaped the economies and employment profiles of global cities. In Singapore, restructuring has not yet reached the stage of comprehensive 'hollowing out' of industry common to advanced western societies, and indeed has succeeded in promoting industrial productivity through technological intensification and human capital upgrades. But decline in the manufacturing share of total employment appears to be relentless, with change over the last decade clearly favouring service industries, as shown in Table 6.1.

Changing shares of total employment over time are clearly significant in identifying longer-term developmental trajectories. But employment data incorporating

Table 6.1 Changing shares of total employment by principal sector for Singapore, 1995–2005

	1995	1996	1997	1998	1999	2000	2001	2002	2003	2004	2005
Employment as at year-end (000)	1,823.2	1,925.8	2,046.1	2,022.7	2,062.6	2,171.1	2,171.0	2,148.1	2,135.2	2,206.6	2,319.9
Manufacturing	24.5	22.8	21.6	20.5	20.3	20.5	19.8	19.8	19.7	20.2	20.5
Construction	12.7	14.7	16.1	16.0	14.9	14.2	13.2	11.8	11.0	10.3	10.1
Services	62.1	61.7	61.5	62.7	64.0	64.5	66.3	67.7	68.6	68.9	68.7
Others	0.8	0.8	0.7	0.8	0.8	0.8	0.7	0.7	0.7	0.6	0.7
Employment change	109.0	102.6	120.3	–23.4	39.9	108.5	–0.1	–22.9	–12.9	71.4	113.3

Notes: 1. Data are compiled from administrative records and are based on SSIC 2005.
2. Change in employment is the difference in the employment level at the end of the reference period compared with the end of the preceding period.

Source: Ministry of Manpower, Labour Market Statistics, Government of Singapore.

absolute change among a more disaggregated industry grouping, while affirming high growth in the services-producing industries, also suggest strength in manufacturing employment formation (Table 6.2). Singapore is, therefore, a highly tertiarized but not 'postindustrial' economy and society.

Manufacturing remains one of the pillars of Singapore's economy, especially industries within the advanced-technology production sector such as telecommunications, but as K.C. Ho has demonstrated, *professionalization* constitutes an important socioeconomic process within the labour force as a whole. As might be anticipated, financial and business services 'are significantly top-heavy in the sense that there is a much higher component of professionals and managers' (Ho 2005: 96). That said, Ho's research clearly shows a rising trend of managers, professionals, and technical and associated professional occupations within the manufacturing sector (Table 6.3). This trend reflects the advanced stage of industrial production and specialization of labour within the city-state's manufacturing sector, as well as the pervasive nature of the professionalization tendency within the economy as a whole.

The last decade and a half has provided further opportunities for Singapore to demonstrate its policy adroitness and resiliency. Pressures now include not only the continuing international division of labour in the manufacturing sector but also increasing competition among the advanced services within which Singapore has long maintained a leading regional position: to illustrate, by 2003, both Malaysia and Thailand aspired to compete with Singapore for larger shares of

Table 6.2 Changes in Singapore's employment by industrial sector, 2003–2006

	2003	2004	2005	2006(p)
Total	−12,900	71,400	113,300	173,300
Goods producing industries	−22,800	16,500	39,500	62,600
Manufacturing	−5,000	27,000	29,100	40,900
Construction	−17,500	−9,100	8,700	20,700
Others	−400	−1,300	1,700	1,000
Services producing industries	9,900	54,900	73,800	110,700
Wholesale and retail trade	−2,300	11,000	12,600	17,900
Transport and storage	−700	2,800	6,400	5,900
Hotels and restaurants	1,900	4,100	5,700	12,100
Information & communication	−2,500	2,800	3,700	5,900
Financial services	2,200	6,200	7,700	11,000
Business services	4,100	15,300	20,200	33,900
Other service industries	7,100	12,800	17,500	24,000

Notes: 1 Industries are classified according to SSIC 2005.
2 'Others' comprise agriculture, fishing, quarrying, sewage and waste management.
3 Business services comprise real estate and leasing, professional services, and administrative and support services.
4 (p) = Preliminary statistics.

Source: Ministry of Manpower, Manpower Research and Statistics Department, Government of Singapore.

Table 6.3 Manufacturing and financial/business services occupational structure for Singapore, 1993–2003

Occupation	1993	1995	1997	1999	2001	2003
Manufacturing						
Management and administration	8.7	11.9	11.4	11.9	14.2	14.3
Professionals	4.8	6.5	9.3	11.1	13.0	14.1
Technical and associate professional workers	12.3	15.4	16.8	16.9	17.2	16.9
Clerical workers	11.3	9.7	10.7	11.3	10.8	11.0
Service workers	1.0	1.6	1.1	1.3	1.1	.09
Production workers, cleaners and labourers	61.9	54.8	50.7	47.5	43.6	42.8
Financial and business services						
Management and administration	12.2	6.6	14.6	13.4	16.5	16.0
Professionals	13.1	16.5	18	18.2	22.8	22.3
Technical and associate professional workers	27.7	27.4	28.5	31.6	27.0	26.9
Clerical workers	29.8	23	27.9	20.8	18.4	18.6
Service workers	5.2	3.9	3.1	5.3	5	5.4
Production workers, cleaners and labourers	12.0	12.5	7.9	10.7	10.3	8.4

Source: Singapore Ministry of Manpower, *Report on the Labour Force in Singapore,* various years.

investment and employment in education, health care, and media services (*The Straits Times,* 24 January 2003).

There are to be sure constants in Singapore's longer-term development policy record, notably the ongoing manipulation of labour markets, fiscal, and monetary policy. But we can readily identify features of innovation in the recent policy record. To illustrate, Singapore is striving to deploy its regional advantages in communications, financial expertise, multiculturalism, and international connections to tap into the growth momentum of Asia's economic giants, China and India.[2] The EDB is thus endeavouring to position Singapore as 'bridge to Asia and the world'. The rhetoric of the 'knowledge-based economy' (KBE) as the latest/next 'big thing' among advanced economies has provided impetus for new investments (Wong and Bunnell 2006), including not only funding for NUS and other institutions of higher learning, but also a substantial program of partnerships with leading international universities. These incorporate institutional development *in situ* as well as exchange programs, described by Nigel Thrift and Kris Olds as Singapore's 'global schoolhouse' initiative (Thrift and Olds 2005).

The growth of creative industries in Singapore

Prospects for new industry formation and employment growth in the cultural economy have generated policy support for Singapore's arts community and creative services sector (Yue 2006). This 'cultural turn' has stimulated new institutional and programmatic initiatives, notably 'Renaissance City 2.0', which aspires to promote Singapore as a global centre of the arts; Design Singapore, which, like similar agencies elsewhere, is intended to encourage excellence in applied design; Media 21, a sector-support strategy underpinning new media industries in Singapore; while a school of art, design, and media has recently been established at NUS (Ho 2007).

While these policies endeavour to expand Singapore's cultural economy as an element of the national production and trade sectors, creative industries already contribute significantly to revenues and employment formation. Table 6.4 shows that creative industries oriented toward the technology and producer services sectors in particular are important according to a number of developmental measures. These include IT and software services (almost S$3 billion in receipts in 2000, over 14,000 employees, and exports of S$312 million, with a value-added of almost S$80,000 per worker), advertising (over S$2 billion in receipts, and 5,584 employees, with value-added of over S$90,000 per worker), broadcasting media (over S$1 billion in receipts, and 3,747 workers), and publishing industries (almost S$1 billion in receipts, and just under 5,000 workers) leading the way (Toh

Table 6.4 Direct economic contributions of Singapore's creative industries, 2000

Creative industry	Receipts (S$ Million)	VA (S$ Million)	Employment (Number)	VA/Worker (S$)	Exports (S$ Million)
IT and Software Services	2,892	1,137	14,290	79,661	312
Advertising	2,010	510	5,584	91,332	85
Broadcasting Media	1,212	229	3,747	61,116	25*
Publishing Industries	925	283	4,972	56,919	68
Interior, Graphics and Fashion Design	653	187	4,863	38,865	NA
Architectural Services	616	433	7,185	60,264	45
Art/Antiques Trade and Crafts	192	36	1,945	18,509	0.5
Performing Arts	125	71	2,003	35,447	NA
Cinema Services	121	53	938	56,503	NA*
Photography	80	27	1,137	23,747	NA
Industrial Design	28	12	186	64,516	NA
All Creative Industries	8,853	2,977	46,850	63,543	536
All Distribution Industries	8,803	2,022	31,868	59,264	3,129
Total	17,656	4,999	78,718	61,740	3,665

Note: *Exports for cinema services are subsumed under figures for broadcasting media in Singapore's Trade Classification.

Source: Singapore Department of Statistics.

et al. 2006). Creative industries imbued with a strong fine arts and design character, including fashion design, architectural services, antiques trade and crafts, performing arts, cinema services, and photography, generate lower revenues, exports and employment. But these industries illustrate the diversity of cultural production in Singapore, and perhaps offer a platform for future development.

In fact, the performance of Singapore's creative industries since the mid-1980s (Table 6.5) demonstrate significant levels of growth (albeit from a low base in most cases), with almost all industries experiencing double-digit growth in the 1986–1990 period, while some (notably IT and software services, and advertising) displaying robust growth over the three quinquennial periods. As Table 6.5 indicates, growth among the creative industries exceeded that for the Singapore economy as a whole in each of the statistical periods.

The new industries of Singapore's twenty-first-century New Economy are situated largely within the familiar strategic-scale terrains of Singapore's space-economy, including the CBD, Jurong, and other industrial and science parks, Changi and NUS. But recent diversification efforts – and more particularly the emergence of the creative economy – have brought new districts and sites into play, including the heritage zones situated in the central area, shaped by distinctive intersections of conservation, culture, and technology. Relative to the global-scale complexes of industries, firms, and labour cited earlier, these New Economy sites situated within the heritage districts are to be sure micro-scale, weighted toward the 'small' end of the scale within the small and medium-sized enterprises (SME) sector. But at the same time these historically-resonant districts

Table 6.5 Compounded annual growth rates of Singapore's creative industries, 1986–1990, 1990–1995, 1995–2000

Creative industry	1986–1990	1990–1995	1995–2000
IT and Software Services	29.58	26.78	24.33
Advertising	23.71	11.50	12.73
Broadcasting Media	15.42	2.16	3.52
Publishing Industries*	–	21.34	7.26
Interior, Graphics and Fashion Design	16.12	31.65	6.45
Architectural Services	19.84	21.64	6.32
Art/Antiques Trade and Crafts	16.74	.18	4.94
Performing Arts	5.66	23.14	13.85
Cinema Services*	–	13.76	13.47
Industrial Design^	–	–	–
Photography	13.54	8.47	3.96
All Creative Industries	23.65	16.64	12.86
All Distribution Industries	21.10	10.57	3.76
Total	22.15	13.31	8.61
Singapore GDP	14.36	12.21	5.80

Notes: *Data for numerous segments are not available in year 1986.
       ^ Industry data only available for year 2000.

Source: Ministry of Trade and Industry, Singapore Department of Statistics.

assuredly encompass industrial ensembles of hundreds of firms; function as crucial sites of experimentation, creativity, and innovation, as well as cultural spectacle and consumption; and represent important signifiers of the global-local development interface in twenty-first-century Singapore.

Cultural production and spectacle in Chinatown

As is well known, most of Singapore's historical built environment, including the colonial era buildings as well as the landscapes produced by the racial and ethnic segregation of the colony's population, was demolished and redeveloped during the 1960s and 1970s.[3] The face value motivation for this program was straightforward enough: the exigent need to clear obsolete sites to make way for the construction of what would later emerge as the Tropical City of Excellence, a classic project of economic modernization and modernistic design values. A platform of rational planning values drove the demolitions, as many of these historical districts occupied prime land resources for the new commercial, institutional, and residential infrastructure required to realize the state's ideals of progress, largely shared, if not formally endorsed, by the population at large (although these demolitions were not achieved without vigorous expressions of dissent, see Yeoh and Huang 1996).[4]

Unpacking motivation and meaning in Singapore's heritage areas

But there were other ideological motivations underpinning the pervasive demolition programs of the first two decades of Singapore's independence, including a not-so-silent repudiation of the past, a desire to eradicate the evidence of congestion, prejudice, segregation, and squalor redolent of the era of colonial settlement. Erasure of a history of deprivation, disease, and poverty would clearly signal a bright new future, with the city as a tabula rasa for a comprehensive program of modernization and progress.

As in the preceding discussion of the larger contours of industrialization and restructuring, the mid-1980s represented a policy watershed for heritage and preservation in Singapore, shaped by changing public attitudes and evolving policy discourses of the state itself. Indeed, it may be that there are crucial if in some respects implicit connections which form the new policy narratives. The emergence of a newly-prosperous urban class of professionals, entrepreneurs, and managers, if not precisely conforming to the precepts of Bell's postindustrial society in a state which after all retains a significant manufacturing base, was by the mid-1980s more aware of the 'existence value' of Singapore's historic environment heritage. At other levels of society too, there was a latent community valuation of Singapore's past as represented by the historic built environment, although the program of residential relocation to the new and outlying HDB estates served to deplete the inner city heritage sites of their population base, and perhaps to weaken the sense of connection between people and place.[5]

The momentum of physical redevelopment and insistent modernist values was, however, resistant to pleas for conservation on purely cultural grounds, so the URA and the larger heritage community resorted to arguments incorporating an economistic rationale. Here the value of the heritage districts of Chinatown (to the west of the CBD, Figure 6.1), and Little India and Kampong Glam (to the east) in attracting international tourists seeking a resonantly 'Asian' experience was invoked. The historic districts offered a contrast to the largely generic modernist landscapes of the CBD, new mega-scale shopping centres, and high-rise hotels of the central districts, landscapes not likely to meet the experiential needs and interests of all visitors, including those positioned within the fast-growing 'cultural tourism' sector.

Given the mix of motivations underpinning the formation of conservation policy in Singapore, explicit and otherwise, it is not surprising that scholarship has disclosed contrasting and in some respects conflicting interpretations and meanings embedded within the city-state's heritage districts. Apart from the original dualism of the heritage areas as repositories of historic signifiers and as economic generators, Yeoh and Kong (1995) have written forcefully about the contradictions of state-constructed heritage identity versus community and individual memory. This divergence can be acknowledged as part of the story-line of many such sites in Asia and elsewhere (see the recent work on Berlin, Trafalgar Square), but the distinctive colonial experience of Singapore and its racialized spatial segregation lend a visceral quality to this dichotomy.

New industry formation and the conservation ethos

Although the conservation of heritage districts implies a bias toward stability at least with respect to physical form, a closer reading of experiences over time discloses new narratives of reconstruction and identity formation, as well as recurrent conflict and tension. The role of heritage areas as theatres of spectacle continues, both for local and visitor consumption. But over the past decade or so, a number of Singapore's heritage areas have also emerged as sites of new industry formation, presenting vistas of specialized production, replete with signifying episodes of innovation and restructuring.

As in the London case addressed in the previous chapter, Singapore's inner city industrial experience exhibits not only change over time, in response to new development conditions and cycles, but also significant spatial variegation, reflecting the micro-scale contingencies of place and space in the global city. To illustrate, K.C. Ho has written expressively about the fortunes of a small film company start-up amid the 'unruly' ambience of Little India, situated about a kilometre or so east and north from the Singapore River and the CBD (Ho 2007). As Ho recounts, the somewhat chaotic street life and jumble of activities in Little India are conducive to the creative imperative at the lower, start-up end of the cultural economy structure, where experimentation (and turnover) are most rampant. Here we can acknowledge the creative synergies between the disorderly quality of the Little India *habitus* and the creative impulses flowing throughout the articulated systems and expressive rubrics of the city's cultural economy.

Figure 6.1 View of Telok Ayer (Chinatown) and the Singapore CBD, circa 1980.

Source: Composite photo image from Urban Redevelopment Authority archive, Singapore.

In contrast to the hurly-burly of Little India, a quite different cultural economy ensemble has emerged within the more coherent spaces of Chinatown, situated immediately to the west of the CBD. Chinatown as a whole maintained a measure of vitality even in the aftermath of the depletion of its residential population base, derived from consumption activity and the arts, as well as from the performance of spectacle and memory played out among the spiritual sites and clan associations of the wider area. The initial stimulus to the rise of design professionals and companies in certain precincts of Chinatown was the familiar combination of aesthetics and artistic production acknowledged as preconditions for the cultural economy more widely (Ley 2003), together with the more distinctive cultural, historical and spiritual resonances of individual sub-areas within the official heritage district. With a number of disruptions and shifts, to be documented below, the cultural production trajectory has inserted a new narrative of development for Singapore's Chinatown.

Setting the scene: Chinatown's textured landscapes

Singapore's Chinatown Historic District comprises four sub-areas, each presenting a distinctive spatiality, built form, and imagery (Figure 6.2). Detailed and more expertly informed scholarship on Chinatown and its constituent sub-areas can be found elsewhere (Yeoh and Kong 1994), but for the purposes of this

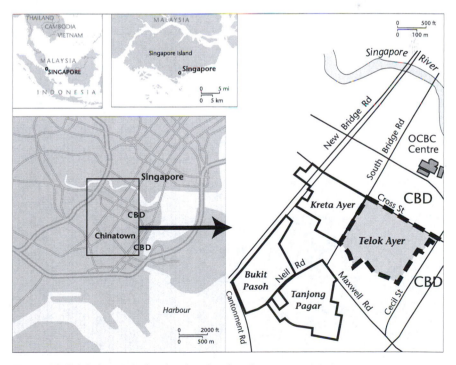

Figure 6.2 Telok Ayer in its local and regional settings.

narrative Tanjong Pagar is widely acknowledged as a critical repository of memory for the earliest Chinese migrants, particularly the Hokkien population; Bukit Pasoh, to the north-west, encompasses a more raffish character, including a red light district and inexpensive hotels; while Kreta Ayer comprises a compact, bustling hotel, and retail activity site on New Bridge Road, a prime locus of the annual Chinese New Year celebrations and other cultural festivals.

While each of these Chinatown sub-areas has attracted a measure of creative activity and labour, Telok Ayer has emerged as the most salient site of new industry formation within Chinatown, and indeed (as its inclusion in this volume attests) represents a classic exemplar of the contemporary inner city cultural economy phenomenon. Telok Ayer encompasses multiple sites of historical and spiritual significance, including the Thian Hock Keng Temple (1840), established as a spiritual commemoration of arriving Chinese immigrants, before early landfill and reclamation deprived Telok Ayer of its shoreline (Figure 6.3). But although the Chinese immigrants were critical to the area's early nineteenth-century provenance, including the formation of numerous Chinese clan associations along Club Street, Telok Ayer also presents a distinctively multicultural identity, including two major mosques and a Tamil shrine, as well as the ethno-cultural meaning embedded in its name ('Telok' is Malay for 'bay', 'Ayer' Malay for 'water').

Telok Ayer consists of attractive landscapes, encompassing a defining built environment of historically and socially-resonant two- and three-storey shophouses dating from the 1840s (Tan 2006), as well as the spiritual sites noted above, and a proliferation of intimate urban spaces (Figure 6.4). There is also a

Figure 6.3 Thian Hock Keng Temple, Telok Ayer, Chinatown.

Figure 6.4 Three-storey shophouse and 5-foot walkway, Telok Ayer.

profusion of consumption amenities, including traditional South-East Asian cafés and hawker stalls, as well as upscale European restaurants, bars, and coffee houses. While these internal attributes are (as we shall see) relevant to Telok Ayer's vocation as site of creative industries and cultural production, this trajectory of new industry formation has also been manifestly shaped by its location, immediately adjacent to Singapore's Central Business District, across Cross Street, and just north of Robinson Road, close to the financial district. Tanjong Pagar and Bukit Pasoh are 'cut off' from the CBD, not only by distance, but also by the street patterns which tend to compartmentalize the sub-areas of Chinatown. Telok Ayer, by way of contrast, encompasses streets (notably Amoy Street and Telok Ayer Street) which provide a direct thoroughfare to the CBD. Indeed, the shop-house built environment extends beyond the official Chinatown Historic District, across Cross Street, encompassing Far East Square.

Landscapes of innovation and restructuring in Telok Ayer

The status of Telok Ayer as a heritage district on the edge of CBD, including the intimate spaces and textured built environment typified by 1840s shop-houses and a mix of spiritual sites and clan associations, provided ideal baseline conditions for the emergence of design activities and other small businesses in the 1990s. These businesses included, notably, small shipping companies and traders which represented a legacy of Telok Ayer's nineteenth-century vocation as waterfront distribution area, as well as a vestigial presence of wholesalers, warehouses, and storage space. The presence of numerous restaurants, cafés, and bars, distributed widely along Club Street, Ann Siang Road, and Amoy and Telok Ayer streets, provided a congenial amenity base for small businesses in the area. These casual consumption spaces included a number of traditional hawker stalls and outdoor eating places, including the long-established Nasi Padang stall in the 121 Eating Centre at the corner of Boon Tat and Telok Ayer Streets, and the Swee Kee Fish-Head Noodle House (Figure 6.5). Rents

Figure 6.5 Swee Kee Fish-Head Noodle House, Amoy Street, Telok Ayer.

were attractive to many small shops and businesses, reflecting the price shadow effect of the proximate CBD, and the earlier migration of former residents to newer HDB estates elsewhere in the city and the suburbs produced a substantial stock of shophouses, protected by heritage legislation, for adaptive re-use.

These defining conditions of Telok Ayer on the eastern margins of Chinatown – central location, distinctive historical resonance, intimate spatiality, adaptable built environment, and modest rent structure – also proved to be spectacularly conducive to new enterprise formation congruent with the emergent developmental trajectories of the late-twentieth and early twenty-first centuries: first, the technology-intensive New Economy, followed in short order by the imprints of the cultural economy and its constituent creative work force, and most recently by the hallmarks of the knowledge economy, which combines features of each. This multi-phase experience stands in contrast to the larger processes of industrial restructuring which typify Singapore's development, both in terms of scale, and also in its more spontaneous (rather than state-induced) origins. That said, Telok Ayer's recent experience vividly illustrates the specific interactions of global and local development factors in the industrial innovation process, as well as the saliency of place (and space) in the New Economy.

Telok Ayer as New Economy site, circa 2000

The final years of the twentieth century saw the dramatic rise of a New Economy of innovative and technologically-advanced industries and firms, with well-known cases including 'Silicon Alley' in Lower Manhattan (see Indergaard 2004, and forthcoming), San Francisco's South of Market Area (SOMA), to be addressed in the following chapter, and Yaletown in Vancouver, to be presented in Chapter 8 of this present volume. At the turn of the millennium, Telok Ayer too became for a time a theatre of industrial innovation, in ways similar to those observed for these other cases, a demonstration of the New Economy experience as a global phenomenon, but in other respects exhibiting more nuanced features derived from localized conditions and contingencies.

An initial site visit to Telok Ayer in the summer of 1999 disclosed the pervasiveness of design-based enterprises within the district, including architects, artists, and graphic designers, but including as well inscriptions of the New Economy, at the apogee of this late twentieth-century development trajectory. These imprints of the New Economy in Telok Ayer took the form of (evidently new) firms specializing in telecommunications, Internet services, and digital photography and reproduction, with firm names typically including as an appendage the signifier '.com' (Figure 6.6).

A year later (site visits in July and December 2000), a more intensive program of research which included interviews as well as mapping and archival work revealed a more comprehensive recasting of Telok Ayer as a New Economy site, expressed in the ubiquity of the dot.coms within the principal streets of the area. As Figure 6.7 shows, the heritage landscapes of Telok Ayer encompassed

Figure 6.6 The New Economy comes to Telok Ayer: '2bSURE.com', Amoy Street, 2000.

numerous firms labelling themselves as dot.coms, or otherwise exhibiting a discernible New Economy identity.

These dot.coms, totalling over 30 firms concentrated within Amoy and Telok Ayer streets and an additional cluster in Ann Siang Road, represented the dominant production enterprise type within Telok Ayer circa 1999–2000, following (on a smaller scale) the experiences of London, New York, San Francisco, and other global cities within developed societies. Within this specific territorial expression of the New Economy in Telok Ayer the structure of enterprise included representative firms in telecommunications, Internet services, digital marketing, digital graphics and art, and software development. Visits to about one-half of these dot.com firms, and informal conversations with workers at cafés, coffee houses, and the area's public spaces, disclosed an employment profile incorporating mostly younger workers (twenties and thirties predominating), and a mix of creative workers (i.e. with formal training in art and/or applied design), technical staff (self-described as 'tekkies'), and entrepreneurs. Relative to the typically

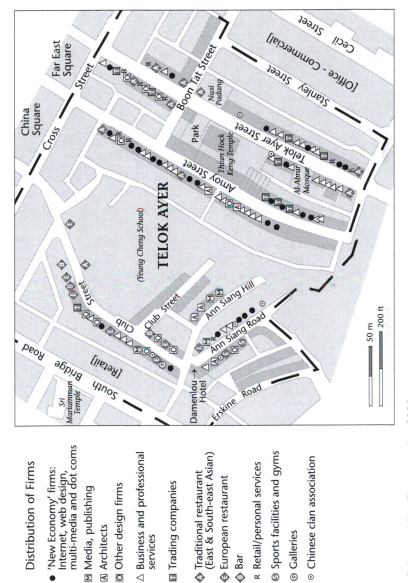

Figure 6.7 Telok Ayer as New Economy site, 2000.

segmented and hierarchical occupational structures of office-based corporations, most of the firms exhibited 'flat' organizational profiles, with relatively few designated executives or managers, and a fairly fluid task orientation in which staff could be deployed anew for multiple (and often coincident) projects.

Aside from the clusters of firms whose New Economy identity conformed to their industrial classification, advanced technology production and communications system capacity, and specialized labour profile, Telok Ayer encompassed in 2003 enterprises of a more traditional sort clearly attempting to capitalize on the cachet of the technology boom. Thus several real estate and personnel firms appended the 'dot.com' descriptor to their company name, in an apparent endeavour to benefit from an association, however tenuous, with the more authentic high-technology enterprises in their midst.

'Root causes' underlying Telok Ayer's New Economy experience

At the time of my first site visits, it seemed natural to ask, in connection with the rise of the dot.coms: 'Why here? Why now?' In retrospect, though, it seems clear enough how Telok Ayer transitioned from a sleepy heritage district backwater in the 1980s, spatially and functionally on the margins of Singapore's central area, to a New Economy site configured by the dot.com landscapes circa 2000. Singapore's leading position as a centre of advanced-technology production and services industries within South-East Asia generated the enabling conditions for the emergence of New Economy firms in a form not dissimilar to that experienced in 'Western' societies and cities. Singapore's labour force also encompassed cohorts of artists, designers, technical workers, and entrepreneurs, key human capital elements of new industry formation at the turn of the last century. Telok Ayer emerged as a prime site for these New Economy firms in Singapore, possessing advantages of intimate spatiality, highly 'textured' built form, adjacency to the concentrations of capital, clients and suppliers in the CBD, inexpensive rents, and high amenity value, following the patterns of new industry formation experienced in other advanced cities. Relative to the overtly corporatist environs of the CBD, Telok Ayer was described to me in conversations as 'cosy' (corporate branding firm, December 2000) or alternatively as 'intimate' (digital design firm), as it afforded the opportunity for small firms to project a distinctive enterprise identity close to, but separate from, the corporate office complex of the CBD. Among the heritage districts, Telok Ayer possessed advantages as a putative site of New Economy formation over Little India and Kampong Glam (more distant from the CBD), and also over other Chinatown sub-areas, which were already highly developed as retail, hotel, and tourist sites (Kreta Ayer), or not as well connected as Telok Ayer to the CBD (Bukit Pasoh).

Paradoxically, though, public policy, at least in the form of state industrial policy as enunciated by the Singapore Economic Development Board, didn't figure directly among the cluster of factors underpinning Telok Ayer's emergence as a New Economy site. In an interview with a staff officer of the Urban Redevelopment Authority in July of 2000, I was informed that the URA (and, to her knowledge, the EDB) had introduced no policies to promote New Economy industries in particular spaces within the broadly-defined metropolitan core (personal communication, 2000). To be sure, the heritage policies promulgated in 1986 and 1988 effectively preserved the spaces and built environment for the New Economy, and certainly the state's emphasis on education and training must be given some of the credit in the formation of human capital essential to the new industries and enterprises observed within Telok Ayer in 1999 and 2000. Indeed, in true Singaporean fashion, the heritage legislation of the late 1980s specified that the conservation districts should in fact generate some economic return, but the precise form of this generative outcome was explicitly allocated to the market, and was largely assumed to favour tourism over production firms and business services (personal communication, URA).

Far East Square: simulacrum or parallel universe?

While Telok Ayer can be interpreted largely as a *spontaneous* New Economy site of creativity and innovation, developed with only indirect influence from the state or corporate interests, directly across Cross Street, an exemplar of *induced* heritage area new industry formation was shaped for Far East Square. In this case, the potential of heritage districts as an environment for creative enterprise was realized by a consortium of state and corporate interests, both local and international, in the reconstruction of Far East Square as 'The Creative Hub' of Singapore's central business district (Figure 6.8). Leading property interest included the Straits Development Corporation, and Keppel, with partners including the Canadian High Commission; (then) Prime Minister Jean Chrétien attended the official opening of the Far East Square Creative Hub in 1998. As in Telok Ayer, the traditional straits settlement shophouse was redeployed for new industries, although with a more 'finished' and coordinated look and feel, incorporating cantilevered roofs to protect pedestrians both from monsoon rains and direct exposure to the equatorial sun (Figure 6.9). The richness of the Telok Ayer heritage landscape was replicated to a degree in Far East Square by the presence of the Fuk Tak Chi Temple, whose restoration and maintenance are supported by the developers.[6]

By mid-1999, near the apogee of the technology-driven New Economy phase, Far East Square domiciled an impressive array of leading companies and institutions, including the Canadian Tourist Board, Nortel Networks, Yahoo!, Leo Burnett, and the BBC. Far East Square therefore projected a more corporate imagery than the dominant SME profile of Telok Ayer, just across the street, although pains were taken to establish an association with conservation values and heritage landscapes in the marketing and sales program. To some extent at least, this site branding exercise was met with a positive market response. David Mickler, director of sales for Yahoo!, was quoted in the weekend edition of *The Straits Times* as follows:

> Being housed in these conserved shophouses with modern office con-
> veniences has its charm. Also, the exotic ambience provides a conducive

THE CREATIVE HUB

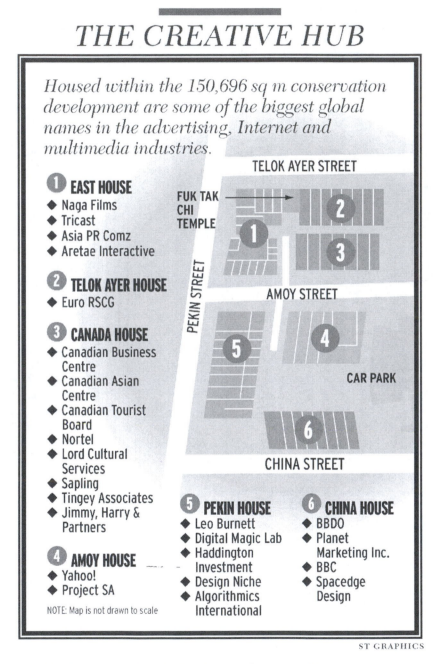

Housed within the 150,696 sq m conservation development are some of the biggest global names in the advertising, Internet and multimedia industries.

① EAST HOUSE
◆ Naga Films
◆ Tricast
◆ Asia PR Comz
◆ Aretae Interactive

② TELOK AYER HOUSE
◆ Euro RSCG

③ CANADA HOUSE
◆ Canadian Business Centre
◆ Canadian Asian Centre
◆ Canadian Tourist Board
◆ Nortel
◆ Lord Cultural Services
◆ Sapling
◆ Tingey Associates
◆ Jimmy, Harry & Partners

④ AMOY HOUSE
◆ Yahoo!
◆ Project SA

NOTE: Map is not drawn to scale

⑤ PEKIN HOUSE
◆ Leo Burnett
◆ Digital Magic Lab
◆ Haddington Investment
◆ Design Niche
◆ Algorithmics International

⑥ CHINA HOUSE
◆ BBDO
◆ Planet Marketing Inc.
◆ BBC
◆ Spacedge Design

TELOK AYER STREET
FUK TAK CHI TEMPLE
PEKIN STREET
AMOY STREET
CAR PARK
CHINA STREET

ST GRAPHICS

Figure 6.8 The 'Creative Hub', Far East Square, China Square Project.
Source: The Straits Times (Singapore), 1999.

Figure 6.9 The crafted landscapes of the 'Creative Hub', Far East Square.

environment for creative work . . . the close proximity to our clients and being part of the creative hub here has also contributed to our decision to locate at Far East Square. As an Internet media company, it helps to be close to people in the creative industry such as advertising agencies, as we work closely with them to provide Internet media solutions.

(*The Sunday Times*, 'Sunday Plus' section, 27 June 1999: 10/11)

These themes were picked up by other tenants surveyed in the same article: Lars Solberg Henriksen, director of Naga Films, affirmed that '[t]he pull factor is that this place is very lively and very happening . . . When you tell people that your office is at Far East Square their reaction is: "Wow . . . cool. That must cost a fortune" (ibid.: 10).

The partnership aspect of business development in a heritage district was expressed by Donna Brinkhaus, regional director of the Canadian Tourist Commission (Asia-Pacific):

We understand how important preserving one's heritage is to Singapore, and we can identify with that as we are looking into preserving our culture and heritage too. Operating from Far East Square allows us to show our support for preserving an important part of Singapore.

(ibid.: 11)

On a less altruistic note, the attraction of Far East Square for the consumption amenities acknowledged as essential elements of the creative milieu was endorsed by John Hastings, managing director of Carnegie's Pub. Hastings disclosed that Far East Square was seen as a 'very strategic location', with a solid local client base (an estimated 75 per cent of the customers situated in Far East Square), including 'a very strong happy-hour trade' (ibid.). This representative of the Far East Square bar scene observed optimistically that the lunch-time trade was stronger than for its outlet in Hong Kong, 'thanks to the *al fresco* dining area that we have here' (ibid.).

At the turn of the last century, then, New Economy firms were presented with two principal locational options on the western CBD fringe in Singapore: Telok Ayer, a spontaneous site, and Far East Square, an induced (and more consciously reconstructed) site. Each was typified by contrasting price points and ambience, with Far East Square rents closer to those of the corporate office spaces of the adjacent CBD than to the appreciably lower accommodation costs of Chinatown. But the development of both sites served to demonstrate the attraction of more intimate heritage districts for creative, New Economy industries and firms, and the importance of consumption amenities and historical markers for creative industries and tenants. Further, both Telok Ayer and Far East Square were by the 2000 survey period evidently thriving, as demonstrated by the strong base of ascendant industries and firms domiciled within each.

Telok Ayer as cultural production site, circa 2003

A program of fieldwork undertaken in January 2003, including a new site mapping exercise, a panel of interviews, and photography, disclosed a fresh cycle of redevelopment in Telok Ayer and its environs, representing in many respects a dramatic departure from the New Economy landscapes and enterprise structure of 2000, but incorporating as well some signifying features of developmental continuity.

Details of this new phase of territorial industrial transformation will be explicated below, but the chief features observed were as follows. First, there was only a residual trace of the 2000-era dot.coms. A vestigial presence of telecomms and digital arts and photography firms was discernible, but the multiple clusters of New Economy firms which largely defined the industrial imagery of Telok Ayer in 2000 had disappeared, as was (for the most part) the dot.com signage which proclaimed the earlier New Economy identity of the site. Second, the dominant trajectory of development in Telok Ayer was now manifestly one of cultural production, with new creative industries augmenting the scattered professional design firms observed in the earlier fieldwork. A program of building renovation and restoration underway in the district in part underscored the aestheticization of Telok Ayer's landscapes (Figure 6.10). Over a period of less than three years, then, Telok Ayer had transitioned from a New Economy landscape of dot.coms, to an aesthetic landscape of the ascendant urban cultural economy. Third, on the eastern side of Cross Street, Far East Square was able to attract a substantial base of firms requiring a more overtly business environment than the more highly textured landscapes of Telok Ayer, but had also lost a number of the larger

Figure 6.10 'Entrepreneurial conservation' in Boon Tat Street, Telok Ayer.

corporations which initially located in the 1998–2000 period. Rents were a factor, as the costs of space in Far East Square were far closer to those of conventional offices in the CBD than for Telok Ayer's shophouses.

Distributions of principal industries, firms and activities in Telok Ayer disclosed by the 2003 surveys are shown in Figure 6.11. First, there are important constants to acknowledge, including the presence of the numerous traditional/regional consumption activities, a number of professional design firms, advertising and corporate branding, media activities (such as the International Herald Tribune located on Club Street, near Ann Siang Hill), and the historical Damenlou Hotel on Ann Siang Road, as well as (presumably permanent) Clan associations and religious/spiritual institutions and landmarks. Even during sequences of rapid industrial restructuring and enterprise 'churning', phenomena identified in the surveys and observations of new industry sites in London presented in the preceding chapter, there are features of continuity as well as transition and succession. The inner city industry site exhibits features of the multi-layered palimpsest for the inscription of new narratives of industrialization.

But the 2003 survey revealed that the New Economy landscapes of ubiquitous dot.coms had largely been eradicated, supplanted by an aestheticized landscape of cultural production, replete with new concentrations of arts, professional design, and creative service firms – an integrated ensemble of design-based creativity and labour formation in this resonant heritage site. This new trajectory was evident in the reconstructed enterprise profile of two principal streets, Amoy Street and Telok Ayer Street, which now accommodated a larger congregation of design and creative services firms. On Ann Siang Road, a conspicuous cluster of dot.coms

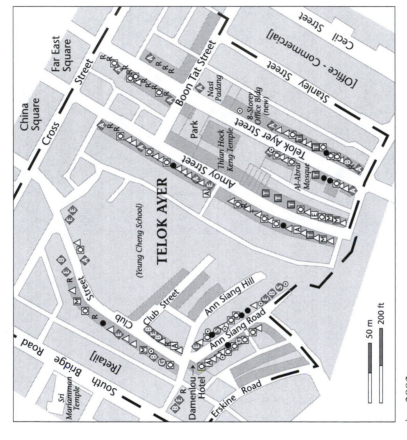

Distribution of Firms

- • 'New Economy' firms: Internet, web design, multi-media and dot coms
- Ⓜ Media, publishing
- Ⓐ Architects
- ◎ Other design firms
- △ Business and professional services
- ▦ Trading companies
- ◈ Traditional restaurant (East & South-east Asian)
- ◈ European restaurant
- ◈ Bar
- ʀ Retail/personal services
- Ⓢ Sports facilities and gyms
- Ⓖ Galleries
- ◎ Chinese clan association

Figure 6.11 Telok Ayer as cultural production site, 2003.

which had shaped the street's enterprise identity in 2000 was now effaced, supplanted by creative design firms, coffee houses, and restaurants. Further, this aesthetic tendency was accompanied by new consumption spaces in the area more generally, including Italian restaurants (a signifier of new industry sites the world over) and coffee houses (but not Starbucks, pervasive elsewhere in Singapore) along Amoy and Telok Ayer Streets, and the eastern end of Club Street.

Reports from the field: expressions of affinity and association

A panel of interviews conducted with a range of creative workers and professionals in January of 2003 confirmed the affinity between arts, creative industries, and the cultural landscapes of Telok Ayer, although the specific mix (and relative weighting) of signifying attributes varied from case to case, as disclosed in the following examples drawn from the diverse product sectors represented within the district. What emerges from the panel of interviews are clusters of attraction, including the general appeal of Telok Ayer's heritage landscapes, and (in some cases) a sense of personal affinity and spiritual connection, as well as more prosaic business factors including location, rents, and access issues.

First, the chief executive of a creative branding firm located on Club Street since 2000 ('Quatise'), endorsed the suitability of Telok Ayer as a site for his firm's operations with respect both to the area's locational attributes and character. Quatise sought to bring together information technology and the creative skills of its staff to produce concepts 'below the sight line' – or outside the box. The mental image of Telok Ayer as a sensuous cultural environment constructed by creative workers was perceived as critical to this business mission. At the same time, more everyday business features of the site, including proximity to the MRT service, and reasonable rents for this relatively central location, underscored the operational advantages of Telok Ayer.

Telok Ayer's emergence as a site of cultural production included firms involved in film and music production. With respect to the latter, a conversation with 'Matt' of Schtüng Music, a company of some ten years' standing on Amoy Street, disclosed that 'for our kinds of people the area is really thriving'. Schtüng's workforce of a dozen professionals included sound engineers, mixers, and technicians, with services including on-site recording and postproduction for 'all kinds of music', perhaps an anomaly in an era of specialization. Matt himself was English, and he noted that Schtüng's workforce exhibited a multinational profile, a fact seen as advantageous for a creative enterprise. An interview with Daen Tay of 'theapostrophe', a film and video company situated since 2001 at 204 Telok Ayer Street, confirmed the district's advantages for industries synthesizing culture and technology in the inner city's recombinant economy. Tay of 'theapostrophe' (with ten staff, half 'creative', and half on the technical and sales side) acknowledged the amenity features of Chinatown as a locational inducement, but also stressed the quality of the business environment: he liked the fact that there were 'lots of similar firms' in the area. Further, he suggested that 'there is competition, but it all helps to contribute to a positive, creative image for the firm'.

Sarah Tham, design director of Cube, offered an evocative appreciation of Chinatown as a site of creative inspiration for the arts and design community. Tham had rejected the idea of locating in a conventional office, with its anonymous identity, conformity, and sense of alienation, and positively selected a shophouse environment for her work. Cube's location on Ann Siang Road was close to her home, which conferred a practical advantage of proximate work place-resident relations, seen as an ideal by many creative workers.

But it was the quality of Telok Ayer as a place and its constituent shophouse built form that represented the most salient feature for her creative enterprise. Tham affirmed that Telok Ayer (and Ann Siang Road in particular) were both 'attractive to clients' and singularly 'conducive to creative work'. These qualities included the historic resonances of Chinatown, the enduring beauty of the shophouse exterior (including ceramic tilework, and shuttered windows which could be opened, unlike those of most office buildings), and highly adaptable interior space. As the sole occupants of the shophouse, Tham and her colleagues could 'turn the music up loud' if they wanted, and could also enjoy easy access to the area's proliferation of restaurants, cafés, and coffee houses, underscoring the links between convivial urban space and creative production.

Corporate branding and design firms, reflecting the increasing importance of creativity in business imagery and marketing, were well represented in Telok Ayer during the January 2003 interview period. An interview with Alan Lim, senior VP of AdXplorer (48A Amoy Street), a digital marketing firm, affirmed that the firm's mainly young staff greatly appreciated both the casual amenities and landscape resonance of Telok Ayer, with 'buzz' explicitly acknowledged as a crucial factor. Lim observed that the historic character of Telok Ayer provided 'creative inspiration' for the staff. Interestingly, perhaps, the firm had initially considered a location in Jurong, but the pull of a substantial client base in the CBD and in Chinatown was in the end decisive for selecting the Amoy Street location. The rent structure of Telok Ayer still favoured this area over the CBD, although Lim did acknowledge that there had been some inflation over the last two years. Through the medium of information technology AdXplorer was 'effectively networked' both locally and internationally, critical to maintaining contact with external clients. Lim noted (echoing the comments of several of the panel of interviewees) that the proximity of the MRT station was a plus for the staff journey to work. But business contact was paramount: Lim emphasized that its Telok Ayer location afforded opportunities for collaboration, as well as immediate access to clients. He observed that some meetings could be arranged at the area's abundant amenities, including restaurants and coffee houses, while the nearby Telok Ayer Green was occasionally used for more informal sessions with staff.

An interview with the senior business manager (Cassandra Wong) of a corporate branding and design firm disclosed perspectives generated by a relatively lengthy tenure in Telok Ayer. Su Yeang Design Pte has been located at 84 Amoy Street since 1990, and so represents one of the pioneering creative services firms in the district, surviving through periodic swings of the business cycle, and the rise and crash of the dot.coms. Su Yeang in fact owns their building (Figure 6.12),

Figure 6.12 Su Yeang Design, Amoy Street, Telok Ayer.

and their staff (of which two-thirds are 'creative', one-third administrative and sales) 'love the building and the area'. Su Yeang has 'no interest' in Far East Square, a two-minute walk to the east, as it is seen as 'too monolithic, lacking in building and design integrity' relative to the more textured and organic landscapes of Telok Ayer. Su Yeang sees itself as a medium-sized business, with clients mainly in the region (including China, Indonesia, and Malaysia, as well as Singapore), large enough to internalize inputs to the design process, and close to printing and other outsourcing operations in Telok Ayer and in the CBD fringe when needed. Although self-positioned within the creative sector, with its implied aesthetic orientation, Wong emphasized that Su Yeang was engaged in a 'tough business environment', with a mission of 'helping companies survive' in increasingly competitive markets through excellence in business design and branding.

Saffron Hill Research,[7] located in Amoy Street, conducted consumer research for companies in Singapore and elsewhere, and (like Su Yeang) saw itself as a business enterprise engaged in creative work. Saffron Hill's locational choice experience points to the importance of micro-scale considerations of spatiality in the New Economy of the inner city: an initial site on Duxton Street in Kreta Ayer was rejected as it was 'just a bit too far out' of the central business district. In contrast, Telok Ayer was and is close to the CBD, with its concentrations of banks and corporate clients, but is not 'sterile and intimidating' like the modernist towers of the central corporate office complex. The aesthetics of the Telok Ayer site were acknowledged by Saffron Hill's workforce as conducive to creative work, while the clients and focus groups invited to the premises enjoyed the heritage ambience of the shophouse: 'a conventional office just wouldn't be the same' (interview with Mr Raymond Ng). At the same time, Ng observed that some 'mainstream' business services had relocated from the area to the CBD or a suburban business park, in some cases motivated by a desire to congregate in a more traditional business environment than that of Chinatown.

Telok Ayer as site of business service activity

If creative industries and cultural production represent the defining trajectory of development for Telok Ayer in the early years of the twenty-first century, then the growth of more prosaic business services in the area suggests a second, parallel pathway, despite the comments above. The 2003 survey disclosed a continued erosion of Telok Ayer's traditional shipping and transportation roles, and a con-comitant growth of more contemporary business services, including legal and accounting firms which can be fairly regarded as the successor businesses to these shipping and transport functions. To some extent at least, the emergence of main-stream business services in Telok Ayer may be attributed to spillover effects of commercial development in the immediately adjacent CBD, reflecting the supply and demand for office space and significant rent differentials between the corporate office complex and the inner city, a phenomenon observed earlier in the London City Fringe case study.

While Telok Ayer's production landscapes and industrial structure in 2003 thus presented a marked contrast to that of 2000, we can also identify some important developmental commonalities. First among these is the clear evidence of enter-prise *clustering* as a locational tendency within this bounded heritage district. As in other inner city new industry sites across a range of city types, Telok Ayer (in 2003) encompasses not merely a discrete set of firms but highly internalized production networks, with dense patterns of backward linkages connecting primary design firms with the accoutrements of specialized cultural production. These include printers and photographers (including digital printing and photog-raphy), Internet services firms, telecomms, and other business services. In add-ition, the relational geographies of production in Telok Ayer included the rich amenity base of restaurants, cafés, bars, open spaces, and fitness centres generic to these inner city New Economy epicentre zones. As might be expected, though, the channels of forward linkages connecting Telok Ayer creative industries and cultural production firms were more spatially extended both in the 2000 and

2003 surveys, reflecting the international marketing reach of some of the area's firms, and the potential of advanced telecommunications for transmitting design products across space.

Second, there is (and was) a defining *design* orientation to industrial activity in Telok Ayer, observed both as a production process in the 'New Economy' phase documented in the two survey exercises in 2000, and in both the 'process' and 'product' orientation of the cultural economy mode predominant in 2003. The design orientation is self-evident and self-defining in Telok Ayer's 2003 vocation, but a substantial portion of the 2000-era dot.coms were also fundamentally engaged in design functions – albeit with a deeper technology base for process, production, marketing, and communications – as exemplified in software design, web-design, advertising and 'branding', and digital arts.

A third developmental commonality derived from the 2000 and 2003 survey exercises relates to the saliency of Telok Ayer as a zone of industrial experimentation, innovation, and restructuring, constituting over the two survey periods a type of *territorial innovation zone* (after Morgan 2004). In each case, the imprints of ascendant development trajectories were immediately legible among Telok Ayer's landscapes and spaces, demonstrating both the global reach of the New Economy and cultural economy modes in its localized manifestations, as well as the volatility of new industry formation processes as exhibited in the experiences of accelerated transition and succession.

Telok Ayer as global village: media and lifestyle services, circa 2006

Attendance at a conference at NUS in December of 2006 afforded an opportunity to observe the latest sequence of industrial transition and its accompanying landscape signifiers in Telok Ayer and adjacent areas. Although the cycle of change observed since the principal 2003 fieldwork exercise in Telok Ayer was not as transformative as that documented for the period following the crash of the dot.coms post-2000, this latest survey disclosed a number of significant shifts. First, the December 2006 observations suggested a continuing densification of land use and activity in Telok Ayer, including new clusters of firms, shops, retail uses, and consumption amenities (Figure 6.13). This intensification was evidently achieved through a limited amount of new building, but more comprehensively through the renovation of existing structures, and the greater utilization of upper floors of the area's shophouses to accommodate new enterprises.

Second, Telok Ayer's emergence as a site of cultural production, a trajectory well established in the 2003 survey phase, was now undergoing a deepening of industry and enterprise representation. There was considerable 'churn' within the base of firms, including the relocation (or closure) of many of the companies included in the earlier survey, but new entrants more than made up for the contractions in the 2003 sample. The area as a whole still sustained a charmingly local feel, but there was also evidence of a larger international presence, including a BBC operation on Club Street (Figure 6.14). There were also several new international shipping and communications firms, evoking Telok Ayer's original

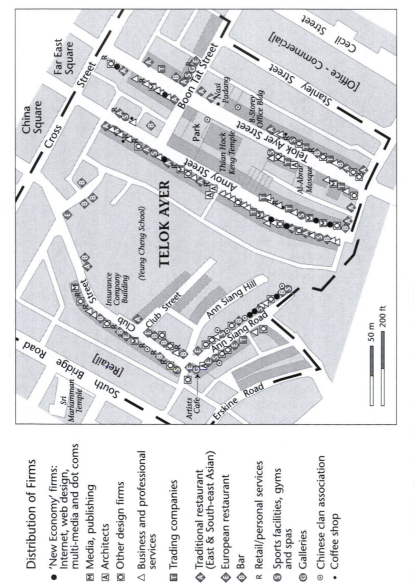

Distribution of Firms

● 'New Economy' firms:
  Internet, web design,
  multi-media and dot coms

Ⓜ Media, publishing

Ⓐ Architects

◎ Other design firms

△ Business and professional
  services

▥ Trading companies

◈ Traditional restaurant
  (East & South-east Asian)

◈ European restaurant

Ⓑ Bar

Ⓡ Retail/personal services

Ⓢ Sports facilities, gyms
  and spas

Ⓖ Galleries

⊙ Chinese clan association

• Coffee shop

Figure 6.13 Telok Ayer as 'global village': media, culture and amenity, 2006.

Figure 6.14 BBC Global Channels, Club Street, Telok Ayer, 2006.

vocation as a distribution centre in the colonial period, but with an updated technological capacity and global sweep.

The cultural production orientation evident in the 2003 survey period was in late 2006 manifestly augmented by a fresh entry of firms in the 'lifestyle' sector – including gyms, spas, massage facilities, and private fitness and lifestyle counseling services throughout Telok Ayer. The street survey of December 2006 disclosed at least a dozen such new establishments in this sector within the area (Figure 6.12). This incursion of lifestyle services may indicate a trend toward 'personal' (as opposed to 'collective' or social) consumption as a new phase of the area's development; while the burgeoning restaurant, bar, and coffee house scene, a defining feature of the last decade of Chinatown's development, indicates a high level of conviviality and sharing of social space in the inner city, the new lifestyle centres are more about catering to the individual and to personal needs and preferences, characteristic of the self-actualization lifestyles of advanced postindustrial societies.

We can conclude our most recent observation of markers of change in Telok Ayer with these three prospectively (re)signifying features. First, the volumes of tourist visitations to the area seemed markedly higher in the 2006 survey than for the previous site visits, suggesting the possibility that Telok Ayer occupies at least a modest niche role within Singapore's cultural tourism sector. Second, the historic Damenlou Hotel, situated at the Club Street end of Ann Siang Road, and an essential part of the raffish charm of early Chinatown, is now being redeveloped as an artists café, punctuating the reorientation of the area's identity and enterprise profile toward art, aesthetics, and consumption. Finally, the intimacy of the area's landscape has been compromised a little by the construction of an eight-storey office building on the south side of Telok Ayer Street. The principal tenant and owner is a long-established Chinese clan association, Singapore Hokkien Huay Kuan, and the building replaces an earlier, smaller office structure, so the URA classifies this as a one-off replacement project. The new office building is, however, a reminder that the CBD and its modernist landscapes are but a short distance away, and may yet place development pressures on this protected heritage area.

Conclusion: Telok Ayer as signifier of innovation and restructuring

At one level, this study of Telok Ayer's evolution as a site of specialized production represents an instructive narrative of localized change within an evocative district of Singapore's Chinatown. What we have observed within the intimate spaces of Telok Ayer over the span of less than a decade is a compressed sequence of new (and reconfigured) industry formations, including initially, pioneering artists and designers, then a spectacular inflow of technology-intensive New Economy firms, followed by apparently more durable ensembles of creative industries, coupled with a more recent lifestyle orientation. Telok Ayer has thus become a 'new industrial district' of sorts. The progression of activities faithfully follows the development sequences and patterns of other inner city heritage districts, as documented in the London cases presented in the preceding chapter, and in the San Francisco and Vancouver examples to follow.

There is, of course, one quite striking point of contrast between Telok Ayer and most of the other cities and sites discussed in this volume: the relocation of the residential population to outlying HDB estates effectively removed the potential for social dislocation accruing from the entry of new industries and firms toward the end of the twentieth century, although the processes of transition and succession outlined above certainly suggest aspects of the industrial gentrification phenomenon described by Andy Pratt in the Hoxton area of London. As the current spatial strategy of encouraging more residential development in Singapore's downtown matures, too, there may be more pressure to accommodate housing in Telok Ayer and other districts of Chinatown, a tendency which in other cities has exerted destabilizing pressures on small firms.

To be sure, there are now expressions of the cultural economy trajectory found within other Chinatown districts, but Telok Ayer's distinctive location, spatiality, and landscapes have shaped its salient position within the creative economy of Chinatown. The area rapidly recovered its industrial vitality and innovation functions following the crash of the dot.coms in 2000, suggesting a localized condition of robustness in the face of change, and the resiliency of a favoured district even amid a context of rapid churning at the level of the firm. In contrast, Far East Square, situated just across the street from Telok Ayer, lost many of its major corporate clients in the aftermath of the crash, and has struggled to recover its competitive position, in part because of a rent structure that excludes many small enterprises and start-ups. Far East Square occupies a kind of 'no man's land', located on the margins of the central business district, but presents a less conventional business imagery than the CBD's corporate complex, and is separated from the heritage landscapes of Telok Ayer both by the major arterial of Cross Street and by a steep rent gradient.

These localized inscriptions of new industry formation on the textured landscapes of Telok Ayer are intrinsically significant, but there are larger implications for Singapore's developmental story-line. The unplanned success of Telok Ayer as a zone of industrial experimentation and restructuring offers one modest example of innovation that lies outside the *dirigiste* traditions of the exemplary development city-state. But the experience outlined here also points to the potential of smaller spaces to contribute to Singapore's aspirations for a creative economy on a larger scale, supplementing the established strategic sectors of finance, intermediate services, tourism, technology, and the knowledge sector. The blurring of industrial categories between hitherto separate arts, design, and corporate branding observed in the enterprise profile of Telok Ayer also suggests the competitive advantage potential of the 'recombinant economy' of the 'new inner city'.[8]

Finally, the experience of Telok Ayer in its larger cultural economy setting might point to the possibilities of capturing new trajectories of development in a context of volatile global restructuring. Jonathan Rigg cites Krugman's well-known dictum that Singapore's commitment to investments in human capital and physical infrastructure has its limits as a growth strategy, in that 'Singapore's growth has been largely based on one-time changes in behaviour that cannot be repeated' (Krugman 1994: 71; quoted in Rigg 1997: 24). That said, the successive refinements of Singapore's development strategy bear witness to the state's determination to extract more value from these investments via rearticulations of policy vision which capture the latest big thing, the developmental Zeitgeist represented by the latest growth trajectory of the most advanced states and urban-regional jurisdictions. In this interpretation, the twenty-first-century development orientation of the exemplary developmental city-state might take (in part) the form of smaller increments of growth and change, absorbed over shorter time periods, in contrast to the sweeping restructuring episodes of the past.

7 The New Economy and its dislocations in San Francisco's South of Market Area

Introduction: from industrial to postindustrial on Route 101

The twenty-five-minute taxi or shuttle ride north from San Francisco Airport (SFO) to the City of San Francisco reveals a sequence of illustrative landscapes and landmarks. The excursion from SFO to the City enables views of the Fordist hillside housing landscapes of Daly City, which achieved a sort of pejorative fame in Pete Seeger's ironic homage to the 'ticky-tacky' houses and lifestyles of postwar suburban America. A roadside archipelago of stand-alone office buildings and acres of surface parking scattered along the highway exemplify the 'edgeless city' phenomenon of ex-urban North America. The journey northward on Route 101 also offers panoramic views of San Francisco Bay, including anchored naval ships which remind us of the more martial identity of the paradigmatic 1960s city of harmonic peace. The East Bay (part of Alameda County) encompasses the persistently industrial Oakland and the University of California at Berkeley, a bastion of intellectual liberalism and radicalism, with the cerulean skies of northern California as backdrop to one of the world's most favoured regions.

Another of the sights to greet the northbound traveller en route to the City is a hillside sign advertising the imminence of 'South San Francisco – the Industrial City', unconsciously mimicking the famous 'HOLLYWOOD' hillside insignia proclaiming Los Angeles as the world's movie capital. In stark contrast to the iconic Hollywood banner 500 miles to the south, however, the 'South San Francisco' signage embodies no suggestion of self-regarding glamour, global media projection, or conspicuous consumption. Instead, the self-designation of an 'industrial city' situated within the famously high-amenity, increasingly postindustrial Bay Area, with its signature wineries (including those of Napa and Sonoma Counties), cook-book restaurants ('Chez Panisse'), and convivial lifestyles, conveys a defiant and retro imagery of place.[1]

Once over the last set of hills the more emblematic features of San Francisco come into view: the point towers of the Central Business District (notably I.M. Pei's pyramidal Transamerica Tower, rising above the more austerely modernist office towers of the CBD), San Francisco Bay (including Alcatraz), and in the far distance the Golden Gate Bridge and the arcadia of Marin County and beyond. But before the shuttle arrives in the downtown, with its four-star hotels, boutiques, heritage attractions, and restaurants, the visitor is treated to fleeting glimpses of physical and social landscapes in the South of Market Area (SOMA), an experience which offers clues to the wrenching episodes of industrial restructuring that have underpinned San Francisco's transformations.

This first reconnaissance of San Francisco's southern inner city and CBD fringe districts provides a vantage point from which to identify evidence of inner city change which defines in part San Francisco's larger saliency as a site of industrial regeneration and social dislocation. Of course, the choice of route through the inner city matters. If the driver selects a western passage through SOMA, a usual route to the central city's hotels and B&Bs, the visitor transits the gentrified Potrero Hill neighbourhood, with its contingents of the new middle class, including successful artists and designers; the condos and recreational amenities of Mission Bay; and the 1,000-foot-long blocks of the old warehouse district, now encompassing big box retail and restaurant uses and artists galleries and co-ops, as well as just a few traditional industrial activities. The San Francisco Giants baseball stadium, AT & T Park, which has already metamorphosed from its earlier SBC Park and Pac Bell corporate designations, is ensconced within China Basin, with the loftier, often steroid-propelled home runs clearing the fence and making a splash in Mission Bay.

If the shuttle continues north on Third Street, the traveller may catch a glimpse of the bellwether site of South Park, initially an elite residential enclave in the mid-nineteenth century, modelled on the London townhouse square, but a terrain of redefining social and economic change over the past century, and a correlate, in some ways, to the Telok Ayer precinct described in the previous chapter (Figure 7.1). Before reaching Market Street, our traveler can hardly fail to notice the arts and gallery precincts of Yerba Buena, a flagship site of San Francisco's cultural economy and tourism sector, superimposed on the marginal residential districts of SOMA in the last decades of the twentieth century.

SOMA as site of restructuring, regeneration, and dislocation

Market Street diagonally bisects the metropolitan core of the City of San Francisco, with the CBD, elite residential areas and hotels and tourist infrastructure concentrated largely to the north of Market, and the old industrial and warehousing districts and historically low-income ethnic neighbourhoods situated to the south. Market Street itself presents a rich (and often troubling) narrative of historical and contemporary change, from the middle-class Twin Peaks district in the extreme south-west, then proceeding north-east to Castro, with its large gay and lesbian communities, and the City's municipal government precinct at the intersection of Mission and Van Ness. Beyond this civic precinct we encounter the Tenderloin with its enclaves of poor black and Hispanic populations, then the Powell Street intersection with its Cable Car junction and the start of the major hotel and retail district, just south of Union Square, and finally the southern fringe of the CBD and the reconstructed Embarcadero site on the Bay. Over time, new groups have imposed themselves along the Market Street corridor, presenting a social

Figure 7.1 Entrance to South Park, off Second Street, South of Market.

relayering of space and the production of a version of Bourdieu's 'mixed space' of diversity and pluralism, with connotations of conflict and tension as well as interaction.

Since the 1980s, SOMA has been transformed by a full-blown cultural economy experience. The centrepieces of this trajectory include major galleries such as the San Francisco Museum of Modern Art and the Yerba Buena Center, and public exhibition and convention space, notably the Moscone Center, as well as art schools and colleges, historical archival centres and similar agencies, collectively enhancing San Francisco's status as a principal American centre of arts and culture, and a magnet of attraction for its huge tourist industry. At the same time SOMA's cultural makeover encompasses a proliferation of artists' studios, workshops, and co-ops, many of which present a more edgy identity, including numerous avant-garde arts enterprises designed to further activist social and environmental agendas, essential features of the much longer bohemian imagery of the City and its liberal traditions and tolerant values. This civic liberalism has been sorely tested in recent years, reflecting government cutbacks at state and local levels, exacerbated by restructuring in the Bay Area's key industries. Certainly there has been no evident diminution of the homeless population or in the proliferation of street people who lend a palpable imagery of deprivation and inequality to the city's social spaces.

Multiple layers of industrial restructuring in the Bay Area

Industrial restructuring in the Bay Area comprises not one but several coincidental shifts in the experiences of growth and decline among key sectors, including the erosion of San Francisco's corporate control and financial activity, following the ascendancy of Los Angeles as the dominant metropolis and international gateway city in the American west. But this saga of shifts in national and regional urban hierarchies tells only part of the story. At one level, San Francisco's experience of restructuring follows the familiar contraction of basic manufacturing, goods distribution, and ancillary industries, and coincidental expansion of specialized services industries and labour, incorporating a large contingent of the new middle class in the City and Bay Area. These stories have been played out along similar lines, of course, in many cities. But in the last decade of the twentieth century, San Francisco was the site for a quite remarkable industrial transformation, emerging as one of the world's largest New Economy agglomerations. Over the latter years of the 1990s global-scale concentrations of the so-called 'dot.coms' – shorthand for a more diverse but technologically-intensive sector including new media, video production, computer software development, and computer graphics and imaging, among many other industries – were established in San Francisco, most notably in SOMA and the Mission, but also in the CBD and the CBD fringe.

Resonances of the New Economy were of course felt in many advanced cities, as the accounts of London and Singapore (as well as New York) in preceding chapters of this volume attest. That said, the scale, growth rates, and intensity of innovation and creativity in late twentieth-century San Francisco were in many respects singular, owing in part to the City's distinctive regional arts traditions and cultural milieu, to the proximity of San Francisco to the leading global clusters of innovative industries and firms clustered in Silicon Valley, forty minutes to the south, and to the environmental qualities of the South of Market. For a time, the technology-driven New Economy comprehensively transformed the landscapes, social spaces, and identity of SOMA, creating an enormous creative energy and 'buzz', but also displacing hundreds of long-established firms and thousands of low-income residents of marginal communities. The quick fortunes to be made from the New Economy also interrupted for a time the artistic and creative design trajectory of redevelopment in SOMA.

Saliency of the San Francisco–SOMA experience

This chapter presents a retrospective of the New Economy experience of San Francisco, half a decade on from the dot.com crash of 2000 and its recessionary aftermath, with a view to placing this experience in a longer-run perspective on contemporary industrial restructuring and its implications for cities and urban communities. The setting for this place-situated industrial phenomenon will be set out in some detail, as there can be little question that the SOMA experience illustrates the social nature of the advanced economy.

Nor is there much room for doubt concerning the harsh effects of the New Economy on the communities in which the new media firms and other ascendant industries imposed themselves in the last half-decade of the twentieth century. This is not a saga of 'normal' gentrification, typified by an insidiously gradual social upgrading of older communities, proceeding incrementally over a period of decades. As Rebecca Solnit, in a bitterly eloquent eulogy for the displaced neighbourhoods of SOMA tells it:

> Gentrification is transforming the city by driving out the poor and working class . . . But gentrification is just the fin above the water. Below is the rest of the shark: a new American economy in which most of us will be poorer, a few will be far richer, and everything will be faster, more homogeneous and more controlled or controllable.
>
> (Solnit 2000: 13–14)

Solnit extends this multiscalar projection of restructuring and its dislocations by suggesting that while many have positioned Los Angeles as urban prototype, with its 'urban decay, open warfare, segregation, despair, injustice and corruption', San Francisco may instead emerge as the 'capital of the twenty-first century' by virtue of its 'frenzy of financial speculation, covert coercions, overt erasures, a barrage of novelty-item restaurants, websites, technologies, and trends, the despair of unemployment replaced by the numbness of incessant work hours and the anxiety of destabilized jobs, homes, and neighborhoods' (ibid.: 14).

The contours of late twentieth-century New Economy development within the city's production districts will follow this introduction, including consideration of the social and local planning contexts for this process. The following section situates the imposition of a distinctive New Economy regime in the South of Market Area, including the earlier patterns of redevelopment, restructuring, and dislocation that served as precursors to the late twentieth-century experience, and SOMA's transitory status as San Francisco's 'Multimedia Gulch'. Next, the chapter presents a profile of South Park and its environs, demonstrating the rich yield of insights to be derived from the imprints of industrial change legible at the localized scale, based on surveys, mapping, and interview programs conducted in 1999–2005 in this evocative site. This narrative includes the emergence of South Park as the epicentre for SOMA's Multimedia Gulch in the late 1990s, its fate as collateral damage in the crash of 2000, and after, the slow and painful recovery of South Park's enterprise base, and some recent evidence for its contemporary recasting as an exemplar of a second generation New Economy site. The concluding section will offer an outline of larger implications of the SOMA New Economy experience, anticipating the more systematic comparisons and retheorization exercise presented as the concluding chapter of this volume.

Contours of economic change in the Bay Area

Viewed from the contemporary vantage point, in the aftermath of the technology boom and bust of the late twentieth and early twenty-first centuries, the defining imagery of San Francisco's economic trajectory is one of accelerated and destabilizing change, exemplified by the rapid rise and even more compressed crash of the dot.coms. These pronounced swings of economic fortune are certainly integral features of San Francisco's storyline of transformation since the 1980s, but tell only part of the tale. San Francisco and the larger Bay Area represent one of the most striking examples of multi-layered industrial restructuring within the developed world, with contour lines drawn at the regional, subregional, district, and localized scales. Processes of redefining change occur at each echelon, but there are also aspects of continuity as well as discontinuity which must be acknowledged as co-present features of contemporary urbanism.

For the period 1985–1996, which encompasses an era of massive restructuring among many city-regions, and which takes us to the advent of the tech-boom in the Bay Area, the record is one of relative employment stability among major industry groups at the regional level, coupled with significant growth in service employment and occupations. To illustrate the former tendency, employment in the key FIRE category (Finance, Insurance and Real Estate) in the larger Bay Area was the same in 1996 (195,000) as it was for 1985, with a peak of 207,000 in 1993 (City of San Francisco 1987). Counter to the trends of other city-regions, too, employment in manufacturing was relatively stable over this period, with a modest decline from 481,000 jobs in 1985 to 471,000 in 1996. The story was similar for other principal employment sectors, including a small increase in government jobs (from 426,000 to 436,000 over the 11-year period), and similar increments for construction and mining and retail trade. The major gainers among the Bay Area's industries included transportation (from 89,000 to 168,000 jobs), and a spectacular growth in service employment – from 666,000 in 1985, to 966,000 in 1996, increasing its share of regional employment from 25 per cent to 31.4 per cent. For the Bay Area as a whole, then, the profile of economic change is one of a marked *tertiarization* of employment, but not one of industrial decline on the scale encountered in other city-regions during this period.

Data for the Bay Area as a whole, however, tend to mask significant changes in industrial employment at the sub-regional scale. Employment growth over the period 1985–1996 for the sub-regional areas outside the city and country of San Francisco was robust overall, especially in the South Bay area which includes Santa Clara and Silicon Valley, one of the principal growth areas of the US economy during this period (Saxenian 1991). But in San Francisco, the central urban place of the Bay Area, most major industry groups exhibited either modest growth or decline, and indeed San Jose is now the largest city in the Bay Area. San Francisco's restructuring experience included, as might be expected, contractions in manufacturing (a loss of 7,000 jobs from its 1985 level of 42,000), wholesale trade (a steep decline from 35,000 to 23,000 over this period), and communications (26,000 jobs in 1985 and 16,000 eleven years later), with smaller decreases in government and construction jobs. As might be anticipated for an important regional, national, and international office centre, employment in services (which includes business or 'producer' services) was fairly robust, increasing from 172,000 in 1985 to 204,000 in 1996.

But employment within the key FIRE category actually declined, from 81,000 in 1985, to 66,000 in 1996, indicating a significant erosion of the banking and financial sector which comprised a centrepiece of San Francisco's economy over its first century of development. Changes in the organization of labour within the financial sector, as well as transitory business cycle effects – and the impacts of the U.S. Savings and Loan fiasco – likely represent components of this trend. But the contraction of employment in the financial sector also reflects the shift of corporate banking activity from San Francisco to Los Angeles, a more structural feature of change (Erie 2004). The record of employment change for San Francisco over this decade suggests a profile of *relative* decline overall, both within its regional setting, but more decisively vis-à-vis Los Angeles, with Tokyo the first-order global city within the Asia-Pacific region.[2]

Profile of San Francisco's industries and employment, 2000

The restructuring processes described succinctly above produced in San Francisco a development profile marked by a pronounced *postindustrial* character in terms of employment categories, as well as a complementary *professionalization* of the city's occupational structure. Table 7.1 shows employment by industry generated by the 2000 US Census.[3] The dominant industry groups were the professional, scientific, management, and administrative category, comprising almost one-fifth of total employment (19.3 per cent), followed by educational, health, and social services, with about one-sixth of San Francisco's employment by industry (16.2 per cent). These figures demonstrate the importance of both private and intermediate services, as well as public/final demand services, to San Francisco's economy at the turn of the twentieth century. The third largest industrial employment group in 2000 was that of arts, entertainment, recreation, accommodation, and food services (11.2 per cent), reflecting both the growth of the cultural economy and the importance of the City's tourism sector. The other large industrial categories, each with just over one-tenth of San Francisco's employment, included retail trade and FIRE, the latter representing a substantial decline from its peak levels in 1991. But manufacturing employment had suffered the largest contraction, and by the 2000 census comprised only 6.6 per cent of San Francisco's industrial employment, a percentage similar to that of London (Chapters 4 and 5), although (as noted earlier) manufacturing retains a significant presence in the larger Bay Area economy.

Employment data classified by occupation provide additional insight to San Francisco's development orientation, and underscores the magnitude of the

Table 7.1 Employment by industry, San Francisco County, 2000

Employed civilian population 16 years and over	Percent	Industry	Number	(%)
427,823	100	Agriculture, forestry, fishing and hunting, and mining	825	0.2
		Construction	14,961	3.5
		Manufacturing	28,228	6.6
		Wholesale trade	10,954	2.6
		Retail trade	43,935	10.3
		Transportation and warehousing, and utilities	19,111	4.5
		Information	30,000	7.0
		Finance, insurance, real estate, and rental and leasing	43,479	10.2
		Professional, scientific, management, administrative, and waste management services	82,573	19.3
		Educational, health and social services	69,461	16.2
		Arts, entertainment, recreation, accommodation and food services	48,079	11.2
		Other services (except public administration)	21,995	5.1
		Public administration	14,222	3.3

Source: Bureau of the Census reports.

City's professionalization trajectory. Table 7.2 shows that employment in management, professional, and related occupations approached one-half of the total for San Francisco City and County (48.3 per cent), with sales and office occupations accounting for a further one-quarter (25.6 per cent), together comprising almost three-quarters of the total employment – high levels by any international standard.[4] By way of contrast, an occupational category including transportation as well as production activity represented just 7.5 per cent of San Francisco's employment, with construction and extraction comprising a further 4.2 per cent.

Growth industries and the New Economy

The policy response to structural change took the form of a vigorous municipal commitment to expanding San Francisco's crucial tourist, visitor and convention sector, based on its enviable combination of high urban amenity and cultural assets, as well as access to the amenity and lifestyle attractions of Marin, Napa, and Sonoma Counties to the north, and Santa Cruz and the Big Sur to the south. With these regional competitive advantages, tourism was widely seen as a prospectively

Table 7.2 Employment by occupation, San Francisco County, 2000

Employed civilian population 16 years and over	Percent	Occupation	Number	(%)
427,823	100	Management, professional, and related occupations	206,804	48.3
		Service occupations	61,364	14.3
		Sales and office occupations	109,316	25.6
		Farming, fishing, and forestry occupations	462	.01
		Construction, extraction, and maintenance occupations	17,990	4.2
		Production, transportation, and material moving occupations	31,887	7.5

Source: Bureau of the Census reports.

more sustainable growth vector than other key industries, notably FIRE, manu-facturing and wholesale trade, each of which appeared to be in decline. This commitment included business investments, for example, in hotels and ancillary enterprises (such as tour operators), but the projected growth of the tourism sector was to a large extent underwritten by the public, an ongoing theme in San Francisco's development history we shall return to later in this chapter.

But amid these concerns about the prospective decline of key sectors, and hopes for an increasingly culture- and amenity-based based tourism and convention sector, new growth industries associated with the tech boom of the late twentieth century appeared as a timely deliverance from the prospects of decline. A dazzling set of new (and technologically reconfigured) industries, including software development, multimedia, Internet services and web design, and film and video production, together with new labour cohorts, were inserted into the CBD fringe and the more textured inner city landscapes and complex social milieu of SOMA. The New Economy of the late twentieth century in San Francisco was also mani-festly a multi-layered experience, as the localized concentrations of new industries were supported in large measure by sub-contracting and commuter linkages with the South Bay, while the amenity value of Marin, Sonoma, and Napa Counties served to attract scientists, entrepreneurs, professionals, and creative workers. Further, the ascendancy of these New Economy industries and firms, however evanescent their rise proved to be, and however problematic their positioning within the vocabulary of industrial urbanism has been, generated a large and distinctive consumption economy, and interacted in complex and forceful ways with housing markets and residential communities.

The cultural economy and its dislocations in SOMA

The 1990s New Economy phenomenon in San Francisco tends to be seen as a highly transitory phenomenon. While it is true that both the take-off phase and collapse of the New Economy occurred within an astonishingly abbreviated time frame, it is also the case that both the developmental conditions and residual effects follow a more extended and complex pathway. A powerfully consistent pattern of redevelopment pressure has been imposed externally upon SOMA both by corporate interests and agencies of the municipal government dating back at least as far as the 1950s, as well as a record of sustained resistance on the part of marginal populations and long-established industries and firms.

The complex and conflicted history of SOMA is described in a number of influential books, notably Chester Hartman's (2002) *City for Sale: The Transformation of San Francisco*. Hartman's masterly critical narrative of the South of Market's postwar history takes in the area's role in the larger redevelopment of San Francisco, as well as its redefining episodes of change, most often involving external pressures for redevelopment and the corresponding internal points of resistance. In his recitation of SOMA's transformations and recurrent episodes of displacement, Hartman also draws on the work of historians such as Kevin Starr as a means of enriching this distinctive urban storyline, which incorporates interdependencies between industrial restructuring and social change.

Starr suggests that SOMA's redevelopment patterns mimic the larger processes of San Francisco's transformative experience in the postwar period, and this presents an especially evocative and exemplary site of urban scholarship:

> This district represents the most comprehensive paradigm of San Francisco. More than any other neighborhood in the city, South of Market is the part that contains the whole: the one matrix that subsumes unto itself every successive layer of urban identity in the history of the city. Here indeed is the anchor district of San Francisco . . . Here is the residential district of its most diverse population . . . The other neighborhoods of the city . . . seem practically empty – or at the least mere occasions for residents – in comparison to the rich life of hotel, union hall, shipping, industrial manufacture, government office, newspaper room, church, school and orphanage, and residential life. South of Market was an urban district containing the full formula of the city.
>
> (Starr 1996; cited in Hartman 2002: 59–60)

Starr's description of SOMA's larger signifiers for San Francisco's development saga echoes Richard Tames's similar designation of Clerkenwell as paradigmatic metropolitan district in the London case presented in Chapter 5. The basis for each claim comprises a unique co-presence of diverse, specialized industries and social groups, and redefining sequences of restructuring and social upheaval.

The 1906 earthquake and subsequent fires devastated SOMA, but the area was comprehensively rebuilt within three years. What emerged from the reconstruction of the South of Market was a rich and variegated mix of populations and industries. The recovery phase included characteristic congregations of small hotels and apartments, which accommodated working men and the increasing flow of immigrants.[5] With regard to the latter, the social reconstruction of the South of Market between 1910 and 1920 was shaped in part by new immigrant groups, including Irish, Germans, and Greeks, many of whom found employment in the area's industries, or, alternatively, established small businesses.

During the twenty years of industrial development and employment growth in SOMA following the 1906 earthquake and fire, life in SOMA for many featured hardship and deprivation, especially for the large numbers of contingent industrial workers reliant upon employment agencies for casual labour. As Hartman observes, the economy of SOMA and much of California as a whole exploited 'this industrial reserve of job-hungry men . . . [who] wore out early, with little provision by society for premature old age and premature retirement' (Hartman 2002: 59). The vibrancy of this dense agglomeration of industries, businesses and workers in the South of Market, as elsewhere in the City and America as a whole, was seriously compromised by the Great Depression of the 1930s. The Second World War brought a revival of sorts to SOMA, as San Francisco 'became a dormitory metropolis housing war industry workers and military personnel . . . newly-arrived workers, seamen, soldiers, and sailors joined the traditional residents in the hotels, boarding houses, bars and restaurants' (ibid.: 59) of the South of Market.

This war-time social reconstruction of the South of Market included waves of migrants: first, mainly black workers seeking jobs in factories, warehouses and the docks; and then Asians, including large numbers of Filipinos, who retain a presence in the residential neighbourhoods of early twenty-first-century SOMA. Hartman notes that SOMA remained essentially a 'workingmen's quarters' in the years following the war, with single men comprising 72 per cent of the population (ibid.).

Cycles of redevelopment in SOMA, 1950–1985

The genesis of redevelopment, transition, and dislocations in the South of Market's modern history, a precursor to the impacts of the New Economy phenomenon of the 1990s, can be traced to the initial urban renewal proposals of the early 1950s and 1960s, with SOMA's obsolescent industries and low-income populations characterized by proponents of redevelopment as comprising a classic derelict zone. As in other cities, this profile of an urban district in irreversible decline presented a vulnerable target to the interests of capital and government seeking opportunities to remake the city in ways more attuned to modernistic visions and ideals.

The four decades of redevelopment in SOMA, culminating in the district's present-day cultural economy and leisure orientation, commenced with the San Francisco Redevelopment Agency's submission of an application for a federal urban renewal survey and planning grant. This application, supported by major corporate interests, 'sailed through' the San Francisco Board of Supervisors in

1966, bypassing the City Planning Commission, establishing a pattern of bypassing the City's planning staff which would be repeated over the following decades (ibid.: 45). The elements of the plan for an 87-acre section of SOMA, the Yerba Buena Center, included offices, an indoor sports complex, and a new convention centre, features in tune with the redevelopment motif of the day, and key underpinnings of the contemporary orientation of SOMA as a centre of culture, tourism, and spectacle.

The postindustrial agenda and points of resistance in SOMA

A redevelopment project on the scale of Yerba Buena could not of course fail to generate significant dislocation, including residential areas comprising a majority of older, poor, and single men. As Hartman attests, the prospects of dislocation underscored the conflicting perception of the residents of SOMA as 'Skid Row' or, alternatively 'community' (2002: 60). Clearly, acceptance of the former designation suited the interests of the redevelopment lobby, in marginalizing the resident population, and thereby diminishing impediments to clearance and reconstruction. As the Yerba Buena planners and developers saw it, they were offering a double benefit, 'providing economic revival through construction jobs and increased tourist and convention business, and they were also helping the city clear out an 'undesirable element' (ibid.: 61) of transients and substance abusers.[6]

Views of SOMA's social groups as minor impediments to the Yerba Buena project had a correlate in attitudes towards the area's traditional production sector. During the public hearings on the Yerba Buena proposals in 1965, labour groups protested against the apparent dismissal of the area's industrial sector and blue-collar jobs. Labour representations to the public hearings included concerns about manufacturing decline in the City and County as a whole. (Indeed, the 1960s saw a loss of almost one-fifth of San Francisco's manufacturing employment, positioning the City as one of the early victims of the industrial decline of large metropolitan cities among advanced societies.) The proponents of the Yerba Buena project responded to these concerns by promising to commit 25 per cent of the lands designated for redevelopment for industrial re-use. But in the event 'the project in its final form contained few industrial uses' (ibid.: 47), an outcome broadly congruent with planning decisions for the reconstruction of Vancouver's inner city, 800 miles to the north of San Francisco, an experience to be elucidated in the following chapter.

SOMA's reconstruction as Multimedia Gulch

The approval and eventual construction of the Yerba Buena Center established a powerful redevelopment orientation for SOMA, reinforced by the subsequent (and much larger) Mission Bay project. The scale, complexity, and conflicted interests of the Mission Bay project entailed an almost two-decade process, undertaken during numerous swings of business and local electoral cycles, but like Yerba Buena implying a discounted valuation of the City's industrial sector and blue-collar labour force.

But while the redevelopment lobby was implicitly prepared to endorse a postindustrial land use agenda for much of SOMA, the City Planning Department doggedly pursued a strategic land-use policy process which endeavoured to balance the interests of the industrial constituency (firms and workers) with the redevelopment imperatives of a City experiencing swings in its economic fortunes, encompassing a district of diverse communities defined by social and industrial configurations as well as space and built form (see Figure 7.2 for map of SOMA planning areas, including Yerba Buena, South Park, and South Beach). City planners hoped to establish land use, zoning, and building regulations which would preserve a balance of traditional industrial functions (including production, repair, warehousing and distribution), with a view to maintaining both an element of industrial diversity, as well as a measure of occupational and social pluralism in a city increasingly dominated by a new middle class of executives, managers and professionals.

Figure 7.2 Planning areas in the South of Market Area (SOMA), San Francisco.

Source: City of San Francisco Planning Department.

Survey work conducted circa 1993–1999 by the City Planning Department disclosed a pattern of industrial specializations within SOMA's terrains and the adjacent waterfront, which included an array of traditional production industries along with some newer activities (Figure 7.3). While the restructuring processes acknowledged earlier had severely undercut the City's manufacturing vocation, SOMA retained a mix of transportation, goods handling, and distributional activities, notably in the Central Waterfront, South Bayshore, and Showplace Square areas, as well as a residual base of Fordist industries such as garment production and food processing in North and South SOMA, and East and West NEMIZ.[7] South Park maintained a presence of printing industry firms, but also encompassed

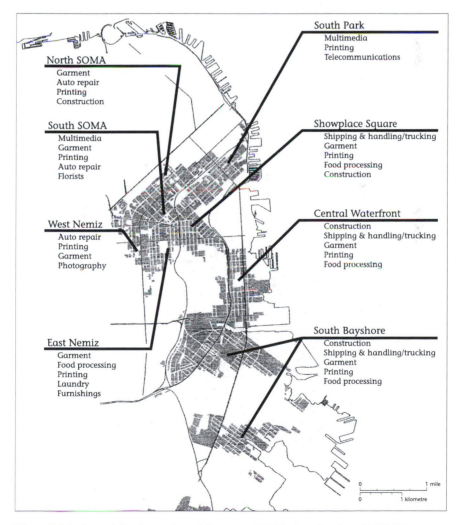

Figure 7.3 Industrial districts and specializations, SOMA, Port Lands and NEMIZ.

Source: City of San Francisco Planning Department, Land Use Planning, 1995–99.

multimedia and telecommunications enterprises, presaging its emergence as epi-centre site of San Francisco's New Economy. As another way of depicting the space-economy of SOMA and adjacent industrial areas, the Planning Department produced a map showing the location of important clusters (Figure 7.4), notably the diverse representation of industries for the SOMA districts identified in Figure 7.2.

Advent of the New Economy in SOMA

In the midst of its review of industries and land use in SOMA, the City Planning Department was obliged to consider the implications of new economic activities which were beginning to transform the production landscapes of the CBD fringe and inner city, juxtaposed among long-standing populations of artists, graphic designers, and students, as well as traditional industries. These new entrants to SOMA (and to North-east Mission, just to the west) comprised an eclectic contingent of ascendant industries and occupations, including computer software designers, Internet firms and web-designers, electronic data processing and storage, technology consultants, digital artists and photographers, film and video producers and editors (including digitalized effects), industrial designers, and fashion designers. The rapid growth of these primary production enterprises was accompanied by a corresponding expansion of complementary industries and companies, including communications, sales, personnel, and (increasingly) technical support companies. There were concentrations of these support firms in SOMA and Mission, as well as in the CBD fringe, where they could cater to corporate clients with outsourcing requirements

The true diversity of industries, firms, and occupations within SOMA and CBD fringe and inner city districts, even following successive rounds of restructuring, should not therefore be downplayed. In San Francisco, as in other cities discussed in this volume, we can affirm the co-presence of 'old economy' and 'new economy' within the terrains of the inner city. But not all industries enjoyed equivalent growth rates, impacts, and imagery, and so a 'New Economy' of multimedia industries and the ubiquitous dot.coms was inserted into media reportage and thus into popular usage, as well as entering the City's planning discourse and lexicons.

In the middle of 1997, the City Planning Department published *Multimedia in San Francisco*, a document which set out the scalar and spatial dimensions of growth, together with a range of policy issues. With regard to the first, the study team (which included Berkeley faculty and students as well as City Planning staff) identified over 400 'core multimedia' firms, specializing in content and service provision, as well as 119 firms specializing in multimedia tools and applications, and a further 200 support service firms, the latter including computer services (training, hardware, consulting), non-digital film and photography, and business services. The study estimated that the San Francisco and New York multimedia sectors were of roughly equivalent size, and comprised the largest clusters of such activity in the US. The 1997 City survey disclosed that significant concentrations of multimedia firms were domiciled within the CBD and CBD fringe, reflecting

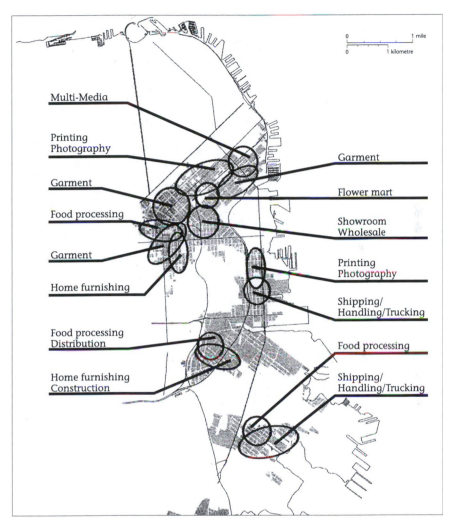

Multi-Media

Printing
Photography

Garment

Food processing

Garment

Home furnishing

Food processing
Distribution

Home furnishing
Construction

Garment

Flower mart

Showroom
Wholesale

Printing
Photography

Shipping/
Handling/Trucking

Food processing

Shipping/
Handling/Trucking

Figure 7.4 Industrial clusters, SOMA, Port lands, South Bayshore, and NEMIZ.

Source: City of San Francisco Planning Department, Land Use Planning, 1995–99.

a positioning of new media as an important business and professional services sector, with smaller distributions in the Mission and Noe Valley (see Figure 7.5 for city-wide distributions of multimedia firms). But the largest clusters of core multimedia firms and support companies were found in the South of Market, as shown in Table 7.3.

The concentrations of multimedia enterprises in SOMA (see Figure 7.6) suggested the formation of a 'new industrial district' of primary producers and proximate networks of suppliers and clients, superimposed upon the obsolescent production landscapes of the inner city. Support for this view was provided in

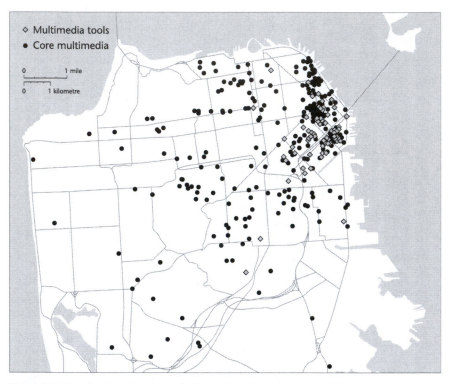

Figure 7.5 Distribution of multimedia firms, San Francisco, circa 1997.

Source: City of San Francisco Planning Department (1997).

Table 7.3 Multimedia businesses in San Francisco: distribution by district and type of business

Districts	Core multimedia	Multimedia tools	Support services	Total
South of Market	165	83	111	359
Downtown-Civic Center	92	30	75	197
Mission-Noe Valley Potrero	53	2	7	62
Marina-Fillmore	42	–	1	43
North Beach-Chinatown Van Ness	24	3	3	30
Rest of the City	31	1	4	36
Total	407	119	202	727

Source: City of San Francisco Planning Department (1997).

City's multimedia document, which reported survey data generated from a panel of interviews indicating that the draw of SOMA for multimedia firms included

[a] dynamic artist community, the need for multimedia businesses to interact

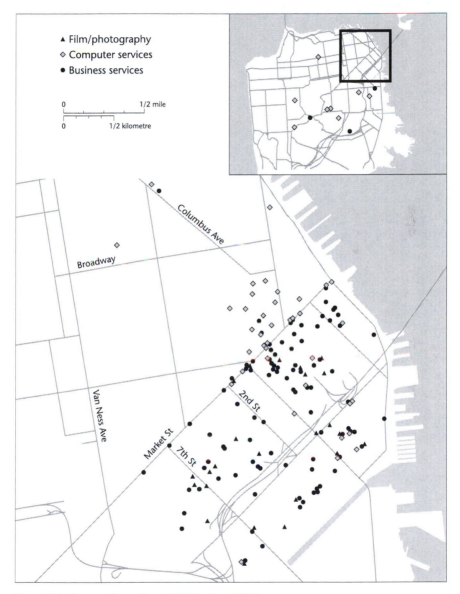

Figure 7.6 New media services, SOMA, circa 1997.

Source: City of San Francisco Planning Department (1997).

and exchange resources in the process of innovation, the availability of warehouse-office space, access to fibre optic cables and servers, and San Francisco's proximity to high technology resources in Silicon Valley and the film industry in Marin County.

(San Francisco Planning Department 1997: 11)

This distinctive spatial profile of San Francisco's multimedia sector comprised both a highly compact formation on the CBD fringe and inner city, as well as a broader regional dimension which included linkages with more peripheral counties within the Bay Area. Together, this construct suggested a highly robust, growth-oriented and durable new industry formation structure, drawing on the considerable human, technological, and environmental assets of the Bay Area.

It must be acknowledged that a sector encompassing fewer than a thousand, mostly small firms necessarily occupies a small niche within a mature metropolitan economy which contains many thousands of enterprises. But the rapid expansion of this sector, attendant media buzz, and the documentation of its growth effects and signifiers by academics, planners, and NGOs seemed to position San Francisco as a capital of the New Economy at the global as well as national scale. Toward the end of the 1990s, the multimedia sector was widely acknowledged as the flagship of a rapidly-growing San Francisco economy, with employment numbers swelled by large contingents of freelance workers engaged primarily in contract work. As Rebecca Solnit observed at the apogee of the Bay Area's technology boom:

> In late 1998 a city survey found nearly as many people were employed in the brand-new Internet/multimedia industry as in the old hotel industry, 17,600 compared to 19,200, and that doesn't count the huge numbers of freelancers working in multimedia who bring the numbers to more than 50,000 in a city whose population is about 800,000.
>
> (2000: 15)

Multimedia (and the New Economy more generally) appeared to offer San Francisco not just a growth industry 'of the moment', but something like a major new trajectory of economic development: a new vocation to complement the long-established banking, business services, manufacturing, and tourism industries, each subject to recurrent restructuring.

Dislocation and displacement in the New Economy

The wave of multimedia firms, dot.coms, and other New Economy enterprises which washed over SOMA in the last years of the twentieth century occurred not benignly in a derelict, postindustrial inner city brownfield site, but rather was imposed upon established, mature communities of residents and businesses. Nor were the New Economy firms the only cause of dislocation. The industrial agency of dislocation in the form of the New Economy firms had a related residential counterpart in the form of live-work studios, an echo of the 'loft living' syndrome chronicled by Sharon Zukin, with the progenitor sites among New York's brownstones. Some of the SOMA live-work occupants followed the familiar pattern of new middle-class entrepreneurs, artists, and professionals inhabiting Manhattan's lofts. But others represented a more distinctive cohort of Bay Area dot.com millionaires, seeking a residential cachet and lifestyle that complemented their high-tech professional identity. Another contrast was in the specificity of built form of the SOMA live-work studio. Manhattan contained a relatively large supply of heritage buildings of good quality construction, eminently suitable for adaptive re-use. In SOMA, the supply of such structures was more limited, so the development sector responded by developing *faux* live-works in the more edgy neighbourhoods south of Market, built to industrial (rather than the more stringent residential) construction standards, generating handsome profits in the residential feeding frenzy of the late 1990s.[8]

Among the many chroniclers of SOMA's dislocation narratives in the New Economy era, Cheryl Parker and Amelita Pascual produced an account which includes a vivid empirical component as well as a critical perspective. Their paper documented the incidence (and location) of displacement, situated in what they termed 'parcel politics': 'the politics of space at the smallest and most complex level . . . grounded in the idea that a great urban place is composed of a complex mix of spaces and places that can accommodate a wide variety of interdependent users' (Parker and Pascual 1999: 55). They acknowledge that the original notion of live-works as a complement to the lively and diverse social landscapes of SOMA, an outcome of an earlier review of industrial lands in the mid-1980s, seemed a good fit with the area, and a way of formalizing the long-established practice of artists and graphic designers deploying old warehouses as studios and workshops.

But the flood of live-works which accompanied the New Economy phenomenon after 1995 represented not an organic development of this mature community, or even a gradual social upgrading of an all-too-familiar kind, but rather a dislocating force which jeopardized the tenancy of both long-established residents and businesses. The pressures generated by local activist groups advocating for a preservation of traditional industries, notably SOS ('Save our Shops') and the Coalition for Jobs, Artists and Housing (CJAH), led to the Planning Department recommending an interim policy for an Industrial Protection Zone (IPZ). The IPZ encompassed much of the production lands within SOMA, with housing and retail uses encouraged beyond the IPZ, with affordable housing to act as a 'buffer' between industrial and these mixed-use areas. As Parker and Pascual concluded, however, the IPZ strategy was a compromise, 'a mixed use plan designed to control gentrification' (ibid.: 63). By 1997, a City Planning survey disclosed the increasingly pervasive infiltration of live-works throughout much of SOMA (Figure 7.7).

South Park as epicentre of the New Economy

If the South of Market Area and the Mission represented exemplary inner city districts of the global technology boom of the late 1990s, then the enclave of South Park can be situated at the epicentre of San Francisco's *fin-de-siècle* New Economy. As the previous section disclosed, advanced-technology firms were liberally distributed among the larger streetscapes of SOMA, typified by the adaptive reuse of warehouses and some building reconstructions along the 1,000-foot street fronts, presenting a vivid transition from old economy to New Economy in a resurrected

Figure 7.7 Current and proposed live-work sites, SOMA, circa 1997.

Source: City of San Francisco Planning Department (1997).

inner city industrial district. This was the New Economy phenomenon writ large upon the landscapes of the inner city, with hundreds of companies, including a representation of large corporations as well as a more numerous SME contingent, forming a world-scale cluster of multimedia, software, and Internet firms.

The concentrations of dot.coms ensconced within the intimate confines of South Park encompassed at best a small fraction of the New Economy enterprise base within SOMA in the waning years of the twentieth century, although these included some successful small- and medium-sized firms. South Park's status as signifier of the New Economy lies not so much in the number of firms situated within the site, but rather in its distinctive mix of production and consumption activities; in its spaces of social interaction, enabling knowledge spillovers and the informal exchange of tacit knowledge; and in its faithful inscription of larger forces of transition and change in this evocative storyline of urbanism. These facets of industrial urbanism and the social ecology of the city, involving continuities as well as jarring ruptures with the past, have in turn comprised the ingredients for the periodic recasting of South Park's identity.

South Park: reconstructions of a New Economy enclave, 2000–2005

Following an initial scoping visit in 1999, a survey program undertaken in the fall of 2000 disclosed South Park and its environs in full bloom as a New Economy site. To be sure a number of residuals of South Park's recent and more distant past could be discerned, including a sign advertising 'John H. Tway & Son Blacksmith' on Brannan Street, a reminder of the horse-drawn fire wagons and mounted police formerly stationed in the area (Figure 7.8). Large warehouses, some still in use but many vacant, lined the streetscapes surrounding South Park, on Second and Third Streets, recalling the more extended South End[9] area's role as storage and distribution area for the Port of San Francisco. The refurbishment of a warehouse on the south-east corner of Third and Bryant for discount fabric sale presented one of the few examples of adaptive re-use of the industrial stock for retail activity (Figure 7.9).

A prominent survivor among the traditional industrial cohorts within the area was Standard Sheet Metal, located on Brannan just west of Jack London Street. Standard Sheet Metal presented initially as a classic metal-bashing concern, a vestige of the first half of the twentieth century, and indeed the company was founded in 1942, in the midst of the Second World War which stimulated a new wave of industrial development within SOMA. The staff comprised mainly middle-aged males, with a few younger workers in evidence. Standard Sheet Metal had survived the protracted industrial decline of SOMA by supplying a traditional market with its products, together with (as we shall see) subcontracting business for growth-oriented industrial designers (Figure 7.10).

Resonances of South Park's contemporary positioning as site of arts and design included a well-stocked and actively-patronized architecture and design book-store, William Stout, on the Park's south side, as well as a number of architects (including Levy & Partners), fashion designers (notably ISDA, close to William

Figure 7.8 Vestiges of the 'old economy' in SOMA, Brannan Street.

Stout), and industrial designers. 'The Ligatures', a large lithography and engraving enterprise located at the corner of Second and Bryant, reinforced the area's pedigree as locus of applied design within SOMA. South Park also encompassed a significant residential population, including two buildings catering to the Filipino community (mostly retirees), although the area offered only a sparse array of retail and personal services (with the exception of eating establishments, as we shall see). South Park itself, however, one of the few green spaces south of Market Street, offered open-air amenity both for area residents and workers.

While South Park in the fall of 2000 displayed aspects of functional continuity with the 'old economy', however, the signifiers of development lie in the representative enterprises of the New Economy. These principally took the form of self-identified multimedia firms and the ubiquitous dot.coms, sprinkled liberally through South Park and its environs (Figure 7.11). South Park's multimedia sector encompassed a diversity of firm types, including graphic arts and design, business applications (including consultancies providing advice on the deployment of

Figure 7.9 Fabric clearance store, circa 2001, Third and Bryant Streets (now a digital art gallery).

Internet technologies); a number of artists combining standard photography with new digital technologies; publishers, including several serving the multimedia community in San Francisco (notably *Wired*, located on Third, close to South Park); educational services firms; and several film and video producers (several of which were concentrated within a new four-storey building at the corner of Brannan and Jack London Streets; see Figure 7.11).

One of the more intriguing enterprises in South Park circa 2000 was The Idea Factory™, situated in a large warehouse building on the south-west corner of Brannan and Third Streets. The corporate brochure proclaims its mission: 'to redesign the practice of innovation for companies seeking competitive advantage and enhanced wealth creation ability in the new economy'. As an elaboration of this mission, the document stressed the need for new thinking on business practices for the New Economy:

> The Idea Factory™ focuses on systematizing disruptive innovation, sustaining it over time, and integrating it into the organizational fabric. We enable our clients to engage in meaningful experimentation with a manageable level of risk in which the future, not the past, becomes the guiding frame of reference.

Beyond this statement of philosophy The Idea Factory™ identified a business toolkit comprised of 'strategic foresight, user-centred design, stagecraft, new media, and digital business', with 'areas of particular interest' including 'new

Figure 7.10 Standard Sheet Metal, South Park (established 1942).

media, e-services, and applications of wireless technology in business and entertainment'. A visit to The Idea Factory™ in the fall of 2000 disclosed a lively workplace, including what resembled a movie set of props and stage equipment as well as computer stations and striking graphic displays, and a decidedly upbeat staff presence. With its operational synergies of art, design, and technology deployed in the service of business innovation, The Idea Factory™ presented in a particularly exuberant form the imagery of an iconic enterprise in SOMA's turn-of-the-century New Economy.

Much of the New Economy in South Park in the fall of 2000 was on display, evidenced in the proliferation of dot.com signage at the street level, including (as in the Telok Ayer case presented in the previous chapter) the appropriation of this insignia of cutting-edge technology by more prosaic concerns, such as real estate companies. But, like the below-the-surface shark trope employed by Solnit earlier in this chapter, a number of larger New Economy firms operated sight unseen. A conversation with a young (twenty-something) software engineer in South Park

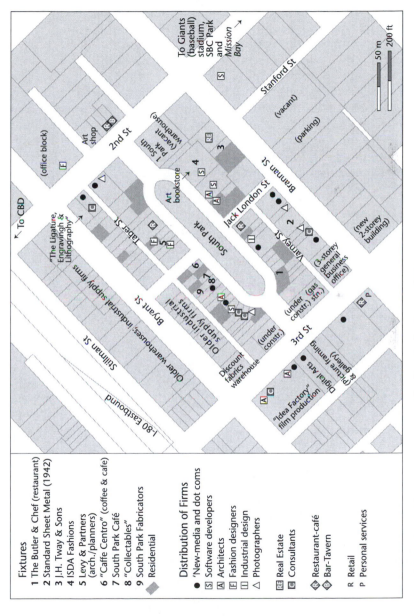

Fixtures

1 The Butler & Chef (restaurant)
2 Standard Sheet Metal (1942)
3 J.H. Tway & Sons
4 ISDA Fashions
5 Levy & Partners (arch./planners)
6 "Caffe Centro" (coffee & cafe)
7 South Park Café
8 "Collectables"
9 South Park Fabricators
◆ Residential

Distribution of Firms

● 'New-media and dot coms
Ⓢ Software developers
Ⓐ Architects
Ⓕ Fashion designers
Ⓘ Industrial design
△ Photographers

Ⓡ Real Estate
Ⓒ Consultants

◈ Restaurant-café
◈ Bar-Tavern

R Retail
P Personal services

Figure 7.11 South Park as epicentre of SOMA's New Economy, 2000.

disclosed the presence of a software development company with 'several hundred' employees, labouring in the upper (and windowless) floors of a warehouse on the south-east quadrant of the park. I was assured that there were other software firms on this scale hidden away in other nondescript buildings, mostly working as subcontractors for larger Silicon Valley Corporations, demonstrating South Park's partial incorporation within this propulsive industrial district.

But these working solitudes aside, the South Park New Economy was in 2000 manifestly a *social economy*, expressed in the patronage of the area's numerous restaurants, cafés, and bars, in the social interaction levels of the streetscapes, and in the palpable 'buzz' of life in the park itself. The rich amenity base of the precinct included an upscale Italian restaurant ('Ecco Restorante') located prominently on the corner of Jack London and South Park Streets (Figure 7.12), a standard accoutrement of New Economy epicentre sites from Clerkenwell to Telok Ayer, and the chic South Park Café. But the liveliest scenes were observed in Pepito's California-Mexican eatery, with cheap and cheerful meals available for

Figure 7.12 Ecco Restorante, South Park, at the height of the New Economy, 2000.

young New Economy workers with limited consumption resources (time and money); Caffé Centro, strategically located on the Park's northern perimeter, and (in the fall of 2000) busy all day serving meals and take-away coffees; and The Butler and the Chef on the south side of the Park, managed by Pierre Chatel. While Caffé Centro and Pepito's embodied the California consumption ambience in full, Chatel's The Butler and the Chef promised a Parisian Bistro experience. That said, Chatel's postcard design showing South Park as a circuit board (Figure 7.13) conveyed in a particularly vivid way the contemporary imagery of South Park as locus of the technology-driven New Economy.

The social nature of the New Economy was also expressed in the spaces of

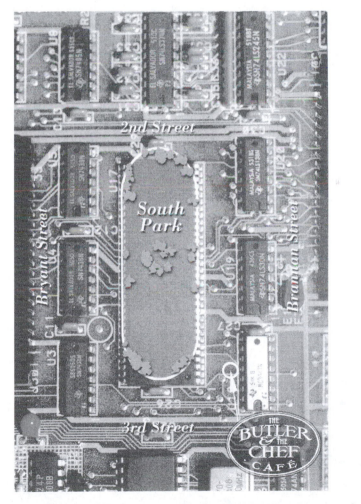

Figure 7.13 South Park as metaphor for SOMA's New Economy: from 'textured place' to 'hard wired', circa 2000.

Source: Pierre Chatel.

interaction within South Park itself. Observations made during different times of the day in the working week demonstrated high levels of park usage, presenting a mixed social space reflected in the diversity of park visitors (Figure 7.14). Certainly during lunch periods and the morning and afternoon break periods, the park was fully patronized by (mostly young) workers, predominantly white and male, but including a representation of women and visible minorities. The New Economy in SOMA exemplified in large part the centrality of social networks to innovation and business practice, principles of the 'new industrial district' described in Chapter 3.

In many cases, the congregations of workers shared documents and papers as well as meals, underscoring the value of the park space as site of knowledge exchange. This function was facilitated by the supply of park tables which allowed the perusal of worksheets and other documents, as well as the more numerous benches of a more conventional sort. Aside from this dominant population of workers, South Park was also actively used by families (for the most part, mothers or care-givers, with children), concentrated within the zone of the park accommodating a limited supply of recreational equipment), more so on week-ends (when the numbers of workers was diminished). But the park's spaces of amenity and conviviality, shared by co-workers and project teams, were accompanied by a sharper image of deprivation, manifested in the homeless African-Americans clustered almost exclusively on the western margins of the grassy area.[10]

'The crash', restructuring, and its imprints in South Park, circa 2001

The technology bubble burst with a vengeance in 2000–2001, demolishing ebullient visions of a technologically-driven and creatively innovative future economy,

Figure 7.14 Enjoying the 'buzz': South Park at lunch, autumn 2000.

and the confident aspirations of the New Economy class. The crash was felt everywhere the New Economy had gained a foothold, and nowhere more than in the leading centres of innovation, including (as we saw in the previous chapter) outposts in South-East Asia as well as in the 'West'.

As the rise of the dot.coms clustered in SOMA represented one of the most globally resonant story-lines of the ascendant New Economy, so too the collapse of these ubiquitous firms symbolized within the public imagination the larger crash of the technology sector. As the proclaimed hub of SOMA's 'Multimedia Gulch', South Park and its environs inevitably shared in this abrupt reversal of fortunes. A return site visit in November of 2001 and a new mapping and interview program disclosed if not an extinction of the dot.coms, then certainly a greatly diminished presence of these emblematic companies, as shown in Figure 7.15. Among the casualties of SOMA's technology crash was The Idea Factory™, evidenced by the vacant and somewhat forlorn building frontage on Third Street. Many of the smaller enterprises within the enclave of South Park itself had disappeared, including the once-thriving Ecco Restorante, testimony to the comprehensiveness of the downturn in SOMA's New Economy.

That said, the configuration of industries in the South Park precinct observed in this return visit registered a measure of change as well as outright decline. In particular, the emergence of what was described to me by local workers as the 'Second Street Tech Corridor' (see Figure 7.15) appeared as a significant development, amid the growing civic pessimism about the area's New Economy vocation. Here the losses of the myriad new media firms which proliferated throughout much of SOMA and the adjacent Mission area were at least partially offset by the growth of larger enterprises along Second Street, south of Brannan Street. These included a number of large software firms, some ensconced in showy new buildings on the west side of Second, while half a dozen medium-sized concerns occupied the former 'South End Warehouse' building on the east side of the street. The Second Street Tech Corridor formed part of a rejuvenated precinct of SOMA which included the Pac Bell baseball stadium at the southern end, together with many new restaurants and bars catering to sports fans as well as to the growing residential population of the adjacent Mission Bay mega-project (Figure 7.16).

The firms occupying the tastefully renovated interior spaces of the South End Warehouse in 2001 included technology consultants and an educational multimedia company ('Look Smart') (Figure 7.17). But perhaps the most intriguing company present was the 'Scale Eight Global Storage Service'. Scale Eight (with 120 employees) specialized in the storage of data, in the form of documents, images and voice mails, thus presenting in its tenancy of the South End Warehouse an exemplary form of functional continuity – from storage of goods in the 'old economy' of the industrial city, to the storage of data in the New Economy. The business concept for Scale Eight was to reduce costs 'by eliminating hardware inefficiencies, removing software and integration expenses, and minimizing staffing requirements'. The 'global service' dimension of Scale Eight's market positioning lies in the capacity for clients to access data from any location, rather than

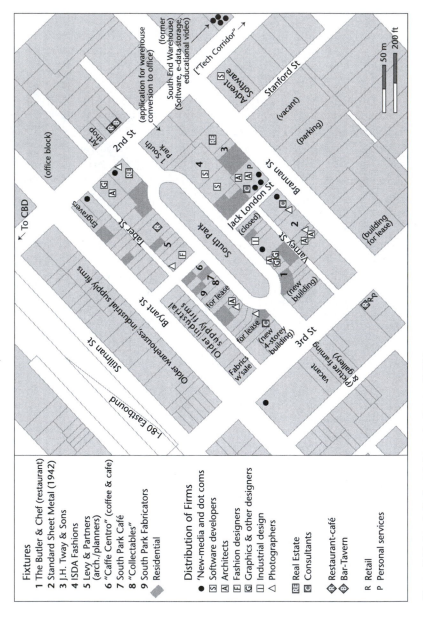

Fixtures
1 The Butler & Chef (restaurant)
2 Standard Sheet Metal (1942)
3 J.H. Tway & Sons
4 ISDA Fashions
5 Levy & Partners (arch./planners)
6 "Caffe Centro" (coffee & cafe)
7 South Park Café
8 "Collectables"
9 South Park Fabricators
   Residential

Distribution of Firms
● 'New-media and dot coms
Ⓢ Software developers
Ⓐ Architects
Ⓕ Fashion designers
Ⓖ Graphics & other designers
▥ Industrial design
△ Photographers

ᴿᴱ Real Estate
Ⓒ Consultants

◈ Restaurant-café
◈ Bar-Tavern

R Retail
P Personal services

Figure 7.15 South Park and the melt-down of 2001.

Figure 7.16 The 'Second Street Tech Corridor', circa 2001, view south to Pac Bell Park (now AT & T Park).

relying upon locally-accessible data storage facilities, thus overcoming the friction of space.

Half a decade on this stored data access concept might not seem a revolutionary breakthrough, but clearly at the time represented a significant innovation. In an article entitled 'Scale Eight Hits the Right Note', George Gilder wrote in the *Gilder Technology Report* that '[u]sing twice as many discs as more "efficient" rivals, Scale Eight's competitive advantage comes from centralizing dumb storage while moving file systems (intelligence) to the edge' (September 2001: 2). Scale Eight's founder was Joshua Coates, a 27-year-old software programmer, consistent with the age cohort and occupational signifiers of the New Economy, while Gilder identified 'computer system titan' Dave Patterson as Scale Eight's chief scientist, underscoring the role of science and innovation in the New Economy milieu. For its part, *The Wall Street Journal* was impressed that Scale Eight managed to raise $31 million from local venture capitalists, 'at a time when the Internet sector was finding the venture spigots drying up' (© *The Wall Street Journal*, 1 March 2001) in the aftermath of the tech crash. Finally, the *Red Herring* report, a business newsletter (combining in their own words 'admiration for entrepreneurs with a larger-than-usual dose of skepticism'), cited Scale Eight as one of 'the 50 private companies most likely to change the world', suggesting an innovative potential (and ambition) well beyond the scope of most technology start-ups (*Red Herring*, 1 May 2001).

A shift from the dense clusters of new media firms and smaller dot.coms to a survivor cohort which featured some larger firms arrayed along Second Street in

Figure 7.17 Directory of New Economy firms, (former) South End Warehouse, 2001.

particular included a sociological dimension. The green space, tables, and benches of South Park at lunch-time were now dominated by cohorts of young males, many favouring iridescent blue shirts and expensive hair-styling, apparently emblematic of the harder technology edge of the software and Internet firms now characterizing the area, as opposed to the social composition of the previous year which included a more balanced gender mix and a more consciously bohemian look. Conversations with a number of these technology workers disclosed that many had walked up to South Park from their corporate spaces along the Second Street Tech Corridor, seeking a slice of amenity as well as noon-hour sunshine. But South Park's distinctive blend of amenities and ambience exerted a larger pull for some: I spoke with three young architects who had driven down to South Park for lunch from their corporate high-rise office, a weekly practice for the group, and much valued as an escape from the concrete landscapes of the CBD. South Park's allure for both nearby and more distant users was one of the most well-defined continuities of my extended field experience in the area.

The party's over: tracking survivors and successors, 2003

A site visit in January 2003 revealed South Park and its environs at perhaps its lowest ebb. The technology wave of the late twentieth century had fully receded, as disclosed by the proliferation of 'for sale' and 'for lease' signage everywhere in the district. South Park (and most of SOMA and the Mission as whole) presented commercial and industrial vacancy rates in the 40 per cent range (Figure 7.18). Within the approximately 60 per cent of premises occupied, a very substantial amount of firm turnover was evident, with new firms including real estate and general business activity, as well as an influx of design firms and artists which portended a prospective return to the area's creative production trajectory, inter-rupted in a forceful way by the New Economy phenomenon of the late 1990s (Figure 7.19).

The profile was particularly bleak in the Second Street Tech Corridor which had exhibited an imagery of cutting edge technology blended with an insistent business 'drive' only a year or so before. The large buildings along Second Street

Figure 7.18 Scenes of dereliction: rampant vacancies in South Park, 2003.

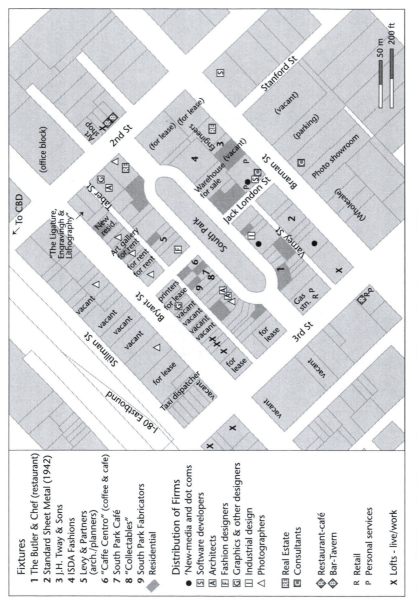

Figure 7.19 South Park in recession: landscapes of live-works and vacancy, 2003.

Fixtures

1 The Butler & Chef (restaurant)
2 Standard Sheet Metal (1942)
3 J.H. Tway & Sons
4 ISDA Fashions
5 Levy & Partners (arch./planners)
6 "Caffe Centro" (coffee & cafe)
7 South Park Café
8 "Collectables"
9 South Park Fabricators
   Residential

Distribution of Firms

● 'New-media and dot coms
Ⓢ Software developers
Ⓐ Architects
Ⓕ Fashion designers
Ⓖ Graphics & other designers
Ⅲ Industrial design
△ Photographers

ⓇⒺ Real Estate
Ⓒ Consultants

◇ Restaurant-café
◇ Bar-Tavern

R Retail
P Personal services

X Lofts - live/work

south of Brannan, including the several new purpose-built structures as well as the renovated warehouses on the eastern side of the street, were now vacant, presenting a bleak landscape of industrial collapse and deep recession. The casualties included Scale Eight, perhaps the most shocking case of business failure, given the high critical regard accorded the company disclosed in the preceding narrative.

A marker of change in South Park was the expansion of live-work redevelopment, reflecting changes in building economics and property values, as well as the abrupt patterns of succession which characterized SOMA as a whole throughout this period of abbreviated restructuring episodes. This expansion of live-works in San Francisco mirrored a trend observed in other cities, including the London (Chapter 5) and Vancouver (Chapter 8) case studies incorporated in this volume.

Enterprise profile: insights from interviews

While the general imagery of South Park and its adjacent territory was one of decline and even distress, the January 2003 survey work disclosed some instructive cases of survival in this particularly bleak market. In the large metropolis there is enough local wealth and resiliency of markets to support a measure of buoyancy for certain enterprises, even during deep downturns.

The mapping, field survey work, and meetings with City Planning staff were augmented by a panel of interviews with representatives of companies, mostly drawn from the evolving mix of design firms and creative services enterprises situated in the South Park district of SOMA. These interviews were undertaken principally to generate insights on the nature of new industry formation in the district (i.e. a probing of the *internal* dynamics of specialized industries), as well as perceptions of the area as an operating field for creative enterprises (i.e. vignettes disclosing an *external* scalar perspective of activity and interdependency within a 'new industrial district').

South Park Fabricators

South Park Fabricators presents an exemplary case of the critical syntheses of culture and technology, place and production, and design and fabrication characteristic of the new inner city industrial district among advanced metropolitan centres. South Park Fabricators, situated in South Park since 1988, are engaged in the design and manufacture of high-value industrial metalwork, for commercial, institutional, and residential clients. Larissa Sand, principal of South Park Fabricators, has worked in this field since 1990, and produces work described in a trade magazine as the synthesis of materials, craftsmanship, and architecture. Sand suggested that South Park Fabricators aspired to achieve 'high style' and high design values in the modernist style, targeted toward clients with good taste. South Park Fabricators was engaged in sales to markets outside the region, but many of its clients were in the Bay Area.

South Park Fabricators survived both the tech boom of the late 1990s in San Francisco, with its inflationary rent cycles and dislocations as well as the dot.com crash of 2000 and its depressed aftermath. With respect to the division of labour, Sand employed (in 2003) four designers, working principally in studio space on the second floor of the building, and six fabricators on the ground floor. That said, Sand emphasized that in a small industrial design enterprise, the exigencies of achieving client needs entailed a fluid interchange of staff between design and production processes, rather than a highly segmented occupational structure typical of Fordist production. The working interface between design, prototype or sample development, and fabrication required high skill, including the capacity to move easily between phases of individual projects.

The success of South Park Fabricators in the market stretched the capacity of its confined premises on South Park, so (at the time the interview was conducted) Sand was engaged in a search for additional space in the area to expand the scope of production. The existing space on South Park was 'functional' rather than elegant, and was a limitation on the expansion of the business. The clear preference was a nearby site, and indeed at the time of interview negotiations were underway to secure new premises on Bryant Street, just behind South Park.[11]

South Park Fabricators' staff configuration, including the multiple skill sets of the workers, enabled a high degree of internalization of inputs in the production process. But there was some scope for outsourcing, which provides a critical insight on the nature of inter-industry and inter-firm linkages in the new industrial district. In this case, the outsourcing patterns of a successful industrial design company evoke a fascinating insight into the co-presence and co-operation of the 'new' and 'old' economy within the production spaces of the new inner city. South Park Fabricators subcontracted for metal fabrication Standard Sheet Metal (see Figure 7.10), a classic industrial era metal-bashing firm located since 1942 in a space on Brannan Street, a three-minute walk across South Park. Standard Sheet Metal presents the gritty imagery of a mid-century machine shop, with an apparent clutter of sheet metal, components and end products distributed promiscuously among the plant, and with a largely older (and exclusively male) staff. But Standard has survived by maintaining a resiliency in the face of recurrent restructuring episodes in the Bay Area as well as recessionary business cycles, achieved in part through producing quality sheet metal products for 'high design value' industrial design firms such as South Park Fabricators. The machine shop, once a mainstay of advanced production systems, is on a long decline in most regions, but is still a valued complement to industrial design.

Fuseproject

If South Park Fabricators represents a successful example of high value industrial design catering to a diverse mix of clients, Fuseproject demonstrates a capacity for success via principal subcontracting for major corporations, both in business and consumer markets.

At the time of the interview (17 October 2003), Fuseproject was located at 123 South Park, directly across from South Park Fabricators. Fuseproject was a start-up in the area (1999), initially serving as a principal design subcontractor for Hewlett-Packard, specializing in the design of personal computer casings and keyboards. The ten staff members were all industrial designers, drawn from a diverse set of nations (and industrial design traditions) including Germany, Sweden, Switzerland, Denmark, and Japan, several of whom had won international design awards: an example of the fusion of cultural production practices characteristic of the transnational city. The mix of industrial design traditions was seen as a distinct competitive advantage for Fuseproject, enabling high levels of productivity and creative diversity in a small company competing in challenging markets.

These company survival skills were put to the test in the aftermath of the collapse of the tech boom in 2000, and attendant oversupply, market shrinkage, and contractions of companies and labour. Fuseproject gained an important new client in Birkenstock, offering an entrée to a relatively robust consumer goods market, offsetting to an extent the temporary downturn in the high-technology hardware producer sector, and underscoring the exigency of subcontracting diversity for industrial design firms.

Like South Park Fabricators, Fuseproject occupied two floors of a building fronting onto South Park, with principal workspaces on the ground level, and a mezzanine area allocated to 'ideas and discussion' use. The mezzanine space also accommodated shelving area for product prototypes and samples, both for Hewlett-Packard and for Birkenstock.

Fuseproject's location underscored the advantages of proximity and affinity within this enclave of SOMA. As all of Fuseproject's employees were industrial designers, there was a clear need for a dependable network of suppliers. These were obtained on occasion from suppliers as distant as Portland and San Diego, but Fuseproject's South Park location also afforded proximate access to local suppliers. These local suppliers included precision cutters and model/prototype builders, who could offer a reliable one-day turnaround. Fuseproject's contingent of designers also valued the amenity attributes of South Park, including views of the green space and its theatre of human activity, as well as the congenial and convenient restaurants, cafés, and coffee shops which abounded in the area.

On the way back? Signs of recovery circa 2005

In the fall of 2005, I returned to San Francisco to observe a new sequence of changes in South Park, two years on from the low-water mark of the area's economy. On a first look, some of the conditions of evident decline seemed not very much different from those witnessed in 2003. Certainly, vacancies and signage advertising sale or leasing opportunities abounded. These included the large former warehouse at 1 South Park, a prominent building occupying a strategic site at the corner of South Park and Second Street, which had remained without tenants during each phase of the area's roller-coaster ride of the late 1990s and early twenty-first century. Conversations with City Planning staff and with local business people indicated that at least a portion of the property owners in the area, especially those with long tenures (and amortized investments) weren't very 'motivated' to reduce prices, even in a slow market. Some property-owners were also apparently waiting for a return of the New Economy, with its potential for greatly inflated land prices and leasing rates.

But on a closer examination it seemed clear that a recovery of sorts was underway, at least on a selective basis. The continued expansion of the adjacent Mission Bay project, and the success of the San Francisco Giants' baseball park in China Basin, contributed to a rejuvenation of the larger area's consumption and amenity sector, and a greater traffic and pedestrian presence than that observed two years earlier. Within South Park itself, new businesses had been established, including fashion design, architects, and artists, as well as creative businesses, such as advertising and marketing. An interview with a fashion designer who had occupied a compact site on South Park's south-east quadrant had expanded to accommodate a larger design studio as well as retail space. South Park Fabricators had also managed a successful expansion within the area, reflecting the strong affinity Larissa Sand and her colleagues felt for South Park as a site of creative inspiration, sentiments expressed by the Telok Ayer designers and entrepreneurs featured in the preceding chapter. Further, there was a discernibly greater social presence and interaction within the spaces of South Park itself, a clear sign of the robustness of the site as a place of conviviality and exchange. Certainly patronage of Caffé Centro and The Butler and the Chef seemed to indicate a revival of South Park at the mid-point of the twenty-first century's first decade.

Like South Park Fabricators, Fuseproject's market success placed growing pressures on the workspaces of the building at 123 South Park, and so the company relocated, although staying within SOMA.[12] In June of 2005, this property was occupied by 'Jumbo Shrimp', a company engaged in business-to-business consulting on technology and advertising.

An interview with Jumbo Shrimp principal Robert Ahearn yielded rich insights into the workings of the creative sector in SOMA, and more particularly the distinctive synergies between work and lifestyle in the new inner city. Jumbo Shrimp had been established for four years, 'one of many creative firms in the Potrero Hill design district'. While Ahearn and his partner enjoyed the working environment of Potrero Hill, they 'had always liked' South Park, like many native San Franciscans. While on a visit to the Park Ahearn had happened to note the 'for lease' sign for 123 South Park, and had jumped at the chance to relocate.

Jumbo Shrimp (with eleven full-time staff) targets its clients among the SME (or 'tier II') business sector, offering personal service to clients, as distinct from major advertising corporations such as McCann Ericsson or Leo Burnett, which tend to use mid-level staff for all but the largest customers. Ahearn and his partner have experience working for large advertising concerns, so aim to bring that knowledge to clients within a small firm setting (hence 'Jumbo Shrimp'). The company intends to grow laterally, rather than vertically, with a view to maintaining direct contact with clients. The company's South Park location offers a convenient base for accessing clients within San Francisco's CBD, half a mile to the north.

For Robert Ahearn and his colleagues, South Park's ambience and amenity serve to enhance personal lifestyle preferences as well as offering a stimulating work environment. The staff enjoys the casual eating establishments and the green spaces of the Park itself; those with young children value South Park's playground equipment. Ahearn is a baseball fan, and takes clients to the Giants' games at AT & T stadium, a ten-minute walk to the south along Second Street.[13]

Like other creative professionals I interviewed, Ahearn suggested that the atmosphere of South Park was 'definitely better' without the frenetic presence of the 'dot.commers'. There was a sense that the New Economy firms of the late 1990s congregated within South Park and its environs in a frenzy, following the locational Zeitgeist, rather than a conscious choice based on a more nuanced appreciation of the Park's history, social quality, and textured landscapes. A conversation with a fashion designer at ISDA (south-east quadrant of the Park; thirteen years in this area of SOMA) affirmed this sentiment, adding that the creative businesses and workers valued a real sense of community which had evolved in the area, as well as an affinity for the Park and its amenities. The dot.com firms of the late 1990s were seen essentially as 'accidental tourists', prepared to draw on the amenity base of South Park to satisfy operational needs, but not really engaged in the life of the community in the same way as the creative firms.

Epilogue: The New Economy redux in South Park?

While the evidence of the 2005 site visit appeared to confirm a recovery based on a perhaps predictable revival of South Park's arts and design vocation, there were also the first signs of a sharper technological edge to the area's businesses, in the form of a number of Internet providers. A year later, in an article published in the *San Francisco Chronicle*, Dan Fost outlined the contours of a prospective 'New Economy II' experience for SOMA, once again centred on South Park and its adjacent areas. This time, the precipitate causality was the so-called Web 2.0 phase of the Internet's development, expressed as a widespread use of high-speed Internet services and mobile telecomms, combined with open-source software.

Fost's account of this prospective second tech wave, entitled 'Web 2.0 has a local address: South Park, the neighborhood that fostered the dot-com boom, is back', included several story-lines of reinvestment and revival:

> Brightest among the new firms are podcasting companies, with at least three near the park. Former MTV veejay and current 'podfather' Adam Curry's Podshow is on Bryant Street, and Evan Williams, who founded Blogger and sold it to Google, runs Odeo in South Park. Odeo moved in January from its office at Second and Brannan to one right on South Park; VideoEgg, a venture-backed startup that makes a tool to edit and share video from camcorders online, moved into Odeo's old digs. 'It is a bit hermit crab-ish', VideoEgg CEO Matt Sanchez said.
>
> (Dan Fost, *San Francisco Chronicle*, 16 April 2006)

Fost quotes *Wired*'s editor Jeffrey O'Brien as observing a renaissance of the area as signified by an increase in the vitality of South Park itself:

On sunny days, the grass fills up during lunch hour, and there always seems to be a line snaking along the sidewalk from the burrito restaurant. Throughout its history South Park has always been an interesting barometer of the local economy. Things must be getting better.

(Jeffrey O'Brien, quoted in Fost 2006)

Fost also quotes Jesse Blout, Director of the Mayor's Office of Economic and Workforce Development, as affirming a connection between the return of the technology sector and the recovery of the San Francisco economy, with a decline in unemployment from 7 per cent in 2002 to 4.7 per cent in February of 2006: 'We're seeing [a] lot of a very particular type of high-tech company in San Francisco that blends the best of the Silicon valley expertise with a very uniquely San Francisco creative niche' (Jesse Blout, quoted in Fost 2006). Even if we delete the redundant 'very's from the commentary, there is some evidence that a more modest (in scale and tone) New Economy phase has been inserted into the landscapes of South Park and the larger SOMA territory that comprised the prime locus of innovation a decade ago.

That said, the crash of the New Economy and the ongoing restructuring of San Francisco's other major industries have left their mark on the city's economy. Employment data generated by the American Community Survey (Table 7.4) in 2005 disclosed significant contractions from the levels reported in the 2000 U.S. Census, even allowing for a larger statistical margin of error. Employment losses were quite substantial in key industry groups, including an approximately 50 per cent contraction in the information employment category associated closely with the New Economy, and smaller but significant declines in FIRE, professional, scientific, and management employment, and in manufacturing. San Francisco's positioning as a favoured city in the Golden State is not proof against the swings of insistent industrial restructuring and deep recessionary cycles.

Conclusion: observations from the SOMA New Economy experience

While the 1990s New Economy boom transformed for a time the landscapes of many advanced and transitional cities, San Francisco's New Economy experience and its aftermath stand out as one of the defining exemplars of this moment in the urban narrative. The growth of multimedia firms and dot.coms within SOMA and Mission, together with the dramatic expansion of specialized, neo-artisanal labour, and the distinctive regional dimensions of the tech boom as represented by the strength of subcontractor linkages with Silicon Valley corporations, represents in many respects the high-water mark of the global New Economy phenomenon.

But the higher and faster the rise, the more precipitous the fall. The technology crash of 2000, shaped by oversupply, grossly inflated technology stock values, and the snowballing effects of corporate debt, layoffs, and losses of sales and incomes, hit the Bay Area particularly hard, following as it did earlier contractions in other key sectors. The frenzied expansion of firms in SOMA, displacing hundreds of

Table 7.4 Employment by industry, San Francisco County, 2005

Employed civilian population 16 years and over	Margin of error	Industry	Estimate	Margin of error
391,953	+/−4,779	Agriculture, forestry, fishing and hunting, and mining	730	+/−495
		Construction	17,284	+/−2,628
		Manufacturing	21,676	+/−2,543
		Wholesale trade	11,088	+/−1,876
		Retail Trade	36,235	+/−3,211
		Transportation and warehousing, and utilities	17,456	+/−2,486
		Information	19,233	+/−2,250
		Finance and insurance, and real estate and rental and leasing	37,196	+/−2,933
		Professional, scientific, and management, and administrative and waste management services	76,440	+/−4,294
		Educational services, and health care, and social assistance	75,151	+/−4,140
		Arts, entertainment, and recreation, and accommodation, and food services	44,039	+/−3,680
		Other services, except public administration	21,787	+/−2,472
		Public administration	12,638	+/−1,687

Source: American Community Survey (2005).

firms and residents in the latter years of the twentieth century, rapidly gave way to a melting away of the dot.com presence which defined the imagery of Multimedia Gulch as one of the principal global bastions of the advanced-technology urban economy. By 2002, the City was in full recession, with knock-on effects including a property market crash in SOMA, a profusion of 'vacancy' and 'for rent or lease' signs along the principal street fronts, and the closure of many of the upscale restaurants and other consumption activities which catered principally to the big spenders created by the evanescent profits of the tech boom. The production-consumption interdependencies characteristic of the New Economy meant that the costs (as well as the gains) would be cast commensurately wide.

The economic pain of the dot.com crash in SOMA and the Mission was compounded by the deep social costs ensuing from, first, the large-scale displacements of pre-existing residents and businesses, brought to life by Rebecca Solnit, Cheryl Parker, and Amelita Pascual, a familiar enough storyline in this urban territory as Chester Hartman and others have chronicled, and, second, the hardships associated with the massive restructuring and fallout of 2000 and afterwards. Even for the survivors of the crash, the increasing pressures of ever-tighter and more competitive markets have made for intensely demanding business experiences and working life for many. The working day for creative workers, professionals, and entrepreneurs encroaches deeper into home and family life, as Andy Pratt and Helen Jarvis have documented (2002), and as the vignettes presented in this chapter have suggested. Some of the glitz and glamour of the New Economy lifestyle, certainly, has been tarnished by the corrective events and trends of the past five or six years' restructuring.

If the crash of the 1990s tech boom and the corollary meltdown of the New Economy were widely acknowledged as classic exemplars of market failure, then the San Francisco experience concentrated within SOMA and the Mission also point to policy (and perhaps more trenchantly *political*) deficiencies. The San Francisco Planning Department has established a reputation for conscientious and progressive planning, including commitments to the City's diverse communities, and to the conduct of diligent policy planning practices. The City Planning industrial land use planning process of the 1990s endeavoured to reconcile the needs of long-established industries and labour, together with attendant social groups and communities, with the pressing demands of growth industries as represented by the multimedia sector and the dot.coms clearly in the ascendant by 1997. But the startling growth of the late 1990s New Economy overran the measured pace of the industrial lands strategy exercise. Further, the planners were severely constrained in preventing wholesale conversions of land use and buildings in SOMA and the Mission, including the transfer of use from industrial to office use which compromised the integrity of the City's zoning regime, and the ubiquitous reconstruction of sites for live-works, by the lack of political will. The Mayor and many of the Board of Supervisors – and influential members of the City's business sector – were entranced by the high-tech vision and its apparent promise of a propulsive New Economy, not subject to the vagaries of business cycle effects which recurrently afflicted the Bay Area's financial, manufacturing, and tourism sectors. When the vision proved ephemeral, the shortcomings of logic and consistency embodied in the political response to the rise of a putative New Economy were exposed, with the costs falling principally on the residents and workers of SOMA.

The San Francisco New Economy story has high intrinsic value, in light of the scale and speed of the growth phase, and the steepness and impacts of its collapse. But the saliency of the SOMA experience in particular can be underscored by comparisons with those of other cities and sites, including the case studies incorporated in this volume. Perhaps the starkest contrast with SOMA is captured in the narrative of new industry formation in Singapore's Chinatown, within which the imprints of the dot.com phenomenon in Telok Ayer faithfully replicated the experiences of South Park and its environs. Unlike the SOMA experience, though, the Telok Ayer New Economy phenomenon generated few social costs, given the previous relocation of the residential population, and the district also recovered from the dot.com collapse much faster than South Park and SOMA. This contrast is due in part to environmental factors, including the greater distance from SOMA to the CBD than is the case for Telok Ayer, immediately adjacent to Singapore's corporate complex, as well as the somewhat forbidding nature of the 1,000-foot blocks of the former warehouse areas of SOMA, as opposed to the highly textured and more intimate spaces of Chinatown in Singapore. It is also the case that the design sector was able to expeditiously recapture Telok Ayer in the aftermath of the dot.com collapse, demonstrating the resiliency of an urban micro-market of fragmented property ownership in accommodating changes in the pattern of demand. Place, scale, and distance are all of critical importance in the economy of the inner city.

8 New industry formation and the transformation of Vancouver's metropolitan core

Introduction: Vancouver as site of recurrent restructuring

Since the 1960s metropolitan Vancouver has experienced a sequence of industrial restructuring processes and events that have underpinned the transformation of its economy, labour force, society, and spatial structure. This compressed period of forty years or so has seen far-reaching shifts, from regional central place at mid-century, to an exemplary twenty-first-century transnational metropolis. As in other city-regions, exogenous factors have been influential, including major investments and policy decisions by senior levels of government, as well as a complex mélange of market and social forces. But as external factors (such as Foreign Direct Investment [FDI], new immigrants, or technology) are absorbed into the structures and systems of the regional economy, the distinction between exogenous and endogenous factors becomes blurred, and the nature of *global-local interaction* assumes greater importance.

Since the 1970s, successive rounds of industrial restructuring, coupled with a transnational urban development trajectory, have produced a distinctive post-Fordist (and post-staples) small- and medium-size enterprise (SME) economy and entrepreneurial labour force. The Vancouver city-region lacks the propulsive-scale corporations (and thus global projection capacity) of cities such as Toronto or Seattle. But its SME economy has, on the evidence, imbued Vancouver with a compensatory adroitness and resiliency.[1]

Vancouver never developed as a major centre of manufacturing on the model of central Canadian metropolises such as Toronto and Montreal, with these latter cities demonstrating path dependency constructed through early industrialization, advantages of market scale and concentration, and access to capital.[2] Vancouver has emerged as a largely post-Fordist production centre, including services as well as goods-producing industries, without first developing as a major site of Fordist manufacturing (Barnes *et al.* 1992: 180).

Consistent with its long-established development trajectory, the structure of Metro Vancouver's economy is increasingly shaped by growth in specialized services, broadly congruent with trends observed in the preceding London, Singapore, and San Francisco cases. In particular, the past decade has seen particularly rapid expansion in professional, scientific and technical services, educational services, and trade, reflecting Vancouver's strength in both intermediate and final demand services, as well as its strategic international gateway functions (Table 8.1). There has also been substantial growth within the 'information, culture and recreation' category, an aggregation which takes in employment generated by Vancouver's cultural industries and creative firms, largely (but not exclusively) concentrated within the City of Vancouver's core districts, as well as the informational industries whose ascendancy was incorporated as an axiom of Daniel Bell's forecast of a postindustrial society (Bell 1973). But while advanced services constitute the cornerstone of Vancouver's economy, the most recent period has included employment growth in the 'goods-producing sector', especially in construction (associated with a sustained boom in residential development throughout the region, and industrial, commercial and institutional development in the suburbs), and in agriculture,[3] while manufacturing employment has been relatively stable.[4]

The comprehensive nature of successive transformative episodes over the last four decades or so has been felt throughout the Vancouver metropolitan region as a whole. The momentum of population growth has shifted to the suburban areas, as the central municipality, the City of Vancouver, now comprises only about one-quarter of the Metro Vancouver (i.e. metropolitan) population (c. 570,000, of a regional population of 2.2 million). Within Vancouver's suburbs, we can identify as important outcomes of restructuring new spatial divisions of production

Table 8.1 Employment by industry for Metro Vancouver, 1996, 2001, 2006

	1996	*2001*	*2006*
Total employed, all industries	946.5	1,039.1	1,187.1
Goods producing sector	182.2	176.2	211.9
Agriculture	5.9	6.6	10.0
Forestry, fishing, mining, oil, and gas	10.0	5.6	8.1
Utilities	5.3	5.5	3.7
Construction	59.4	53.5	85.3
Manufacturing	101.6	104.9	104.7
Services producing sector	764.3	862.9	975.2
Trade	152.0	165.7	191.8
Transportation and warehousing	58.1	66.8	67.6
Finance, insurance, and real estate	78.2	77.8	88.0
Professional, scientific, and technical	74.1	95.8	112.0
Business, building, and support services	37.6	42.7	54.5
Educational services	55.8	72.5	92.4
Health case and social assistance	88.6	96.1	115.8
Information, culture, and recreation	50.4	66.3	70.3
Accommodation and food services	74.6	84.9	86.9
Other services	45.3	52.8	52.7
Public administration	49.5	41.4	43.3

Note: Annual averages (000's of employees).

Source: Statistics Canada.

labour, notably growth in manufacturing and allied industries. There has been significant growth in advanced-technology industrial enterprises, notably within the inner suburban municipalities of Richmond, Burnaby, and North Vancouver, especially in telecommunications, electronics, aerospace and marine industries, life sciences, and alternative energy research. A number of regional town centres (RTCs) designated as growth concentration zones in the Metro Vancouver *Livable Region Strategic Plan*, particularly Burnaby Metrotown and Richmond Centre, have emerged as specialized service poles within a polycentric space economy which includes the metropolitan core, central waterfront (CWF), University of British Columbia (UBC), Simon Fraser University (SFU), and the Vancouver International Airport (YVR) (Figure 8.1).[5]

International immigration has underpinned the formation of major new multi-cultural communities within the region as a whole, constituting an important sphere of global-local interaction. Over the metropolitan area, immigration is reshaping industries and broadening the region's base of entrepreneurship, and is transforming Metro Vancouver's labour market, educational system, and urban landscapes. Increasingly this new immigrant community formation has found expression in the suburban municipalities, notably ethnic Chinese in Richmond, and South Asians in Surrey, as documented in Daniel Hiebert's (2005) research.[6] Table 8.2 shows changes in the distribution and degree of concentration of principal ethnic groups in Metro Vancouver. The data show large concentrations of major ethnic populations in the City of Vancouver, especially Chinese, by far the

Table 8.2 Population group by geographical distribution and degree of concentration, 1996 and 2001, Vancouver Census Metropolitan Area

	1996			2001		
	Population	% in City	Index of Seg.	Population	% in City	Index of Seg.
Total – all groups	1,813,935	28.1		1,967,520	27.5	
Total visible minority pop.	564,595	40.3	39.5	725,700	36.5	41.1
Chinese	279,040	50.2	49.3	342,620	47.0	50.0
South Asian	120,140	21.7	48.9	164,320	18.7	52.8
Black	16,400	30.2	31.0	18,460	25.9	32.8
Filipino	40,715	40.7	33.2	57,045	38.7	37.8
Latin American	13,830	40.5	36.6	18,765	34.6	36.4
South-east Asian	20,370	61.2	52.0	28,550	51.5	48.4
Arab/West Asian	18,155	20.6	41.6	27,270	17.0	47.0
Korean	17,080	24.0	42.2	28,880	21.3	44.5
Japanese	21,880	36.9	30.4	24,025	34.4	32.7
Visible minority, n.i.e.	6,775	31.6	48.6	3,290	35.1	56.3
Multiple visible minority	10,215	35.7	35.9	12,450	36.5	36.3
All others	1,249,340	22.5	39.5	1,241,815	22.2	41.1
Average, weighted			41.5			43.5

Source: Statistics Canada, 1996 Census and 2001 Census; Hiebert (2005).

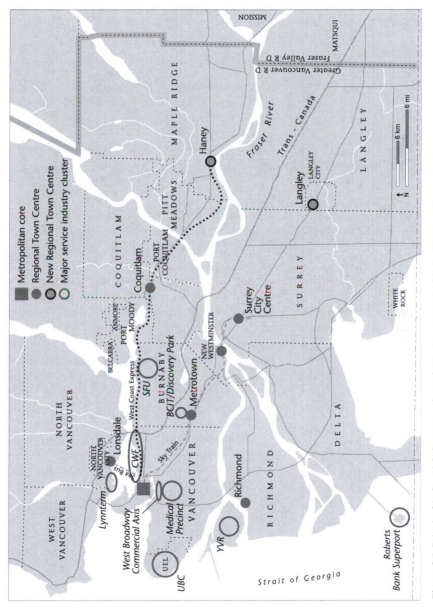

Figure 8.1 Principal industrial clusters and employment centres in Metro Vancouver.

most numerous group, and significant numbers of Filipinos, South Asians, and Southeast Asians. But Table 8.2 also shows a shift in concentrations of ethnic populations to suburban areas. Vancouver's experience is thus broadly consistent with a socio-spatial shift from 'urban enclave' to 'ethnic suburbs' (Li 2006) observed in other North American metropolitan cities such as Los Angeles and Toronto, where multiculturalism is an important dynamic of urban growth and change.[7]

The developmental saliency of the metropolitan core

We can forcefully demonstrate that the *metropolitan core* (central business district [CBD], CBD fringe, and inner city) has represented a major terrain of Vancouver's transformative experiences since the mid-twentieth century, continues to function as a highly salient zone of industrial innovation and change, and has produced in many ways Vancouver's most instructive policy lessons and exemplary planning experiences (see Punter 2003). To illustrate: the defining features of the City's *postindustrial* transformation of the 1970s included the growth of the Central Business District's corporate office complex (Hutton and Ley 1987); the conversion of obsolescent industrial lands and the formation of a mixed-income, medium density residential community on City-owned land in False Creek South (Cybriwsky *et al.* 1986); and sustained gentrification pressures on core area communities (Ley 1980). The most spectacular expressions of the *global processes* of the 1980s were observed in the 'hallmark event' of Expo 86, staged on the former industrial lands of False Creek North, and the subsequent purchase of these properties in 1988 by Li Ka-Shing, accelerating the integration of Vancouver into the global property markets (Olds 2001).

These restructuring experiences are associated with influential theoretical expressions (notably postindustrialism, post-Fordism) and with exigent normative outcomes (such as the emergence of a 'new middle class' of elite managers and professionals, and attendant social polarization). Over the 1996–2001 quinquennial periods, occupations experiencing growth included those in management, the natural and social sciences, government and education, sales and services, and art, culture, recreation and sport (Table 8.3). Polarization as shaped by occupational trends is evident in Vancouver, as is professionalization, but the latter comprises the larger tendency, as Table 8.4 shows. Within the Vancouver region, there is a pronounced concentration of professional workers within the central city census tracts, following trends observed in London, Toronto, and other metropolitan cities where professionalization constitutes a principal trajectory of socioeconomic change.

We can also acknowledge the significance of planning and other policy factors in the reproduction of Vancouver's central area since the 1970s, affirming a commitment on the part of the City and other public agencies to participate in the reshaping of the core's spatial structure, land use, environment, economy, and social morphology. The City's planning model entails a sequence of interventions at the more localized scale, following the strategic reconfiguration of the core as a whole embodied within the seminal *Central Area Plan* of 1991, which

Table 8.3 Changes in Metro Vancouver's occupational structure, 1996–2001

	1996	*2001*
Management occupations	95,300	98,500
Business, finance, and administrative occupations	205,000	194,300
Natural and applied sciences and related occupations	60,700	72,400
Health occupations	45,000	46,300
Occupations in social science, education, government service, and religion	76,200	90,700
Occupations in art, culture, recreation, and sport	33,800	36,400
Sales and service occupations	225,700	300,100
Trades, transport and equipment operators, and related occupations	142,300	134,200
Occupations unique to primary industry	16,400	10,900
Occupations unique to processing, manufacturing, and utilities	42,200	42,500

Source: Statistics Canada.

Table 8.4 Metro Vancouver's workforce employed as professionals and low-level service workers, 1971 and 2001

	1971	*2001*	*Change*
Professionals*			
As a proportion of the workforce	0.170	0.300	0.130
Number of workers	81,190	320,695	239,505
Low-level service workers*			
As a proportion of the workforce	0.131	0.145	0.014
Number of workers	57,845	156,190	98,345

Note: *Occupational definitions based on Walks' (2001) grouping of census variables.

Source: Markus Moos (2007) calculations of Statistics Canada census data (1971, 2001).

consolidated the CBD, favouring housing and a mix of new uses (public amenities as well as new industries) within the inner city. Local planning in Vancouver's central area includes the mediation of social conflicts and contested claims for urban space in the 'new inner city', filtered in part through a series of district-level community plans, as well as urban design site planning.[8] But the Vancouver 2010 Winter Olympic Games are likely to generate a new global narrative for the City's metropolitan core, inserting an international inflection to local policy discourses and agendas, and contributing to the reshaping of space within a densely developed urban setting and inflationary property market.[9]

Reindustrialization and the 'new inner city'

For many urban scholars the master story-line of Vancouver's core area redevelopment since the approval of the *Central Area Plan* concerns the remarkable growth of residential development, consistent with the dominant policy motif and programmatic orientation of the 1991 strategy. This residential trajectory has largely taken the form of precincts of podium-mounted point towers, as seen in the Concord Pacific, Downtown South, and Coal Harbour projects, but also includes loft conversions, live-work and work-live studios, and the adaptive re-use of older commercial buildings. These residential projects have been situated within what some observers have described as North America's highest quality public realm of amenity and environmental attributes (Sandercock 2005), a progressive model of urban design planning John Punter has termed the 'Vancouver Achievement' (Punter 2003). A second, and manifestly darker, urban narrative concerns the acute deprivation of older, marginal communities in the Downtown Eastside (DTES), a territory characterized by endemic poverty, substance abuse, and crime, exacerbated by the relentless encroachment of high-end housing, upscale consumption activities, and spectacle. These produce severe inequality and dislocation, as documented in Heather Smith's research on polarization in the Downtown Eastside (Smith 2000). The DTES, the locus of Vancouver's original townsite in the 1880s, has been the subject and site of numerous experiments in mitigation and in community planning, with the current policy rhetoric one of 'revitalization without displacement'.[10] But the amelioration of acute socioeconomic conditions has to date been largely piecemeal, although there are diverse communities struggling for tenure within the area. The City of Vancouver is therefore in some quarters as infamous for the apparently spiralling descent and misery of the DTES as it is famous for the emergence of a paradigmatic residential central city, presenting a distinctly dichotomous profile of urbanism.[11]

But amid (or adjacent to) these new and established residential communities, an economy of ascendant industries, employment, and workstyles has emerged. These design-oriented, technology-intensive, and knowledge-based industries are situated within new (or reconstructed) production spaces in the core, and exhibit distinctive agglomerative behaviours. They represent a recent and important manifestation of what Scott has termed the internal specialization of production spaces in the metropolis (Scott 1988). These new industries and their constituent labour are associated with the expansion of inner city residential communities and innovative forms of housing style and tenure, including the 'loft-living' lifestyles described by Sharon Zukin, as well as live-work (and work-live) studios. At the same time inflationary rents in the central city housing market have driven many of these New Economy workers to suburban neighbourhoods. The relationship between new industry formation and residential development in the core is therefore by no means entirely symbiotic. The operation of the central city property market, ever sensitive to shifts in the balance of returns on investment (ROI), creates competition between high-end housing and employment-generating land uses (and therefore a measure of instability for many enterprises), a trend observed in London, San Francisco, and other world cities as well as in Vancouver.[12]

The CBD still encompasses the largest complex of industries and employment within the City and the region as a whole, but its dominance has been vitiated by a flight (or takeover) of numerous resource company head offices as a local outcome of globalization.[13] The contraction of head office functions and employment in Vancouver's Central Business District has been in part offset by new uses, including private education colleges and institutions, business consultancies, and New Economy firms which have recolonized the CBD's northern crescent and fringe. There is also a presence in the core of remnants of heavy industry (such as a large cement plant on Granville Island), and a residual 'factory world' of sorts on the periphery of the central city, including an admixture of food and beverage production, garment and fashions firms, automotive repair, light engineering and machining, rendering plants, and materials recycling operations. Further, the metropolitan core contains clusters of commercial and industrial support firms which cater to the subcontracting needs of the central city's lead sectors and industries. These support firms are typically situated in mixed-use areas on the periphery of the central city, sites which afford some relief from the steep rent gradients of the downtown proper, while still allowing convenient interaction between clients and suppliers.

While acknowledging the complexity of industrial organization, labour formation, and production practices in the economy of Vancouver's metropolitan core, there is a defining emphasis in this study of Vancouver's evolving central city economy on industries representative of the dominant trajectory of contemporary industrial innovation and restructuring. These include, for the purposes of illustration, new media industries, film and video production and postproduction, Internet services and web-design, computer software development, and computer graphics and imaging, as well as established but technologically retooled design-based industries, such as industrial design, architecture, and graphic design. These industries, together with the ancillary activities that perform supporting roles in the intricate production systems of the twenty-first-century urban core, and the range of consumption amenities that make up essential elements of the relational geographies of specialized production, have significantly influenced the reconfiguration of the core's space-economy.

Restructuring sequences in the central city

The late twentieth-century crisis of industrial restructuring has left its mark on Vancouver's social and economic landscapes. But the scale of impact of global processes and industrial restructuring in Vancouver has been less wrenching than in many other cities, reflecting the comparatively modest role of traditional industry within Vancouver's economy, and the resiliency imparted by the region's SME economic structure. Episodes of industrial change in the Vancouver case have rarely occurred in isolation from local planning and policy initiatives, suggesting a dialectical process of urban transformation, rather than one directed entirely by the market.[14]

From 'regional central place' to 'international stage'

The baseline conditions from which we can undertake an analysis of recent industrial change in Vancouver, and its important role in reshaping urban space, are embedded within the familiar lineaments of the mid-twentieth-century regional central place. The spatial imprints of the mid-twentieth-century city are depicted in structural terms in Figure 8.2, which shows the basic template of industry, commerce, and housing in Vancouver's metropolitan core. The downtown (zone I in Figure 8.2) included relatively specialized financial and commercial services, but also accommodated a diverse spectrum of region-serving functions which have now largely vacated the metropolitan core, including department stores and other general retail services, automotive dealerships, printing and publishing, and other quasi-industrial uses in the CBD fringe. Service industries and employment led labour force growth in Vancouver during the first quarter century of the postwar period, 1945–1970 (Barnes *et al.* 1992), a precursor of the central city's hyper-specialization experience of the 1970s and 1980s.

Vancouver at mid-century, then, was already a 'service city', but was by no means 'postindustrial'. A substantial inner city manufacturing and industrial belt (zone II) followed in some respects the formation of industrial districts in many other North American cities, which included heavy industry (foundries, concrete plants, chemicals), and light manufacturing (breweries and other food and beverage production), as well as ancillary warehousing and distribution functions. But what defined the distinctive nature of Vancouver's inner city industrial districts was, first, the extensive complex of resource processing and manufacturing operations around False Creek, including sawmills and lumber yards, pulp and paper manufacture, barrel-making, and other secondary wood products, demonstrating the tight bonding between urban core and resource periphery in the classic staple economic structure; as well as, second, the rail yards and major port and shipping installations and services of the inner city and central waterfront which underscored Vancouver's role as terminus of the national rail system, and as Canada's principal Pacific port (Figure 8.3).

As a final structural element of Vancouver's mid-century urban core we can identify a number of older residential neighbourhoods (Figure 8.2, zone III), including single-family houses (largely of wood construction), rooming houses, small apartments, and single-room occupancy hotels. For many of the residents of the core, a central city location yielded benefits in terms of proximity to work, both for the industrial labour situated in False Creek and the Central Waterfront, and also for the service workforce of the downtown. For others the attraction of a central city residence was linked to the environmental features of the core, principally English Bay, Stanley Park and other green spaces, and views of the North Shore Mountains, as well as the central area's retail and consumption services.

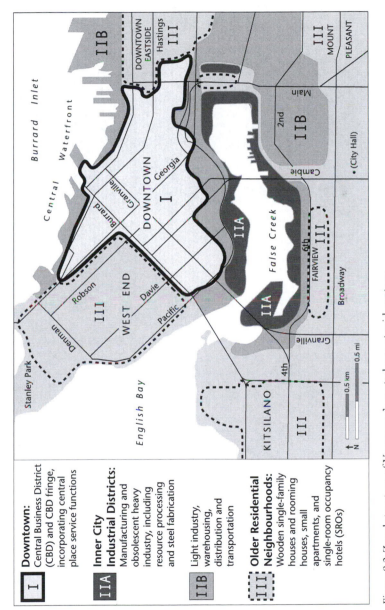

Figure 8.2 Zonal structure of Vancouver's central area at mid-century.

Source: Hutton (2004b).

Figure 8.3 Vestiges of the industrial city: Opsal Steel site, Second Avenue, Mount Pleasant Industrial District.

Restructuring in the 1970s: the rise of postindustrialism

The 1970s saw a fundamental restructuring of industry, urban landscapes, and social class in Vancouver, situated within a context of changing political attitudes and affinities. Growing public disaffection with the problematic features of the City's inner city industrial sector, both in tangible and symbolic terms, led in 1972 to the election of a progressive, reformist Mayor and Council committed to a redefining postindustrial agenda (Hardwick 1974). The centrepiece of this commitment was the conversion of False Creek South, one of the most noxious and obsolete of the inner city resource processing sites, to a mixed-income, medium density, residential community. As David Ley presciently observed a quarter century ago, False Creek South represented 'the most dramatic metaphor of liberal ideology, of the land use implications of the transition from industrial to post-industrial society, from an ethic of growth and the production of goods to an ethic of amenity and the consumption of services' (Ley 1980: 252). Certainly, the rezoning of False Creek South has proven to be deeply consequential, both in terms of the immediate acceleration of industrial decline within the inner city, as well as the establishment of a longer-term policy commitment to privileging housing, consumption, and public amenity in the urban core.

A second major restructuring process for Vancouver's central area in the 1970s was the spectacular growth of the CBD's corporate office complex, impelled largely by the control functions (including head office and financial services) the

City exercised over a vast provincial resource economy, and by local-regional demand for specialized services. As a measure of the core's functional specialization during this period, the number of professional business service firms increased by 140 per cent between 1966 and 1982, while office floorspace in the downtown more than doubled over the period 1966–1982, representing a compound growth rate of 6 per cent per annum (City of Vancouver 1982a). Thirty-six new office buildings were constructed in the core over the same period. At the same time, most other service industries experienced relative decline (retail and wholesale trade) or even absolute contractions (personal, health, and educational services) within the central area over this period, underscoring the dimensions of the hyper-specialization of Vancouver's urban core.

Over the 1970s, then, the rapid growth of intermediate, office-based services, combined with the decline of non-office and final demand services, and the erosion of traditional manufacturing and processing industries, produced a 'hyper-specialized' central city economy and a fundamental reconfiguration of the core's space-economy, as shown in Figure 8.4. The relayering of capital in Vancouver's urban core underpinned both the rapid growth of the CBD's corporate office complex described above, and also the formation of a new high-rise apartment district in the West End, the latter responding both to important zoning changes (enacted in 1956) and the social demand for downtown living in an area of high environmental amenity (Figure 8.4). Further, the rise of a new professional and managerial class produced by the growth of the office district, and the accelerated decline of traditional industries and employment, produced the classic preconditions for gentrification, transforming the social morphology of older residential neighbourhoods within Vancouver's urban core (Ley 1994; Smith 2000)

Restructuring in the 1980s: global processes and hallmark events

The restructuring of Vancouver's core area economy during the 1970s was shaped largely by factors operating at the regional and provincial scales, but a new trajectory of development over the following decade was driven increasingly by global processes. First, the City's hosting of an international exposition, Expo '86, provided a major catalyst to the reconstruction of space in the core, and to the formation of a new global identity for Vancouver. The purchase of the False Creek North Lands for the exposition site (see Figure 8.5) by the Government of British Columbia precipitated the eviction of constituent industries. The City's industrial planner observed in a staff report that this conversion would mean that 'virtually all industrial establishments on the north shore of False Creek will have to relocate' (City of Vancouver 1982b: 55), including rail yards, warehouses, trucking operations, and a sawmill, the last vestige of the inner city's historical resource processing and manufacturing role within a staple economy system. As in the case of the False Creek South experience described previously, the state acted as a facilitator of postindustrialism, although in this case the provincial Government, rather than the City, represented the key agency in the hollowing-out of the inner city industrial sector.

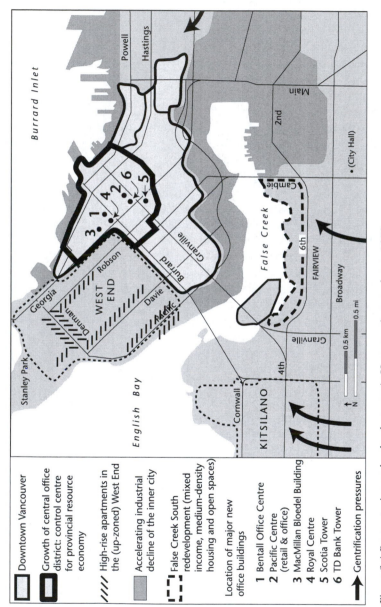

Figure 8.4 Restructuring and redevelopment in Vancouver's central area, 1970s.

Source: Hutton (2004b).

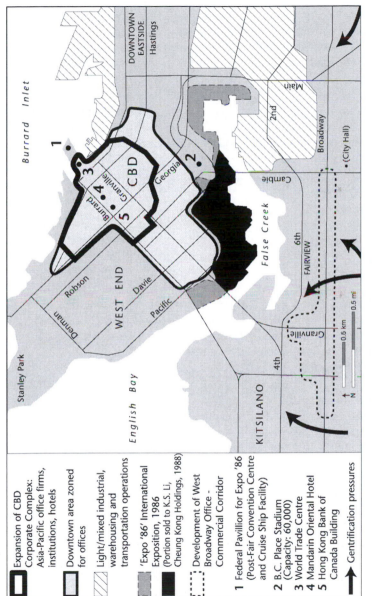

Key / Legend:

- Expansion of CBD Corporate Complex: Asia-Pacific office firms, institutions, hotels
- Downtown area zoned for offices
- Light/mixed industrial, warehousing and transportation operations
- 'Expo '86' International Exposition, 1986 (Portion sold to K.S. Li, Cheung Kong Holdings, 1988)
- Development of West Broadway Office - Commercial Corridor

1 Federal Pavillion for Expo '86 (Post-Fair Convention Centre and Cruise Ship Facility)
2 B.C. Place Stadium (Capacity: 60,000)
3 World Trade Centre
4 Mandarin Oriental Hotel
5 Hong Kong Bank of Canada Building

→ Gentrification pressures

Figure 8.5 Global processes in the reproduction of Vancouver's central area, 1980s.

Source: Hutton (2004b).

Second, a larger international projection for the City's core and the region as a whole was leveraged by new capital investment. The infrastructure development for the 1986 Vancouver Exposition included the construction of new international-class hotels, a federal Government pavilion to be converted to a major convention centre and cruise ship terminal, and a new fixed-rail rapid transit system. Coupled with an intensive international marketing effort supported by the provincial and federal governments and by the private sector (estimated by some to be in the $100 million range), which raised global awareness of Vancouver as an attractive visitor destination site, these major investments in the City's tourism infrastructure contributed to the recasting of the central area as a site of spectacle and consumption, supplanting in short order its traditional resource processing and industrial vocations.

Other important globalization processes implicated in the reconfiguration of Vancouver's central area were in part corollaries of the 1986 international exposition, although their provenance was also to be found in more distant domains. In 1988, the Government of British Columbia sold the bulk of the former Expo '86 lands (Figure 8.5) to Cheung Kong Holdings of Hong Kong, with Li Ka-Shing, the (then) Crown Colony's most influential magnate, as principal director and shareholder. This sale conveyed a signal to other foreign investors that Vancouver was now to be regarded as a prime (attractive, profitable, and 'safe') investment site, expeditiously projecting the City into global property markets. Li's manifest confidence in Vancouver, which included assigning his son Victor Li to manage the development of the former Expo lands, served to add momentum to the flow of Hong Kong immigrants to Vancouver, associated with concerns about the 1984 Sino-British agreement on the reversion of the Crown Colony to China in 1997.

These global processes and events unequivocally repositioned Vancouver's development within a new trajectory of *transnational urbanism*, reinforced in no small measure by an evolving market reorientation of the City's service industries and firms. By the end of the 1980s, the multiple linkages connecting the City to the provincial resource economy, which represented the initial stimulus for the accelerated development of the CBD's corporate office complex in the previous decade, were clearly weakening, as a consequence of global mergers, acquisitions, and takeovers which tended to favour first-order global cities (Hutton 1997). Many of the intermediate service firms ensconced within Vancouver's CBD sought reflexively to explore new, more robust markets, unencumbered by the limitations of a resource sector afflicted by boom and bust fluctuations typical of staple economies, exemplified by a deep recession in the early 1980s that produced unemployment levels of 13–14 per cent in Vancouver. This recession, the most severe downturn since the 1930s, generated permanent losses of capacity as well as more transient effects. By the late 1980s, markets for Vancouver's producer services included the growth economies of the Asia-Pacific, bolstered by marketing supports from government and public agencies (including an economic strategy for the City of Vancouver which for the first time emphasized the potential of service exports and gateway functions), complemented by the increasingly multicultural workforce in the central area's economy and residential neighbourhoods.

Restructuring in the 1990s: 'Living First' and the reassertion
of production

Following the industrial restructuring and globalization processes of the 1970s
and 1980s, the last fifteen years have seen the comprehensive redevelopment of
Vancouver's central area, constituting yet another cycle of signifying change.
Against a backdrop of sustained high growth in the metropolitan economy as a
whole, the fortunes of the central area have taken a marked 'residential turn',
endorsed by the strategic land use reallocations of the 1991 *Central Area Plan*,
and typified by the spectacular development of high-rise residential communities on the CBD fringe and inner city. The *Central Area Plan* included
provisions for numerous policy fields, but the decisive elements were in the
form of major changes to urban structure and land use: first, the consolidation
of the CBD within a smaller territory (Figure 8.6); and, second, the privileging
of new residential communities (Figure 8.7) in the areas of the central city
beyond this more compact CBD. Exemplars include the Concord Pacific
Project and Granville Slopes development on the north shore of False Creek
(Figure 8.8), as well as the residential redevelopment of the old wholesaling
and warehouse district of the Downtown South, and a strand of luxury condominium point towers in Coal Harbour, situated within the central waterfront.
While these represent the cornerstone projects of the City's 'Living First' program for the urban core, Vancouver's leitmotif for a new style of urbanism
since the early 1990s, we can also include loft conversions and live-work
studios, notably in heritage districts and also in certain industrial zones; market
and social housing in the DTES; and mixed-use towers in the CBD whose
vertical extensions far exceed those of the 1970s- and 1980s-generation office
buildings.

The City of Vancouver's 'Living First' model of contemporary planning for the
central city must be viewed as a considerable success, and indeed there has been a
quite remarkable national and international acclaim for this record of high-
amenity and 'livable' development. As an empirical demonstration of the 'Living
First' planning agenda, residential construction in the core exceeds that for commercial office development and other non-residential classes by several orders of
magnitude, effectively reversing the housing: office development ratio of the early
1980s (Figure 8.9).

The 2001 Census disclosed that the residential population of downtown
Vancouver had reached 70,091, compared to 18,983 for Seattle, 12,902 for Portland, and 43,531 for San Francisco (City of Vancouver 2005). Certainly, the
increasing momentum of residential development since the rezoning of False
Creek South and the construction of the West End apartment district in the 1970s,
demonstrated by the scale of new housing built in the core since the approval of
the *Central Area Plan* in 1991, offers compelling evidence of a distinctive and
progressive trajectory of urban growth and change. The relentless encroachment
of this affluent, high-amenity, residential development upon the marginal communities of the inner city, however, has exacerbated the imageries (and realities)

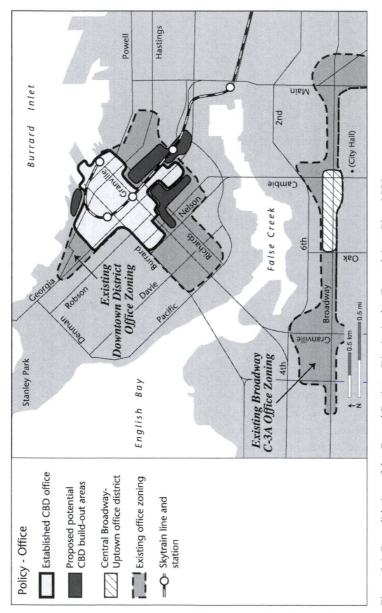

Figure 8.6 Consolidation of the Central Business District in the *Central Area Plan*, 1991.

Source: City of Vancouver (1991).

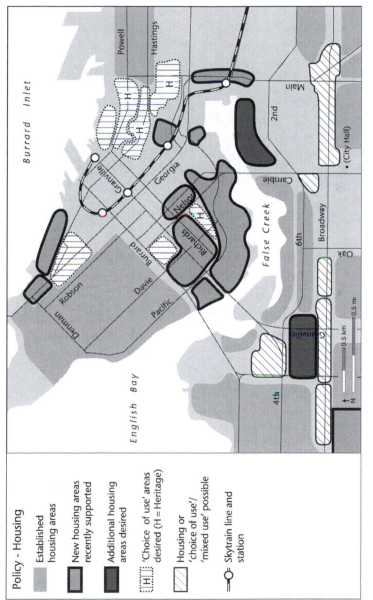

Figure 8.7 Designation of new residential districts in the *Central Area Plan*.

Source: City of Vancouver (1991).

Figure 8.8 Expressions of the 'Living First' planning model: residential point towers, False Creek North.

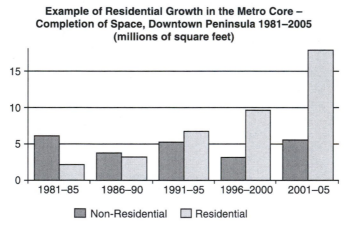

**Example of Residential Growth in the Metro Core –
Completion of Space, Downtown Peninsula 1981–2005
(millions of square feet)**

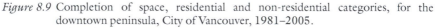

■ Non-Residential   ☐ Residential

Figure 8.9 Completion of space, residential and non-residential categories, for the downtown peninsula, City of Vancouver, 1981–2005.

Source: City of Vancouver Planning Department.

of deprivation in the Downtown Eastside, presenting a blunt corrective to the more euphoric accounts of the Vancouver story-line (Blomley 2004).

Residential development thus leads the reconstruction of Vancouver's central area, but we can readily identify a new phase of industrial urbanism as an important element of the core's contemporary development. Vancouver's metropolitan core continues to function as a critical component of the City's and region's space-economy, including new production and consumption spaces in the recon-structed inner city (Figure 8.10). The office complex of the CBD remains the largest and densest employment centre, but the space-economy of the metro-politan core now comprises a more diverse and complex suite of production sites and ensembles, including cultural industry precincts in the downtown (Yaletown, Victory Square, and Gastown), creative industries and biotech firms in False Creek Flats, and a major cluster of artists studios and galleries in the historic district of Strathcona. Areas such as Mount Pleasant, Granville Island, and Burrard Slopes include creative firms as well as a mix of business and industry technical support firms. While these areas have experienced a measure of fluctuation over the past fifteen years, like others in the case study cities and sites presented in this volume, they also exhibit substantial resiliency and capacity to accommodate new episodes of innovation and restructuring.

The approximately 200,000 jobs situated in the core represent about two-thirds of the City's employment base, and just under one-quarter of the metropolitan total. Within the metropolitan core, employment exceeds the residential popula-tion by a ratio of approximately 5:2, notwithstanding the rapid growth of housing, and loss of corporate head office functions in the CBD since 1991. There has been robust employment formation across a more diversified spectrum of industries, suggesting a departure from the monocultural office economy of the postindus-trial era. Within the key Professional, Scientific, and Technical industry aggrega-tion (33,105 jobs in the core in 2001), for example, legal services (8,195 jobs) still leads, but the computer systems and design category (with 6,800 jobs) now ranks second among constituent industries within this major sector. The important Accommodation and Food Services sector in the core, a key measure of the City's tourism role and consumption activity, numbered 20,490 workers in 2001. Education employment in the core has tended to exhibit lower shares of the regional job base, reflecting in part the suburban location of the major campuses of the University and British Columbia and Simon Fraser University. But there has been robust growth in the central city's education sector, with almost 7,000 jobs in 2001, a significant expansion over this last Census period.

Contours of the core's twenty-first-century economy

The new script for the core's economic development includes features of the more widely observed sequences of restructuring in the cities included in previous chapters of this volume, including the technology-driven 'New Economy' and the transient dot.com episode of the mid- to late-1990s noted in Singapore and San Francisco especially (Chapters 6 and 7, respectively). The technology crash of 2000 and after seriously eroded the base of New Economy firms. But, by 2003, the number of multimedia firms in Vancouver approached 600, compared with about 200 in 1998 (*Vancouver Sun*, 10 July 2003). There is also a more durable imprint of the creative industries which largely comprise (together with related institu-tions and NGOs) the cultural economy of the city theorized by Allen Scott,

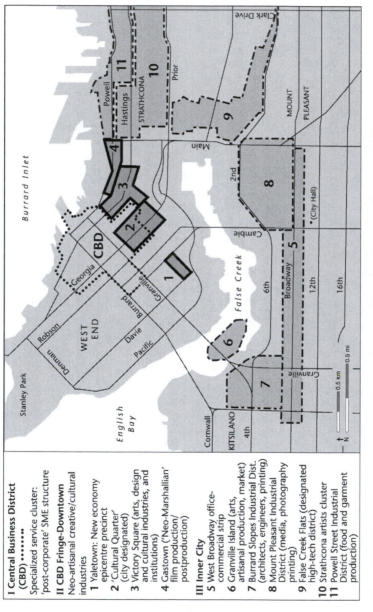

I Central Business District (CBD) ········

Specialized service cluster: 'post-corporate' SME structure

II CBD Fringe-Downtown

Neo-artisanal creative/cultural industries

 1 Yaletown: New economy epicentre precinct
 2 'Cultural Quarter' (city designated)
 3 Victory Square (arts, design and cultural industries, and institutions)
 4 Gastown ('Neo-Marshallian' film production/ postproduction)

III Inner City

 5 West Broadway office-commercial strip
 6 Granville Island (arts, artisanal production, market)
 7 Burrard Slopes Industrial Dist. (architects, engineers, printing)
 8 Mount Pleasant Industrial District (media, photography printing)
 9 False Creek Flats (designated high-tech district)
 10 Strathcona artists cluster
 11 Powell Street Industrial District (food and garment production)

Figure 8.10 The space-economy of specialized production in Vancouver's metropolitan core, 2007.

Bianchini, Landry and others. Creative services were strongly in evidence within the inner city by the late 1980s (Shaughnessy 1988), and the same period saw the City of Vancouver's inaugural 'Design Week', a festival (including juried competitions as well as public exhibitions) established to celebrate the City's distinctive applied design cultures. The 2001 Census demonstrated significant concentrations of creative and cultural workers (broadly defined) in the metropolitan core, including architecture and engineering (just under 5,000 jobs); arts, entertainment, and recreation (just under 4,000); motion picture and sound recording (2,235); advertising (1,600); and newspaper, periodical, and database publishers, and graphic and interior design (1,470 workers each), validating to some extent at least the notion of a 'cultural turn' in the economy of the twenty-first-century metropolitan core.

We can view the spatial dimensions of economic change in Vancouver's core by means of a series of maps which (in sequence) depict distributions of office space, firms within representative industries, and patterns of specialized labour. First, Figure 8.11 shows the geography of office space in the downtown. This map shows the expected concentration of office space in the CBD, but also demonstrates a measure of diffusion and new concentrations beyond the CBD, notably in Yaletown.

Next, the distribution of certain industries comprises another layer of change in the redevelopment of the city. By the early years of the twenty-first century, key inner city districts, such as Yaletown, Victory Square, and Gastown were established as sites of New Economy industries such as computer graphics and new media (Figure 8.12). Figure 8.12 shows that new economy industries have recolonized the northern crescent of the CBD, occupying spaces vacated by 1970s and 1980s generation resource economy firms. Figure 8.12 also shows major concentrations of advertising and architectural firms – formerly concentrated within the CBD – within the CBD fringe and inner city, demonstrating the power of these areas to attract clusters of creative industries.

The maturation (or stagnation, as some would have it) of the downtown office economy in Vancouver can also be seen as part of a larger process of urban industrial change. Speculative office development has slowed very appreciably within Canadian cities as a whole since the mid-1990s (with only Calgary as an outlier in this regard), not only in Vancouver; to illustrate, there has been only one new major office tower constructed over the last decade in downtown Toronto, the sixth largest office market in North America. (Several more towers were approved by 2007, but condominium construction is far outstripping commercial development in Toronto's downtown, as it is in Vancouver.)

There has also been a shift from the more generic middle-management structure of employment that characterized the occupational profile of the classic postindustrial core economy, toward professional and scientific employment (Figure 8.13), not just in the high-rise office environment of the CBD, but also within the new inner city production territories. This evocation of the new spatial division of specialized labour in the core offers perhaps a further vindication of

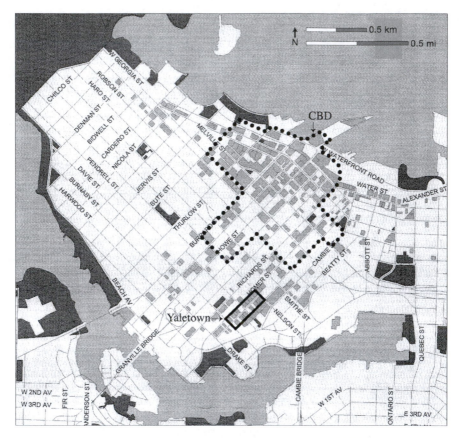

Figure 8.11 Distribution of office space in Vancouver's downtown, 2006.

Source: City of Vancouver Planning Department (2006).

Daniel Bell's forecast of the primacy of scientific knowledge and specialized information as an axial development principle of advanced societies (Bell 1973).

The technology-intensive sectors are at the vanguard of new industry formation in Vancouver's inner city. But changing patterns of location for established industries, such as architecture and advertising, offer another dimension of the changing space-economy of the metropolitan core. Architectural practices in Vancouver experienced rapid growth from mid-century to the early 1990s, with a growth in company listings from 52 in 1961 to 238 in 1991. Over this period the CBD sustained a dominant position, accommodating 53 per cent of architectural firms in 1961, and 46 per cent as recently as 1991. By 2002, the list of architectural practices had expanded to 258 firms, but by that date the proportion of firms situated in the CBD had diminished to approximately 15 per cent. New clusters of architectural practices had established within the inner city, including Yaletown, Gastown, West Pender, and Hastings Street on the CBD fringe, Victory Square (Cambie Street), Granville Island (Creekside Drive), and the Burrard

Figure 8.12 Distribution of firms for selected industries, Vancouver's central area.

Source: Author's survey (2004/05).

Slopes. This shift of architectural practices can be seen as illustrative of a larger new spatial division of production labour in the metropolitan core, which increasingly favours the inner city over the CBD proper for creative industries and firms (including advertising firms as well as architects; Figure 8.12), and for institutions and enterprises of the knowledge-based economy.

Finally, the cultural inflection of the core's economy is vividly shown in the emergence of a significant artists cluster in Strathcona (Figure 8.14). The cluster is comprised of studios and galleries, and is situated within the historically

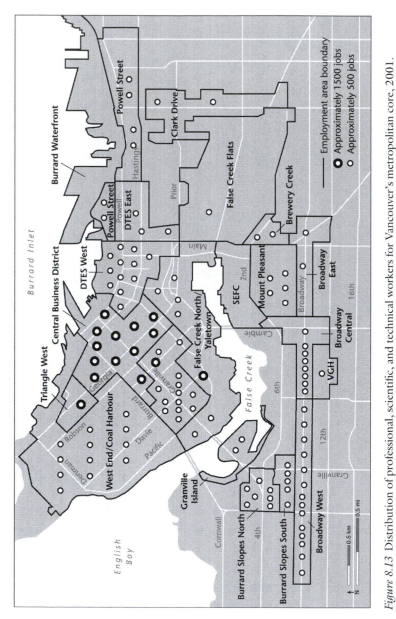

Figure 8.13 Distribution of professional, scientific, and technical workers for Vancouver's metropolitan core, 2001.

Source: City of Vancouver Planning Department (2006), data derived from 2001 Census of Canada.

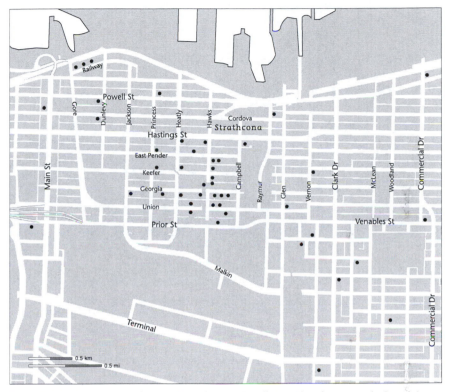

Figure 8.14 Distribution of artists' studios and galleries, Strathcona and Grandview–
Woodland.

Source: Sacco (2007).

textured residential landscapes of Strathcona (Figure 8.15), long established as a
low-income community, part of the high crime incidence and social disorder of
the Downtown Eastside, but now experiencing a substantial gentrification process with attendant inflation in housing prices.

The New Economy in the new inner city: vignettes of restructuring and dislocation

The larger changes observed at the level of the metropolitan core constitute defining elements of the new story-line of economic transformation in the Vancouver
case. At this scale, too, we can readily acknowledge that residential development
and social change have accompanied (and not *replaced*) industrial restructuring
and labour market change as motive forces in the reproduction of the Vancouver
core. The increased 'social density' (and diversity) of Vancouver's core can be
viewed as a factor in the generation of new employment, while the 'buzz' of new
industry formation represents part of the attraction for many new residents of

Figure 8.15 Heritage landscapes of the Strathcona community and artists cluster.

inner city communities, suggesting a pattern of reciprocal socioeconomic development. The growth of consumption amenity observed within inner city districts, as well as the examples of spectacle and allied institutional development, and the proliferation of new industrial spaces, marks something of a recovery of functional diversity in the inner city, however problematic, contrasting with the industrial over-specialization and monochromatic landscapes of the earlier industrial era.

But perhaps the most compelling feature of this most recent round of restructuring concerns the complex and often volatile relationships between industries, firms, and space within the metropolitan core. The last decade and a half or so has seen a significant reassertion of production within the inner city, in some respects resembling a precarious reindustrialization experience, seen most spectacularly in the dot.com boom and bust, observed in our London and San Francisco case studies. But in other regards the larger Vancouver experience suggests a recovery of the inner city's role as incubator of experimentation, innovation, and entrepreneurship. Further, there is, as explicated at a different scale in the London case (Chapter 5), a very marked variation in the particular reindustrialization processes and experiences of individual districts and sites, which militate against a tendency to over-generalize the inner city new industry experience. 'Place matters', not only for the causalities of reindustrialization at the local scale, but also for the nature of spatial engagement and experienced impacts between these clusters and proximate communities.

A more nuanced appreciation of processes of industrial innovation and restructuring in Vancouver's inner city requires a deeper probing into the specificities of change in terms of the core's space-economy, spatial divisions of labour, and land use. What follows, then, are two illustrative vignettes of new industry formation drawn from the instructive Vancouver setting, embedded within the distinctive social environments, histories and landscapes of the City's CBD fringe and inner city: Victory Square (CBD fringe – DTES), and False Creek Flats, situated within the 'new frontier' of Vancouver's metropolitan core. The chapter concludes with a more detailed case study of the evocative Yaletown site.

Victory Square: new industry formation on the CBD fringe

At the advent of the comprehensive reconstruction of Vancouver's core in the late 1980s, the City Planning Department recognized no fewer than thirty-three distinctive planning sub-areas of varying size in the central area, including the Central Business District (CBD), reflecting the complexity of spatial organization in the heart of the city. Victory Square constitutes one of the smaller districts in the core, and in some ways can be constructed as 'interstitial urban space', a fragmented territory inserted between larger zones of the central city, positioned on the CBD fringe, adjacent to the 'true' inner city (Figure 8.16).

But Victory Square's development experience embodies features of both historical and contemporary relevance. At the turn of the nineteenth century, Victory Square constituted the City's principal commercial zone, and throughout the first half of the last century developed as a major office and retail district, with a particularly vibrant street life of workers, shoppers, and visitors. Prominent within the area's landscapes were iconic office towers, including the World Building on Beatty Street, for a time the tallest structure within the extended dominions of the British Empire, and the Art Deco Dominion Bank Building on Hastings Street. The construction of the monument to Canada's war dead at the corner of Hastings and Beatty Streets, Victory Square, gave its name to the larger district

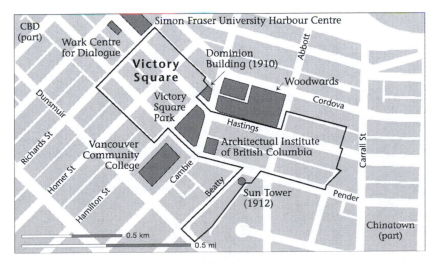

Figure 8.16 Victory Square heritage area and creative industry site.

in which it is situated. Victory Square also contained hotels which provided accommodation for the City's many resource sector workers, exemplifying another dimension of the tight bonding between Vancouver and the provincial staple sector.

Victory Square's once-prime location as the nexus of retail and office activities suffered, however, from the reshaping of space flowing from the shift in the City's commercial centre of gravity over the last decades of the twentieth century. The rapid expansion of the CBD office complex in the 1970s, closer to the centre of the downtown, marginalized the smaller and more isolated office district of Victory Square, while the formation of a new shopping core arranged along the Georgia Street and Robson Street axes seriously compromised Victory Square's critical retail sector. The area's decline was punctuated by the closure in 1993 of the Woodward's Department Store. Over the last decade many of Hastings Street's storefronts were boarded over, presenting an image of decline and disinvestment more usually associated with the most derelict of American rust-belt cities, than with the progressive and prosperous Vancouver urban development model (Figure 8.17).

Victory Square's increasingly precarious economic and social conditions were reflected in the problematic planning process for the area. The Draft Victory Square Plan (1995) acknowledged the divergent needs and interests of three principal sets of constituents in the area: (1) the low-income residents of the single-room occupancy hotels (SROs); (2) the business- and property-owners clamouring for redevelopment opportunities; and (3) the City's active heritage community, which sought official heritage designation for the area's distinctive

Figure 8.17 Street front in the 'interstitial zone': West Hastings Street, Victory Square.

built environment. So a defining issue for Victory Square was the management through an area planning process of the area's revitalization, to be achieved through a careful calibration of reinvestment, and a process for mediation of conflicts among the area's stakeholders.

Over the past decade or so an incremental process of transition incorporating new industry formation, institutional development, and allied social change has inexorably recast Victory Square as a key site of Vancouver's evolution as a centre of creative activity and cultural production. This experience has lacked the immediate visibility and impact of the new modernist high-rise landscapes of False Creek and the Central Waterfront, and has failed to capture the public's imagination in the way that Yaletown's transformations (see below) have. But the growth of knowledge-intensive, creative industries and institutions in Victory Square constitutes an important element of the larger rearticulation of economic and social space in the central area, as well as new episodes of succession and dislocation.

In important ways Victory Square's interstitial siting and marginal location relative to the CBD, factors critical to the area's late twentieth-century decline, have actually worked in its favour in the most recent experience of transition. The state of the built environment, and associated lower rent structures relative to the more glitzy inner city districts to the west, have proved a positive inducement to some businesses and the arts. Real estate agencies and developers are also endeavouring to capitalize on Victory Square's location, promoting a shift in vision from Victory Square as a residual space in an otherwise high-growth central area, to an alluring 'Crosstown' site, favoured by a 'gateway' situation linking the city districts of Chinatown and Gastown with the downtown proper.

We can therefore characterize Victory Square's development profile as including a significant presence of artists and designers, as well as small businesses and consultancies in the realms of film, music, architecture, and environmental systems, enhanced by edgier (non-Starbucks) coffee houses and cafés. This grittier quality is also projected by the raffish look of the area's storefronts, and the offices of the British Columbia Marijuana Party, although the long-standing Enver Hoxha Bookstore (compare with the Marx Memorial Library in Clerkenwell, Chapter 5: p. 129), a magnet for those who prefer their political ideology pure, uncontaminated by careerist backsliders and remnants of European Communist parties, is gone, a victim of the political upheavals of the late twentieth century. Apart from the street-level presence of creative enterprises, some of the area's heritage office towers have undergone complementary processes of adaptive re-use. These include the former Toronto-Dominion building on Hastings across from the Woodward's site (Figure 8.16), a structure which now accommodates a diverse range of design firms, communications companies, and environmental consultancies, among others.

There is also in Victory Square a variant of creeping institutionalism, with the establishment of offices of Simon Fraser University, the University of British Columbia, the British Columbia Institute of Technology (BCIT), and the Architectural Institute of British Columbia, among others, taking advantage of relatively low rents in the area, pointing to the area's apparently growing role in the knowledge-based economy. These institutions provide support for the area's artists and designers, but also signal inexorable processes of upgrading and professionalization within Victory Square and its environs.

Victory Square's distinctively incremental development path as an inner city site of creative and cultural production, however, may well be on the threshold of a more exigent growth cycle. Following several abortive attempts to redevelop the Woodward's department store site over the last decade, a comprehensive project is now underway which will include a significant measure of heritage retention and restoration, as well as some 700 residential units, substantial retail capacity, and a major new site for Simon Fraser University's Centre for Contemporary Arts. The City and the project developers have taken considerable pains to build into the project design significant public benefits, including 200 units of social housing, as well as social services for tenants and residents of the local community. It seems likely, too, that the Woodward's redevelopment will also contribute materially to the revitalization of the rundown and even derelict storefronts along Hastings Street and in adjacent precincts, restoring at least a measure of the area's traditional, vibrant street life.

But a project of this scale, a major departure from the more incremental development of arts, culture, and creative industry of the last decade and a half, will inevitably accelerate the transformative momentum of the area. The real estate sales and marketing approach involved rebranding Woodward's as 'Intellectual Property' (Rennie Marketing Systems 2005), appealing to young professionals prepared to locate in an area of significant crime and public disorder. Astonishingly, all 500 market housing units sold out in a single day, a not uncommon experience in the overheated downtown Vancouver property market, but a portentous event given the site's marginal location. The scale and apparent success of the Woodward's redevelopment have also emboldened other developers in the area, so there are now multiple residential projects underway or proposed for the Victory Square area and its environs. These projects will place greater pressures of dislocation on the residents of the area's SROs, both in terms of spillover rent inflation tendencies, as well as the increasing demands of a more affluent resident population for the police and private security personnel to more stringently manage (and sanitize) the streets and public spaces of the area.

Further, the knock-on effects of property market appreciation are likely to undermine the tenure of Victory Square's active community of artists, designers, and cultural workers, with a very real prospect of driving these cohorts further and further east, rupturing complex and intimate inter-firm linkages and social networks embedded within the area, following (on a much smaller scale) the encroachments and displacements of artists in East London (Chapter 5). Our profile of Victory Square, then, vividly illustrates both the saliency of inner city districts as sites for new industry formation, as well as the somewhat precarious nature of industrialization in the face of accelerated reinvestment and redevelopment in the Vancouver case.

False Creek Flats: high-tech visions on the core's 'new frontier'

False Creek Flats, a large industrial territory located on the eastern margins of Vancouver's inner city, represents in many ways a sharply contrasting case study to Victory Square. Created during the Great War by means of an infill of False Creek east of Main Street, a major north–south thoroughfare, False Creek Flats has experienced a number of industrialization sequences, including manufacturing and processing in its initial development, but its primary function has been one of a major transportation and warehousing district. False Creek Flats has accommodated principal rail facilities, both for passengers and freight, as well as a postwar development as an extensive trucking, warehousing, and wholesaling district, benefiting from the area's proximity to major arterials, the Port of Vancouver, and the downtown.[15]

False Creek continues to perform these transportation and warehousing roles, but they have been in relative decline for some years, in part because of the shift of such uses to suburban areas, closer to the Vancouver region's new industries, and to major new port installations and highways; and also owing to the increasing traffic congestion of the metropolitan core. Over the past two decades, some of the major land owners and principals of large corporations situated in False Creek have approached the City concerning redevelopment possibilities, with a view to more fully realizing appreciating property values, either through a new development venture *in situ*, or sale to a developer or client.

Toward the end of the 1990s, new industrial development trajectories appeared to stimulate a more urgent rethinking of the future of False Creek and its role within Vancouver's larger economy. This new exigency was closely allied to the extraordinary boom of the technology-driven New Economy, seen widely as a dominant influence on the transformation of advanced economies at a global level, and to the possibilities of Vancouver emerging as a player in this burgeoning trajectory of industrialization. A report prepared for the Government of British Columbia in 1996 estimated that the province's high-technology sector, largely concentrated in Vancouver and the adjacent Lower Mainland, was experiencing annual growth of 20 per cent, and called for a more active public policy role in supporting the expansion of this new propulsive sector, in a context of generally slow provincial economic growth.

In 1998, near the apogee of the tech boom, Vancouver's City Council, responding to concerns that information technology firms and other New Economy enterprises were facing constraints of land supply and suitable floor space, authorized staff to proceed with the preparation of a new I-3 High-Tech District Schedule, distinct from the long-standing I-2 (light industry) and M-2 (heavy industry) designations governing land use in industrial areas of the City. False Creek Flats, with its locational advantages, and supply of major landowners eager to expedite redevelopment to more advanced (and profitable) activities, represented it seemed, the ideal space for the realization of the City's high-tech aspirations. Further, False Creek Flats was large enough to accommodate the biotech and information technology industries that might be candidates for a new high-growth, high-tech industrial park in the City. In short, False Creek was seen as potentially the best of both worlds: a campus-like environment for advanced industrial clusters, with proximity to the specialized information services, capital and skilled labour force of the urban core.

In February of 1999, staff submitted to Vancouver City Council a comprehensive Urban Structure Policy Report for a 'Proposed High-Technology Zone' (I-3), setting out the rationale for a new policy model only four years after a major industrial lands review, as well as a detailed zoning schedule, together with the comments and recommendations of the City Manager and General Manager of Community Services (see Figure 8.10 for location: area 9). The latter position entailed high-level supervision of Planning Department staff, and the overseeing of strategic policy proposals for City Council.

The City of Vancouver Planning Department, by most accounts, one of the most progressive and high-powered in North America, has produced a steady stream of successful planning reports which have shaped the City's environment and morphology, but the I-3 High-Technology report proved problematic in a number of ways. There was, for example, no consideration given to an acknowledgement of externality issues for adjacent low-income communities, including the historic Strathcona neighbourhood just to the north across Prior Street.

Preparation of the draft staff report also disclosed a serious division among the City's senior staff. In particular, the General Manager of Community Services identified major concerns about transportation issues and important regional planning conflicts associated with the report. Planning staff had made provisions for larger office space allowances than normally permitted in the City's industrial districts, to accommodate the special needs and flexibility requirements of advanced-technology firms seen as the principal candidates for False Creek Flats, while at the same time placing restrictions and conditions on this office space. But the General Manager of Community Services, while acknowledging that '[t]he objective of encouraging high technology industries to locate in Metro Vancouver is a worthy one' (City of Vancouver 1999: 3), made the following objection to the report:

> The definition of information technology creates a use which looks like, feels like, behaves like and for all intents and purposes is 'office', with attendant high employment densities and high transportation impacts particularly if parking is provided without limit and there is not ready access to transit . . . Potential tenants and the real estate industry are not going to appreciate the subtle distinction between 'information technology' and general office, and our inspectors are going to find this distinction impossible to enforce.
>
> (City of Vancouver 1999: 3)

The General Manager predicted that should the Council approve of the proposed policy plan, '[t]he I-3 districts will become large office precincts' (ibid.).

In his covering memo to the report, the City Manager agreed that 'many of the comments of the General Manager of Community Services should be considered' (ibid.), but his overall guidance facilitated approval of the new I-3 High Tech zoning for False Creek Flats. In any case, some members of Council were of the view that a number of major companies and landowners in the area had made commitments to redevelopment in expectation of a rezoning, and that to withdraw the proposed policy plan would be to demonstrate bad faith. Among some members of Council, too, there was at least a sense that the rezoning of False Creek Flats would enable Vancouver to emerge as a 'player' (rather than a bystander) in the rapidly expanding New Economy that seemed to be the motive force for urban growth and development, the leading edge of urban industrial innovation at the threshold of the new century.

As the planning process evolved during 1999, differences among the Mayor and Council and senior staff intensified, with the protagonists hardening their positions on the key issues of transportation and regional planning impacts, and on the capacity of the new zoning schedules to shape a genuine high-tech cluster in False Creek Flats, as opposed to a de facto new office district which would simply compete with existing commercial centres in Vancouver and the inner suburbs. In the end, the General Manager of Community Services, a long-standing public official with a deep planning policy background and exemplary record of professional integrity, was dismissed, demonstrating that city planning can occasionally take the form of a blood sport, in contrast to the public perception of the bureaucracy as a privileged elite not subject to the vagaries of the private sector. The re-visioning of False Creek Flats from prosaic transportation and warehousing district, to a potentially propulsive high-tech cluster, was duly endorsed by the Council.

The City's I-3 High-Technology Zone for False Creek seemed to capture the developmental Zeitgeist of forward-looking planning among advanced city-regions. But as has been well documented, the crash of 2000 and afterwards marked the end of the technology boom (or bubble) that had been building with almost unprecedented rapidity since the mid-1990s.

This is not to say that False Creek Flats has been in all respects a failure, a rare setback in the annals of Vancouver's planning history. The pace of development has been appreciably slower than anticipated, but there are now important enterprises and institutions situated within False Creek Flats, almost a decade from the initial City staff work on new zoning for the site (Figure 8.18). These include Quadralogic, a major bio-technology corporation, Radical Entertainment, a highly successful video games company for Los Angeles publishers (Figure 8.19), and the Great Northern Way university campus (Figure 8.20), a consortium of four tertiary education institutions created specifically to foster synergies between design and technology widely acknowledged as underpinning forces of the knowledge-based economy. There are plans to locate at least the principal bio-medical research component of St Paul's Hospital within False Creek Flats in the near future.

These isolated developments, however, really don't amount to a cluster of interactive enterprises of a truly regionally-propulsive nature, as envisioned in 1999, and instead represent isolated 'islands of development' within the inner city's advanced production archipelago centred on the CBD, CBD fringe, and 'near' inner city. It may be said that (the cost of at least one professional career

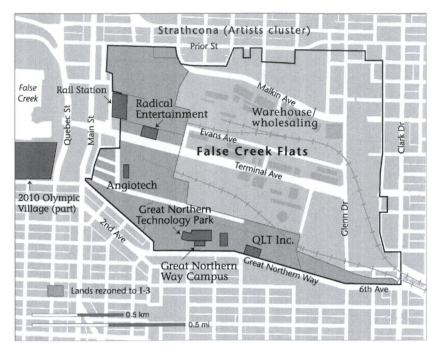

Figure 8.18 False Creek Flats 'New Economy' site, Vancouver metropolitan core.

Figure 8.19 Radical Entertainment, Terminal Avenue, False Creek Flats.

Figure 8.20 From metal-bashing to digital arts: Great Northern Way Campus.

aside) that the sharply truncated nature of False Creek Flat's vocation as twenty-first-century beacon of high-technology has been no bad thing, as the collapse of the tech-boom in 2000 and its aftermath may allow for a more organic pace of development that enhances possibilities of community engagement, social mitigation, and amenity provision.

Yaletown: signifiers of transition in an epicentre site

Although contemporary reindustrialization can generate a proliferation of new production spaces in the inner city, encompassing diverse ensembles of constituent industries, in many cities we can identify a signifying *epicentre* of innovation. These typically compact sites are situated within highly textured, often historically significant landscapes, with high 'imagery value', social resonance, and marketing cachet. Within (and proximate to) these epicentres of the New Economy we find complements of environmental amenity, reflexive consumption, supportive institutions (including NGOs and CBOs), innovative housing styles and tenures, socially interactive spaces, and clusters of leading-edge industries and firms. Further, these epicentre sites function as bellwethers of change, reflecting transitions in taste, style, and cultural expression, new phases of industrial experimentation, leading margins of change in urban social class formation, and, more viscerally, the relentless behaviour of real estate markets and the predations of developers. In the preceding city case studies, representative New Economy epicentres were identified for London (Hoxton), Singapore (Telok Ayer), and San Francisco (South Park).

Yaletown, a tightly-bounded heritage district juxtaposed between the new residential high-rises of Downtown South and the Concord Pacific mega-project on the North Shore of False Creek, is unequivocally Vancouver's New Economy epicentre, robustly meeting each of the criteria cited above. Yaletown's development history can be referenced to gain an overview of its functional evolution, and more particularly the acceleration of change over the last fifteen years or so. Broadly, Yaletown experienced several stages of development over its first century of development, from an inner city industrial area in the 1880s, to the headquarters of the Canadian Pacific Railway, then to a general warehousing, transportation, and distribution district in the first half of the twentieth century, and then a long period of gradual decline in the postwar era. By the 1970s, like other districts within the CBD fringe and inner city, Yaletown was very much in the shadow of the central city's office complex, a peripheral district in a metropolitan core increasingly skewed toward investment and development in the central city office complex. Since the 1980s, however, Yaletown has undergone a sequence of transitional experiences, underscoring its status as the City's most evocative site of signifying change, and encompassing a fluid mix of production, consumption, and housing activities.

As observed in other cases of inner city renewal included in this volume, the initial phases of new industry formation can be traced to the mid- to late-1980s, arising from a context of secular decline in traditional manufacturing and ancillary industry, and underpinned by interdependencies of arts, design, heritage aesthetics, and property market dynamics. Yaletown had shared in this decline, and by the 1980s was populated by low-value wholesaling and warehouse activities, residuals of the area's former vocation. But in a prescient article written in 1988, at the advent of its reconstruction as a new industry site, Robert Jankiewicz wrote that

> [Yaletown] once on the way to becoming a seedy collection of decaying brick warehouses and light industrial plants, is catching the eye of creative professionals looking for alternative workplaces, as well as the real estate developers needed to prepare the space for them.
>
> (Jankiewicz mimeo, supplied by author, 1988: 24)

Reflecting its marginal location within Vancouver's downtown, rents in the $7.50–14 per square foot were considered 'affordable' (ibid.), and highly attractive to start-up firms in the creative industries sector seeking to keep rents under control.

Beyond this urban land economics factor, in the same article Jankiewicz quotes Robert Lemon, a leading heritage preservation advocate, as making a case for preserving Yaletown as the 'only area of Vancouver that has a character and has buildings worth saving' (ibid.: 26) (Figure 8.21). Less fastidious aesthetes would likely affirm as well the heritage value of Gastown and Chinatown in Vancouver's original townsite within the DTES, perhaps, but it is the case that Yaletown's built environment presents a uniquely high-integrity and coherent heritage landscape, constituting a critical supply side factor in the area's contemporary development. In 1986, the City of Vancouver had designated Yaletown's built environment as an HA-3 Heritage area, with a separate HA-4 designation for the district's distinctive loading docks, affording another example of the importance of local policy (and more particularly heritage preservation initiatives) in the redevelopment of the inner city. Looking ahead from the vantage point of the late 1980s, Lemon envisaged a niche for Yaletown as the highly-textured 'pâté' in the City's inner city 'sandwich' (ibid.), ensconced between the incipient high-rise neighbourhoods of the Downtown South and Concord Pacific Place, although the contemporary consumption analogue might well be one which positions Yaletown as the froth on the new inner city's 'latté culture'.

By the late 1980s, Yaletown's new identity as a favoured site for the design community, including high-end retail uses as well as creative design professionals, was well established, providing a timely affirmation of the forecasts of Robert Jankiewizc, Robert Lemon, and others. Although Yaletown's heritage brick

Figure 8.21 High integrity landscapes of the New Economy: Yaletown.

warehouse buildings would be dwarfed over the following decade by the construction of 30-storey residential point towers both to the north and south, the district offered relatively high density for its CBD fringe location, allowing for retail, restaurants, and other consumption uses on the ground floor, and two or more storeys above this ground level for offices, studios, or housing. It may be worth noting that Yaletown's developmental conditions of building configuration, high integrity heritage value, and spatial boundedness closely resemble the profile of Telok Ayer described in Chapter 6.

From its takeoff stage of redevelopment in the late 1980s, Yaletown has undergone an accelerated sequence of transitions, in contrast to the more sedentary rhythms of its earlier industrial history. By the mid-1990s, the initial pioneering cohort of designers was accompanied (and in part supplanted by) new residential uses, including loft conversions and live-work studios, as well as more conventional apartments and condominium tenures. Just a little later, though, the imprints of the technology-driven New Economy were imposed upon Yaletown's landscapes in the form of the ubiquitous (if transient) dot.coms, temporarily at least displacing some of the earlier artists and design professionals responsible for the first phase of commercial gentrification. As we saw in the immediately preceding chapters, this incursion of the dot.coms and attendant dislocations in Vancouver's Yaletown were strikingly similar to the experience in Telok Ayer (Singapore) and South Park (San Francisco), although the severity of social dislocation was markedly higher in the latter case.

The turn of the century saw a precipitous crash of New Economy firms in Yaletown, again closely analogous to the experiences of Telok Ayer and South Park. But by this time Yaletown was firmly established as the central area's epicentre site for leading-edge firms. Vacancies rose for a time immediately post-dot.com, but Yaletown was promptly recast as a district of hybridized, technology-intensive, creative firms, upscale housing, and high-amenity consumption, with the highest leasing values in the City. Indeed, rents for prime Yaletown properties exceeded those for much of the CBD, demonstrating the insistent revalorization associated with social demand and the inexorable workings of the Vancouver property market, which, as we saw earlier, include a substantial cohort of overseas investors and transnationals as well as local players.

The social reconstruction of Yaletown

As a residential neighbourhood, Yaletown has benefited from its unique, highly-textured heritage built environment and interstitial situation between the high-end condominiums of Concord Pacific Place, located across Pacific Boulevard from Yaletown's southern margins, and the more mixed-income and diverse tenancy of the Downtown South area to the north. Yaletown offers authentic heritage imagery, but unlike the lofts and live-work studios of the DTES and core area industrial districts, lacks a palpable sense of 'grit' (and the higher local crime rates of the DTES). For a privileged minority, Yaletown offers the best of all urban worlds, a neighbourhood positioned as Vancouver's ultimate 'live, work and play' site, a paragon of the high-amenity, 24-hour city. About 60 per cent of the Yaletown workforce lives in the City, and about quarter of the workforce reside within the metropolitan core, reinforcing a sense of the 'local' in the district's evolution, and a further demonstration of linkage between New Economy sites and the housing markets of the central city.

The social reconstruction of Yaletown is also evidenced in the dense concentration of high-end consumption outlets in the district and in adjacent areas attempting to capitalize on the market cachet of Vancouver's New Economy epicentre. These include not only the usual 'new inner city' panoply of Italian restaurants, latté bars and Internet cafés, and luxury fitness gyms with personal trainers and spas, the standard accoutrements of such territories from Singapore to Shoreditch, but also a boutique Mini dealership, and the Opus Hotel, a boutique hotel tailored to the requirements of the experiential urban tourist. In the words of Daniel Craig, the hotel's General Manager, a sojourn at the Opus Hotel 'is like staying at your friend's – your really cool friend's' (*Vancouver Sun*, 10 May 2005). The penthouses, with a starting price point of $795 per night, come complete with a $12,000 Bang & Olufsen audio system, while dining and drinking at the Opus Bar, 'the hippest of hangouts' (ibid.), confers at least a temporary emblematic lifestyle status upon the affluent cosmopolite. The washrooms in this latter facility are equipped with video monitors of the Bar, allowing those temporarily indisposed to maintain a watching brief on the action and actors performing in the Bar itself. And, when one tires of the 'luxury' and 'cool' of the Opus Hotel and the Yaletown environs, the waterfront walkway of False Creek Park and its views of the yacht marinas and Granville Island Market, and the quotidian conviviality of the City's public realm, are barely a five-minute walk away, offering an environmental option not available to the habitués of Hoxton or Clerkenwell in London's City Fringe, or even those of Toronto's Liberty Village.

Yaletown's upward mobility as a privileged space of work, living, and recreation has been achieved at a cost some have come to lament. Evelyn Lau, a leading West Coast poet and astute observer of the urban social scene, expresses a measure of this ambivalence in her 'reluctant love letter to Yaletown' (*Vancouver Sun*, 17 April 2004). Lau recounts her purchase of a Yaletown condo in 1995, assuming a $100,000 mortgage that caused her some distress, but a sum that would now hardly qualify as a down-payment for the most modest of accommodation in the area. In observing the remarkable sequence of transitions of the area over the last decade, she observes that '[a]gainst all odds, the area that I remember as a cluster of run-down and brick warehouses has become a neighbourhood' (ibid.: F4). Lau recalls the strikingly low-amenity Yaletown of the mid-1990s, having to drive some distance even for everyday basics of living, and the quiet, almost urban backwater quality of the place. She admits to enjoying the amenity of Urban Fare, with its $100 *poilâne* loaves flown in daily from France, convenient access to the waterfront, and even the boisterous quality of the area's night-life. But Lau acknowledges that for her Yaletown was 'never a comfortable fit', and was, rather, a place 'too undefined' for personal comfort or sense of affinity.

Part of the discomfort Evelyn Lau and others experience amid the pleasure domes of the new inner city is no doubt attributable to the accelerated nature of transition in Yaletown since the late 1980s, and its associative sensory dislocations, as well as episodes of physical displacement. Although Yaletown occupies the peak of conspicuous consumption in the new inner city, its very success as a crucible of industrial innovation and leading-edge socioeconomic transition embeds its own destabilization tendencies. The beer parlours and nightclubs that constituted part of Yaletown's consumption and social infrastructure in the earlier phase of redevelopment, and both female and male sex-trade 'strolls' – activities consistent with the classic CBD fringe economy – have given way to upwardly mobile restaurants, bars, and coffee houses. Further, the urban forests of condominiums in the proximate Downtown South and Concord Pacific Place projects, while perhaps not exactly intimidating, produce an effect of 'closing in' the textured micro-scale spaces of Yaletown, and have certainly eradicated once and for all Yaletown's backwater status within the inner city.

The industrial reconstruction of Yaletown

Yaletown's contemporary identity as an urban place is now intimately linked to its residential role, high-end 'amenity clusters', and attractions for *flâneurs*, local and otherwise. So powerful are the socially reconstructed imageries and cultural resonances of Yaletown that its primary role as a site of intense industrial innovation and restructuring can easily be masked or discounted.

That said, some 70 per cent of Yaletown's floorspace is given over to employment-generating uses, and throughout the sequence of transitional experiences of the last twenty years a substantial quantum of this space has been allocated to *production*. As noted previously artists, designers, architects, heritage advocates, and other professionals were prominent among those in the vanguard of the district's redevelopment two decades ago. Members of these cohorts have maintained a presence even through the most vigorous of the restructuring episodes of the recent past. There is after all an essential robustness about these individuals that enables at least some to maintain a tenacious tenancy both in good times and bad.

So the production spaces of the new inner city are characterized by a measure of continuity, as well as rupture, but the dominant developmental motif of the epicentre site is one of a near-constant churning of industries, firms, and labour, driven by the pressure of the 'next big thing', the intense competition among the most successful of firms for occupancy of the finite space of the district, and the brutally intensifying effect of property market inflation. Survey work conducted by the author and the City of Vancouver Planning Department in the past few years, including interviews and mapping techniques, has disclosed key trends and broad contours of industrial change in Yaletown, as well as the district's status as locus of innovation.

Surveys of land use and businesses in Yaletown provide a useful empirical dimension of the district's salience as a uniquely dense site of New Economy industries and creative firms. At the level of the City's urban structure, there has been a centrifugal movement of firms from the formal confines of the CBD and its modernist office towers, to the new production spaces of the CBD fringe and inner city. Within the inner city, Figure 8.12 (see above) shows the pronounced clustering of creative and New Economy firms in Yaletown and prox- imate areas. Here we find significant representations of many of the key New Economy industries of the early twenty-first century. In the category of Internet Services ('other than access'), a group which comprises firms engaged in web- hosting, domain services, web-design, and e-commerce, Yaletown ranks as the leading cluster within Vancouver's inner city. For Computer Graphics and Imag- ing, another of the defining industries of the New Economy, a major cluster is situated in 'New Yaletown', on Homer and Hamilton streets, between Davie and Nelson. New media firms, engaged in the integration of sound, video, and text for clients, are also strongly represented in Yaletown, as are the related graphic design firms, and video game producers (Figure 8.22). There are smaller but important numbers of other flagship New Economy industries in Yaletown, including Inter- net access providers and computer software firms.

Interviews with specialized, professional workers provide verification of Yale- town's status as signifying inner city new industry site. Brail, for example, con- ducted a set of interviews with creative workers in a number of inner city sites which confirmed Yaletown's preeminence by the mid-1990s as a location for a number of design-based industries, including architecture and graphic design (Brail 1994). In a study of multimedia firms in Vancouver and Seattle in 2002, in the aftermath of the dot.com crash of 2000, Pope's panel of interviews in the Yaletown area strongly supported its status as a highly desirable place to work. Further, Pope's interviewees consistently emphasized the value of Yaletown as a 'creative milieu' for cultural industries and workers, both in terms of the highly legible heritage qualities of the built environment, and also the possibilities of social interaction and exchange of tacit knowledge among the micro-spaces and amenity sites of the area (Pope 2002).

A third set of interviews with design and New Economy workers in Vancouver's inner city in 2003 and 2004 generates further insights into the perception of Yaletown as a site of creative production within a comparative spatial context. A location in Yaletown was clearly seen as the most desirable address, relative either to the CBD, or to other CBD fringe – inner city sites, for many of the respond- ents. For these specialized workers and their firms, the high rents that characterize Yaletown properties reflect both the utility value and prestige level of this district. A web-designer, for example, endorsed the qualities of Yaletown's social and physical landscapes and ambience, in preference to the more impersonal scale and austere topographies of the CBD, as a justification for incurring inflationary leas- ing rates. A representative of a Yaletown graphic design company acknowledged that the firm had originally established in Gastown, in the distinctly grittier and more 'mixed space' environment of the DTES, but had always looked forward to a Yaletown address as a defining business goal, one of the signifying privileges of success in the market. A web-designer based in Gastown, on the other hand, cast

Figure 8.22 'Next Level' Games, Homer Street, Yaletown.

an envious eye toward Yaletown and its unique landscape values and amenities, with Gastown and the DTES constructed as a kind of 'consolation prize' in the contest for the most valuable places and properties in the reconstructed urban core. These observations suggest the workings of a spatial filtering process motivated both by preference and price, and having the effect of continually re-sorting space in the new inner city.

Epilogue: The spatial reconstruction of Yaletown

Yaletown has transitioned from 'outlier' postindustrial district in the late twentieth century to a 'new industrial enclave' in the new millennium, ensconced between the high-rise communities to the north and south. But Yaletown has in effect transcended its tightly localized site via a classic reterritorialization process, colonizing new spaces within the core, and capturing areas of the CBD fringe imbued with weaker identities and more generic morphologies. The area's locally unique combination of popular imagery ('cool'), social energy ('buzz'), and avant-garde industrial identity has projected Yaletown's 'label' well beyond the original heritage zone, underpinned by the relentless marketing of the real estate sector. This extra-territorial dimension of Yaletown's development over the past decade or so has been at least tacitly endorsed by the City, and incorporated within the public imagination. There hasn't been enough of Yaletown to go around, so the obvious solution has been to 'make more of it', with the hope of not diluting beyond recognition the integrity of the brand.

The 'Yaletown beyond Yaletown' experience, a defining process of the respatialization of Vancouver's metropolitan core, can be traced back at least as far as 1991, in the early years of the district's transformation as a site of industrial innovation and cultural production. The City's *Downtown South Zoning Bulletin* (August 1991) recognized among the Downtown South's sub-areas a 'New Yaletown' district, an area several times as large as the compact Yaletown heritage district situated to the east and south. The New Yaletown of the early 1990s was comprised of wholesaling, small businesses, and a promiscuous assortment of quasi-industrial activities. The presence of six heritage buildings in the New Yaletown sub-area, and its adjacency to the official Yaletown heritage area designated in 1986, lent a measure of validity to the extension of the Yaletown identity. The 1991 residential population of New Yaletown was a mere 175, but the area was assigned a central role in the recasting of Vancouver's metropolitan core, with a target population of 5,300, to be accommodated in 3,840 units over twenty-five years.

The quarter-century build-out horizon proved too modest a goal, and the New Yaletown residential community is largely complete, including the new Emery Barnes Park and other amenities. The Yaletown marque has been vigorously deployed in the marketing of residential and commercial properties in the sub-area, clearly to good effect, as the tactic of identity appropriation has been recycled to add lustre to condominium development, sales and marketing in sites progressively distant from the original Yaletown heritage precinct. The City's current Metropolitan Core Jobs and Economy Land Use Plan exercise, the first comprehensive policy review of the central area since the 1991 *Central Area Plan*, now recognizes for planning purposes a 'Yaletown' that extends as far north as Pender Street, and eastward to Main Street. Yaletown has thus metastasized from its original six square blocks to a territory comprising about one-quarter of the downtown peninsula (Figure 8.23).

Yaletown, then, represents a true exemplar of the epicentre site of the New Economy of the 'new inner city'. In keeping with the seismic connotation of 'epicentre', too, Yaletown's boom has been accompanied by a pronounced spread effect on proximate territories, a phenomenon observed in the Telok Ayer and South Park case studies described in previous chapters. Clearly implicit in this process of episodic transition and reconstruction is the complex mix of industrial, sociocultural, and policy forces articulated above, reproducing the City's most distinctive urban landscape multiple times over the past two decades.

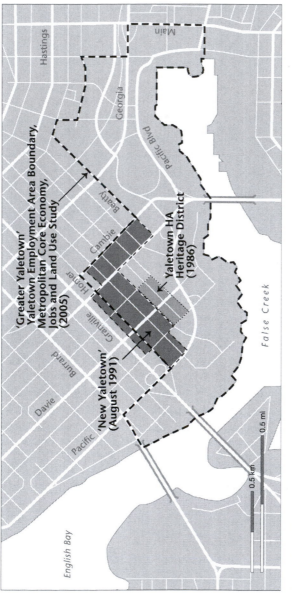

Figure 8.23 The respatialization of a New Economy site: Yaletown, 1986–2005.

Source: City of Vancouver Planning Department, 1986, 1991, 2005.

Paradoxically, one of the City's oldest and most historically resonant sites is simultaneously the district which most truly 'lives in the moment', Vancouver's most evocative site of industrial restructuring, social reformation, and multi-scalar urban place-making.

Conclusion: industrialization vs the social reconstruction of the core?

This chapter has articulated connections between shifts in Vancouver's economic vocation and larger sequences of transformation since the 1960s, with special emphasis on the episodes of industrial restructuring played out in the central area. Over the City's first century of development since its establishment in 1886, industrial development and restructuring acted as salient forces in sequences of transformative change, including the establishment of service industries associated with the City's central place functions in a regional context and the development of staple processing within False Creek and the Central Waterfront, followed by the ascendancy of the corporate office complex of the CBD over the 1970s and 1980s. Since the late 1980s, industrialization, in the form of New Economy clusters and creative industry quarters, has again shaped the configuration of space, land use, and built form in the extended metropolitan core, generating a 'recombinant economy' of integrating factors of culture, technology, and place. Arguably, though, residential development and the social reconstruction of the central area, underpinned by shifts in property market values and building economics which increasingly favour condominium construction and residential conversions, have emerged as decisive processes of the core's transformation.

The advent of Vancouver's post-staples development trajectory was signalled by the City's rezoning of False Creek South from obsolescent resource-processing to a medium-density, mixed-income residential community in the early 1970s, effectively accelerating the deindustrialization of the central area. But the most decisive policy decisions in the reshaping of the metropolitan core were embodied in the seminal *Central Area Plan*, which consolidated the core's zoned office capacity within a smaller CBD, and privileged high-density housing throughout much of the metropolitan core external to this truncated CBD. This new policy orientation incorporated programs for enhancing urban design, amenity, and the quality of the public realm within the core, and coincided both with growing local interest in downtown living and with overseas investment preferences, producing an astonishingly rapid build-out phase for new housing.

The *Central Area Plan* also avoided overly prescriptive guidance for heritage areas, industrial lands, and mixed-use zones on the periphery of the central area, thus providing space (land and buildings) for the rise of new industries following the trajectories of the technology-driven New Economy, the creative industries of the cultural economy, and the enterprises and institutions associated with the knowledge-based economy. Together these forces represent a substantial new production economy on the CBD fringe and inner city, including emblematic creative economy industries such as computer graphics and imaging, new media, software development, Internet services, and digital art and design as well as the technologically deepened, established professional design firms in architecture, graphic arts and design, and industrial design. These production ensembles represent exemplars of the most recent phase of post-Fordism, the relative importance of which has been appreciably enhanced by the erosion of Vancouver's head office functions and the emergence of a post-corporate CBD. In this scenario, small New Economy firms and educational enterprises (including ESL as well as tertiary education institutions) have been inserted in the CBD fringe, and indeed have made inroads within the CBD's office complex itself as space is vacated by corporate entities.

As we saw in the vignettes of new industry formation in Yaletown, Victory Square, and False Creek Flats, however, there is nothing either immutable or generic about these experiences within districts of Vancouver's inner city. There are instead signifying localized contingencies that shape both trajectories and outcomes of new industry formation in the twenty-first-century metropolis. The Yaletown saga just presented can be constructed largely as a 'good news' story of contemporary urbanism, ensconced as it is amid the new forests of modernist high-rise condominiums that represent the most spectacular products of the City's 'Living First' planning program, thus avoiding a significant community dislocation impact. There is also the high exemplary value of Yaletown as a marker of experimentation, innovation, and restructuring in the (post)modern metropolis. Even in Yaletown, though, there has been recurrent enterprise churning and employment change driven by what Andy Pratt described as 'industrial gentrification' in the Hoxton case discussed in Chapter 5, as well as a blurring of localized identities ensuing from the relentless encroachment of 'Greater Yaletown' within proximate districts of the CBD fringe and inner city, so there are some rough edges even to this glittering story.

In comparison with Yaletown, the Victory Square and False Creek Flats narratives embody far more visceral tendencies. These latter cases cannot be conveniently conflated, though, as they too exhibit significant contrasts, both in terms of environmental conditions, and generative processes of change. Victory Square presents as a classic interstitial CBD fringe district, suffering from increasing marginalization as a commercial district as the heart of Vancouver's CBD and retail activity shifted further west, but experiencing a renaissance of sorts as a contemporary site of creative industries, cultural production, and knowledge economy institutions. The ascendancy of these emergent sectors exerted greater pressure, both on the long-established SRO populations, and the residual base of activities in the area. But evidently superseding processes of industrial succession and transition is the recent acceleration of residential development in Victory Square, most dramatically expressed in the Woodward's redevelopment in the heart of the area, and increasingly augmented by smaller (but in the aggregate significant) housing projects. Victory Square is being rapidly transformed from an interstitial space squeezed between the downtown proper and marginal residential communities, to an integral site of new housing and amenity, as the high-momentum, high-externality 'new inner city' creeps ever closer to the City's

Downtown Eastside. The larger, public institutions domiciled within the area can likely thrive in these conditions, as can the most profitable high-end professional design firms, but the smaller and more marginal enterprises will almost inevitably be squeezed toward the lower-rent districts further east. Pressures exerted by new social groups to impose order on the streets and spaces will act to efface the grit and edginess that define the area's identity.

False Creek Flats represents the largest brownfield site in Vancouver's core, and can be interpreted as a 'new frontier' of the inner city with respect to development cycles and potential roles in the reindustrialization process, in contrast to the spatially confined nature of both Yaletown and Victory Square. The exigent revisioning and rezoning of the late 1990s which sought to reconstruct the area as a flagship New Economy site proved to be a false dawn, launched as it was on the threshold of the dot.com collapse in the following year. Certainly the abrupt derailing of the projected postmillennial destiny of the site represents another cautionary tale in the saga of volatile reindustrialization, while the more leisurely pace of economic development in False Creek Flats offers a stark contrast to the spectacular growth of the central area's new residential communities. Almost a decade after the initial high-tech rezoning initiative, it may be that False Creek Flats is belatedly emerging as a site of the knowledge economy, evidenced by the establishment of the Great Northern Way Campus, although the redevelopment experience overall is still piecemeal. The development potential of the site, though, situated on the eastern margins of the inner city, is one of the factors in the City's decision to plan for an expanded 'metropolitan core', supplanting the more restricted Central Area territory deployed since the 1980s, demonstrating the reterritorialization effects of contemporary urban change.

At a conceptual level the Vancouver case study may present both a challenge to foundational models of urban development, as well as an entrée to theoretical refurbishment and new conjecture. Vancouver has, after all, acted recurrently as an instructive site for urban theoretical positioning, notably as a theatre of globalization, transnationalism, industrial restructuring, and allied social change. These will be addressed more fully in the concluding chapter of this volume, but there is scope for an initial sketch of the theoretical possibilities associated with the Vancouver case study. At the broadest level, the emergence and co-presence of new industries and the residential communities which underpin the social reconstruction of Vancouver's core suggest the possibility of synthesizing narratives of industrial urbanism and the social ecology of the city, incorporating the rearticulation of space within the reconstructed metropolitan core. These processes have been replete with disruptions, and reflect to some extent developmental dualisms. But it does seem clear that a compelling retheorization of the twenty-first century will require a thorough probing of the interdependencies of coincident employment and population growth and change in the heart of the city-region, as well as the tensions between these forces played out in the spaces and property markets of the core.

Of equal exigency is the clear need to reflect upon the implications of industrial restructuring and allied labour force and social change for the late twentieth-century foundational theories of urban transformation. Postindustrialism was proposed initially by Daniel Bell as a forecast of redefining social change, a process shaped by the decline of the industrial classes, and the rise of a corollary post-industrial society, implying the ascendancy of scientific knowledge as the axial principle of development, and new cultural values, practices, and behaviours.

Subsequent empirical research by David Ley, Chris Hamnett, and others produced the contours of an urban 'new middle class' comprised principally of professionals, managers, and entrepreneurs, a cohort principally drawn from the segmented labour force of the CBD's office sector, but including public sector elites as well. This new middle class came to represent the advanced city's hegemonic group, implicated in the comprehensive social upgrading of inner city (and, selectively, suburban) neighbourhoods, and associated with the recasting of ideas of citizenship and identity. Ley and Hamnett argue that the new cultural economy workforce can be comfortably accommodated within the new middle class, rather than (as Florida proposes) representing a new class in itself. It appears, then, that the conceptual boundaries of the postindustrial city and new middle class have been stretched by the reassertion of production in the inner city, and more rigorous comparative work will be required to test the resiliency of these foundational theories of the twenty-first-century city.

9 The New Economy of the inner city

An essay in theoretical synthesis

This volume has presented a study of new industry formation within the spaces and landscapes of the metropolitan inner city. The broad intention has been to place sequences of innovation and restructuring observed since the 1990s within broader narratives of contemporary urbanism, with a view to investigating the theoretical implications of these recurrent processes of change. Primary data for the study were generated from a program of field research among exemplary sites in four instructive cities – London, Singapore, San Francisco, and Vancouver – substantially augmented by insights derived from the scholarship of many colleagues working across a broader range of cities.

The study disclosed that processes of new industry formation within the inner city present an exceptionally rich field for scholarly investigation, construed both as *terrain* of inquiry and as research *genre*. While these new industry experiences have been at times volatile, it does seem tenable that the cultural inflection of both production and consumption represents a significant trajectory of development among advanced (and indeed some 'transitional') societies. It is also clear that the metropolitan core, with its unique concentrations of cultural institutions, industries, labour, landscapes, and markets, constitutes the most salient zone of creative industries within the regional space-economy. The locational tendencies of the cultural economy favouring the central city run against the tide of decentralization of manufacturing, retail activity, and other services, underscoring the role of the metropolitan core as (once more) a critical staging area for innovation and restructuring.

In this concluding chapter I propose to set out in some detail the implications of the fieldwork and the engagement with the literature presented in the volume, recognizing both intrinsic and wider signifiers of change. This chapter offers, first, reflections on the four principal case studies, drawing out some of the empirical richness of each; second, insertion of new industries within a model of the 'recombinant economy' of the metropolitan core, acknowledging key developmental synergies and interdependencies in this latest evocation of post-Fordism; third, a venture in retheorization entailing the synthesis of 'industrial urbanism' with the social reconstruction of the metropolitan core; and, finally, some suggestions for new research directions which can fruitfully extend this inquiry.

Reflections on the case studies

The study approach entailed presentation of case studies which combined elements of economic geography (description and analysis of production systems, firms and labour inserted within the 'new industrial district' construct), situated within an urban geography and urban studies setting (emphasizing attributes of place, in historical and contemporary settings), as a means of drawing out the complex interdependencies of economic and social development in the 'new inner city'.

Observations derived from the four case study cities, encompassing multiple sites, provide a robust analytical platform both in theoretical and normative terms. While the debate concerning new urban theory in the early twenty-first-century city remains at an incipient stage, we are dealing with cities which present contrasts to defining features of the 'classical' postindustrial city, circa 1965–1990, with respect to urban structure and land use, industrial development, and divisions of labour. Further, there are corollary implications of these new tendencies for the reshaping the metropolitan space-economy, and for new experiences of urban place-making. The nature and consequences of these changes stretch the limits of received urban theory, conceived in earlier periods of restructuring, and justify exercises in conceptual reflection and proposals for new typologies of change.

Framing the analysis: the New Economy as developmental signifier

The four cities under consideration in this volume – London, Singapore, San Francisco, and Vancouver – of course, differ significantly in scale, their larger regional and national setting, magnitude of global projection, and governance structures and policy systems, among other attributes. London occupies the peak of the global urban hierarchy with New York; Singapore and San Francisco are second-order global cities, underpinned by different bundles of specialization; while Vancouver exemplifies the transnational city, defined principally by gateway functions and by diasporic linkages rather than by corporate or industrial power. That said, each has followed a postindustrial development path over the last decades of the twentieth century, characterized by the secular contraction of manufacturing capacity and labour, the rise of a specialized (but not autonomous) service economy, and the reformation of social class favouring the interests and preferences of a hegemonic 'new middle class' of professionals, managers, and entrepreneurs: an elaboration of Daniel Bell's original formulation and forecast of a dominant postindustrial society, with its distinctive cultural signifiers. The four metropolitan cities in the sample are also transformed by ethno-cultural diversity, with profound consequences for housing and labour markets, among other impacts.

Each of these metropolitan cities has experienced a significant resurgence of specialized industrial production within certain inner city districts, signalling a new chapter in the long-running saga of industrial urbanism, and its connections with broader processes of metropolitan transformation. Again, the specific narratives exhibit significant contrast in scale and impacts. London and San Francisco, in their own ways, represent 'big stories': London because of the strategic dimensions of new industry formation within the City Fringe, and interdependencies between creative industries, the arts, and property markets; and San Francisco because of the global-scale magnitude of its technology-led New Economy phase of the late 1990s. Both in London and San Francisco, the new industries formed in certain inner city districts over the 1990s drew upon the arts and cultural resources of those areas as inputs to production, while at the same time reflexively displacing many of the artists who had initially revalorized these territories. The new industry experiences documented in Singapore and Vancouver are calibrated at a smaller scale, but nonetheless embody larger consequences, including demonstrations of the global reach of the cultural turn, the power of local contingency, and implications for planning.

London: the New Economy in the global metropolis

The larger significance of London's City Fringe includes the reassertion of industrial production within many of the same territories of East London's traditional light manufacturing economy which had thrived (not without wrenching social costs) from the early nineteenth to the mid-twentieth centuries; but which had suffered a collapse of catastrophic proportions following the early 1960s. In Shoreditch, Clerkenwell, and Bermondsey, the emergence of 'new industrial districts', incorporating ensembles of industries and production networks characterized by localized backward linkages as well as the increasing use of digital production and telecommunications technologies, and a rich base of amenity, was documented. These new industries were inserted within the exceptionally diverse production landscapes of inner London, which included pre-Fordist 'relics' and Fordist 'residuals'. Some of the latter, such as publishing, have been effectively retooled through the introduction of new digital technologies and successful penetration of international markets, while others, notably printing, appear to be in secular decline, overtaken and displaced by new production methods which deprivilege traditional advantages of proximity in favour of leading-edge technological capacity.

The dominant industrial aggregations of inner London comprise distinctive sectors of the contemporary post-Fordist regime: first, an enormously powerful, global-scale financial-commercial sector which asserts itself almost wherever it chooses, not only in the mega-scale Canary Wharf financial district, but also in locations such as Paddington Basin, in Broadgate on the periphery of the City, and in Southwark; and, second, a cultural economy of creative industries, institutions, and labour. Each of these distinctive sectors has engendered new policy discourses as well as developmental consequences, with the uniquely lucrative financial and producer services established as the defining platform for London's global economy. The expansion of London's global financial and commercial services is now robustly endorsed not only by the usual advocates of big capital in the City, but also by New Labour in Parliament and by the socialist Mayor of London, underscoring Britain's post-Barlow policy values and political economy. In contrast, we can acknowledge a more humble role for the creative sectors, cast primarily as agencies of local regeneration in the postindustrial inner city, although central government agencies have also acknowledged the national importance of creative industries.

The local regeneration mission in London is evidently fraught with difficulty, in light of gaps between new industry occupations and local labour market conditions, and pressures of competition which tend to promote rapid turnover of firms. Accelerated firm turnover or 'churn' is seen as an operational feature of the 'managed workspaces' of Bermondsey and elsewhere within London's inner city. This business strategy likely yields benefits in terms of industrial experimentation and innovation. But the pace of churn can limit local regeneration potential, which may require a more sustained engagement between a stable base of enterprises and local labour and suppliers.

Beyond the industry-specific dynamics of competition and restructuring, perhaps an even more problematic feature of the 'new inner city' for firms is the relentless encroachment of high-end residential development, particularly within the City Fringe districts which have accommodated many of the new cultural industry firms and institutions. This residential development is driven by a long-running social upgrading trajectory in the metropolis, augmented by the periodic 'wall of money' at the disposal of the 'new gentrifiers'. To some extent at least the development sequence within inner London can be viewed as one of 'precarious reindustrialization', subject to recurrent cycles of innovation and restructuring, as well as the destabilizing effects of the London property market. That said, the scale and quality of creative industries, institutions, and labour concentrated within the inner city comprise important features of London's (and Britain's) cultural economy, diversifying its portfolio of global city functions, as well as enhancing the capital's economic primacy at the national level.

San Francisco: vicissitudes of the New Economy in SOMA

While the dimensions of London's inner city industry reflect the scale of the first-order global metropolis, the New Economy phenomenon concentrated within the South of Market and Mission may have played a relatively larger role in San Francisco's transformation over the last decade of the twentieth century. While the 1990s as a whole saw a resurgence of population and economic growth in London, this period was one of concern for the prospects of major industries such as finance, manufacturing, and tourism in San Francisco, including the relocation of back offices noted by Allen Scott (1988). To some, at least, the technology-driven 'New Economy' appeared as *deus ex machina* for an urban economy experiencing pressures of globalization and restructuring that favoured other cities, notably Los Angeles. With New York, San Francisco was a global epicentre of the New Economy, shaped by its proximity to Silicon Valley, by long-standing communities of artists and designers who were early pioneers in multimedia, by the Bay Area's leading universities and research parks, and by the distinctive spatiality and built environment of SOMA.

While the inner city New Economy episode in most cities featured mostly small firms, San Francisco's technology boom in the late 1990s included a higher share of medium-sized and larger corporations, as well as an army of freelancers who comprised a contingent workforce for the Bay Area's New Economy enterprises. This world-scale complex of technology-intensive firms situated in San Francisco's southern inner city, notably new media, software development programmers, and Internet services, proved unsustainable in the face of the oversupply conditions and drastic market correction manifested in 2000 however, and the SOMA dot.coms disappeared even more quickly than they had appeared. What distinguishes the San Francisco story-line, then, is the rapidity and scale of the tech-boom, the dimensions of social dislocation in SOMA and the Mission, and the correspondingly precipitous gradient of its collapse in 2000 and 2001.

That much is well known. But the nature of SOMA's slow recovery since the tech-crash of 2000 also offers insights into the contemporary restructuring of the urban economy both in its more particularized form as well as its general expression. Successive site visits, mapping, and interviews with key informants since the crash have generated a profile of incremental redevelopment. Continuing high vacancy rates within SOMA's property market, evidenced by the proliferation of 'for sale/lease' signs within the streetscapes of the area, provide proxy evidence for the scale of 'what was'. The barometer site of South Park, on a visit in the fall of 2005, showed signs of a partial recovery, as new design companies and other professional offices and studios have been inserted into the area, and South Park now exhibits a semblance of its pre-dot.com life, if not the frenetic congestion and density of social interactions that marked the New Economy phase at its height, circa 2000. Interviews and conversations with companies in South Park, including new creative firms and several long-established industrial design companies, point to a gradual (and selective) recovery of the area's trajectory as site of applied design and cultural production. Some observers note a renewed influx of technology firms in South Park which could herald a New Economy II experience, likely an inflated prognosis, but one which recalls the volatile nature of new industry formation in the spaces of the inner city.

The SOMA New Economy experience offers particularly acute lessons for planning systems operating in contexts of rapid restructuring. The San Francisco City Planning Department, a conscientious and talented civic agency, has the complex task of assessing the land-use implications of the late twentieth-century tech boom in the Mission and SOMA, as well as working with low-income communities which bore the largest costs of the New Economy phenomenon. The natural tendency of a sophisticated and progressive planning agency is to commit to the necessarily protracted period of consultation and survey work required for strategic policy development in this complex and conflicted domain of the city. The rapid onset of the New Economy phenomenon and the severity of its dislocations compromised this planning approach, inserting a sense of crisis and exigency into the long-running land-use and social planning process within

SOMA. Overall, the San Francisco New Economy experience demonstrates the extraordinary pressures that rapid restructuring places on city planning systems and on the marginal communities which play unwilling hosts to these new industries, highlighting the need for policies which enhance the resiliency of the city and its constituent systems.

The New Economy and its signifiers in Singapore's heritage areas

The sequences of new industry formation in the highly-textured heritage landscapes of Telok Ayer, a micro-scale unit of Singapore's space-economy dominated by the global-scapes of finance, international gateway functions, and knowledge-economy institutions, faithfully reflect the abbreviated restructuring episodes experienced among advanced cities since the mid-1990s. The early phase of Telok Ayer's development as new industry site was largely spontaneous, achieved without conscious policy direction, constituting an anomaly in the classic developmental state, in comparison with the corporate- and state-directed 'creative hub' of Far East Square. Telok Ayer has demonstrated a resilient quality in the face of restructuring pressures, as its individual property tenures, diverse site choices, abundant amenities, and authentic built environment are conducive to accommodating a rapid turnover of firms. In contrast, Far East Square, with its significantly higher rent structure and more corporate ambience, has experienced difficulty in securing a stable enterprise base to support the longer leaseholds required in the business plan. Far East Square is also ambivalently inserted at the crossroads of the CBD and Chinatown, not indelibly embedded in either – a problematic condition, given the fine-grained locational tendencies of creative industries within the micro-spaces of the inner city.

The robustness of Telok Ayer's development as a site of cultural production experienced over a sequence of restructuring episodes from 2000 to 2006 demonstrates a realization of the state's intention to generate an economic (as well as cultural) return from the designated heritage districts. Further, a site visit in 2006 disclosed not only a renewal of creative firms in Telok Ayer, but also a spread effect of such firms in other Chinatown districts, notably Bukit Pasoh. This most recent site visit also revealed a proliferation of lifestyle amenities in Telok Ayer consistent with the pronounced self-actualization tendencies of affluent societies, a stark departure from Lee Kuan Yew's original model of Singapore as a 'rugged' society. These observations suggest a modest contribution on the part of firms embedded within inner city micro-spaces toward Singapore's vision of a global-scale cultural economy, not least in affirming possibilities of 'creative spontaneity' amid the mega-structures and systems of the developmental state.

Vancouver: new industries and the social reconstruction of the core

The Vancouver experience of new industry formation, situated within the metropolitan core of the smallest city in our sample, demonstrates both the pervasive features of global processes of industrial innovation and restructuring, as well as more distinctive aspects. Local factors included an innovative central area plan-ning model which eschewed growth management tactics deployed in cities such as San Francisco and Seattle in favour of a bolder reconfiguration of urban structure, form, and land use.

As for each of our case studies, we can readily identify a sequence of new activities which developed in the aftermath of the restructuring of the 1970s and 1980s. In Vancouver restructuring took the form of a distinctive post-staples trajectory, incorporating a protracted decline in the inner city's traditional resource processing, manufacturing, and distribution roles, accelerated by a postindustrial planning agenda. This post-staples experience generated a legacy of warehouses, rail loading docks and other industrial infrastructure within the inner city, occu-pied initially by artists, and then steadily appropriated by professional designers by the late 1980s. The invasion of the dot.coms a decade later for a time interrupted the dominant trajectory favouring creative industries and labour. But the principal site of cultural production, Yaletown in the Downtown South, recovered quickly following the crash of 2000, reflecting its privileged location as new industry site, high-integrity built environment, and cachet.

At this broad narrative level the Vancouver experience appears to conform to the sequence of shifts in economic activity and land use in the other three cities and reference cases. But more distinctive features of the Vancouver case must be acknowledged. First, we can readily identify a local variant of Scott's concept of the internal industrial specialization of the metropolis, shaped by spatiality, location, and the filtering effects of amenity and the property market among Vancouver's inner city production sites. Yaletown functions as the epicentre of Vancouver's New Economy, with major concentrations of new media, computer graphics and imaging, and architecture and other creative industries, as well as sumptuous local amenities and upscale loft conversions; Victory Square and Gastown on the CBD fringe–Downtown Eastside comprise grittier terrains which accommodate large contingents of artists and designers, as well as film production and postproduction, forming what Coe has termed a 'neo-Marshallian' satellite industrial district; while False Creek Flats on the 'new frontier' of the inner city, the subject of the City's high-tech visioning at the apogee of the New Economy phase in 1999, now accommodates a number of larger creative industries and institutions which cannot be domiciled within the more intimate spaces of Yaletown and Victory Square.

A second distinctive feature of the Vancouver case is the stripping of much of the City's head office firms driven by rounds of globalization, producing a largely 'post-corporate' CBD and a stock of offices which have been recolonized in part by New Economy firms. Third, the influence of the seminal *Central Area Plan* (1991) on new industry formation includes a comprehensive program of residen-tial development, which has generated a substantially enlarged labour market in the core, and a rich platform of amenities, as well as a greatly increased 'social density' which enhances interaction and exchange. At the same time, the extraordinary condominium development over much of the inner city has squeezed employment uses in these critical domains, while producing an insistent

revalorization of land values which further destabilizes the core's base of industries and firms. In this regard, Vancouver's experience increasingly resembles
London, where the relentless social reconstruction of the inner city, especially in
the former East End, places exigent pressures on all but the most profitable firms.

Transnational urbanism and the fusion of design cultures

As a final reflection on the case studies, we can acknowledge the increasingly
diverse social capital in each city which reshapes the cultural bases of specialized
production. In his influential 1997 article on the 'cultural economy of the city' in
IJURR, Allen Scott acknowledged the durability of historical centres of design
and creativity, and their capacity to shape markets for cultural products. But
unlike orthogenetic cities such as Florence, Paris, Tokyo, and Beijing, each associated with a master narrative of foundational high culture, the metropolitan cities
in our sample are defined in large part by a transnational identity.

At a larger scale this transnationalism shapes a fluid synthesis of design influences, symbols, and work practices within the creative industries of the cultural
economy, concentrated mainly (but not exclusively) within the metropolitan
core. At the level of industries and firms, we readily see the influence of multiple
design influences, for example, in the industrial design firms interviewed in South
Park, and in the successful architectural practices in Vancouver. In the London
case, international immigration is seen as contributing to new energies and design
ideas for the capital's creative industries, continuing a tradition which included
Flemish and Huguenot craftsmen and a later immigration of Italians which
helped shape Clerkenwell's precision trades. In Singapore, multiculturalism has
produced a palette of design influences, with a century and a half of resonant
Chinese (notably Hokkien) culture providing creative inspiration for artists and
designers in the textured but orderly spaces of Telok Ayer. But Singapore's inner
city also includes the 'unruly spaces' of Little India, an area conducive to the
hurly-burly creative enterprise characteristic of what K. C. Ho acknowledges as
the high-risk, high-turnover base of the cultural production pyramid. The
experience of the cities incorporated in this volume points to the emergence of
multicultural cities as basing points of inter-cultural exchange, fusion, production,
and transmission: a defining aspect of economic development and labour formation in the early twenty-first century.

Dimensions of the recombinant economy

As a provisional venture into new theoretical vocabulary, we might describe
the complex ensembles of industries, firms, and institutions situated somewhat
tenuously within the evolving twenty-first-century metropolitan core as the
'recombinant economy of the inner city'. This concept is derived from genetics, in
which *recombination* involves multiple processes by which genetic material of
different origins becomes combined. In this interpretation, we can propose a
multilayered construction of the inner city's economy as *recombinant*, shaped by

complex synergies, syntheses, and interdependencies, in the following ways. First, we can readily discern, within the contemporary economy of the metropolitan core, industries, and labour associated with each of the restructuring episodes of the past two decades: i.e. mainstream intermediate services associated with the classic postindustrial era, artists and cultural production, technology-intensive industries of the short-lived New Economy, a large base of creative industries and enterprises, and, finally, representative enterprises of the 'knowledge-based economy'. The economy of the inner city, as demonstrated in the program of fieldwork, comprises a range of each of the principal industrial regimes: pre-Fordist, Fordist, and post-Fordist (Figure 9.1). The London case study exhibited the greatest richness and diversity of industry representation, reflecting its scale, early provenance in the industrialization process, and first-order global city specializations; but each of the four cities demonstrated this in some measure.

Second, there is in many medium and large metropolitan inner cities the co-presence of 'old' and 'new' economy industries, which exhibit a range of relationships, ranging from *collaboration* (seen in the case of South Park Fabricating's subcontracting relationships with Standard Sheet Metal, in South Park, SOMA); *competition*, observed in the 'industrial gentrification' episodes documented by Andy Pratt in Hoxton; and *co-existence*. Co-existence in this spatial context is interpreted as the co-presence of 'new' and 'old' industries, the former through leading-edge product development and delivery; and the latter by means of catering to distinctive markets, sustained competitiveness achieved though specialization and achieving higher value-added, and (in some cases) security of tenure through favourable leaseholds or outright fee simple property ownership. Prominent examples of co-existence include the three narratives of Manhattan's industrial districts presented in Chapter 3, and the diversity of industries situated in Clerkenwell described in the London case study.

Third, the emergence of hybridized occupations embodying high-level design skills with increasingly advanced technical capacity (and in many cases entrepreneurial instincts as well) offers an additional aspect of *recombination* within the inner city. In the larger metropolis, notably in the London case addressed in this volume, there is a sustained presence of neo-artisanal labour engaged in the production of high-value cultural products, as described by Glen Norcliffe (Norcliffe and Eberts 1999). The most salient aspect of labour specialization lies in the high-level design and artistic skills essential for such activities as fine arts, commercial graphics, and video game production. Firms in each of these industries also seek to achieve higher productivity through the deployment of new techniques and technologies. But the *recombination* of design and technology is most vividly exhibited in 'New Economy' industries such as computer graphics and imaging and web-design.

Fourth, the recombinant economy of the metropolitan core is defined in part by the synthesis of service functions with goods production, as noted by Allen Scott, demonstrated in the increasing prominence of cultural products in advanced industrial production systems and consumer markets, in contrast to the systems of the classic post-industrial period in which service production ensconced in the

---

1. 'Sublation' of restructuring episodes and production regimes
   - pre-Fordist production ('relics')
   - Fordist production ('residuals')
   - post-Fordist production
     I. office-based intermediate services
     II. cultural production and creative industries

---

2. Co-presence of 'old' and 'new' economy industries
   - collaboration (subcontracting)
   - competition (dislocation and succession)
   - co-existence

---

3. 'Hybridized' occupations in specialized production
   - synthesis of arts, design, and technology (computer graphics and imaging, other 'new media')

---

4. Synthesis of service-type and manufacturing processes
   - design and fabrication of 'cultural products'
   - (industrial design, architecture, fashion)

---

5. Space and industrialization
   - 'industry-shaping power of space' and 'space-shaping power of industrialization'

---

6. Traditional production networks and digital systems
   - localized input-sourcing in the 'new industrial district'
   - increasing use of Internet and digital technologies for sourcing

---

7. Production-consumption interface in the new industrial district
   - 'amenity-seeking' behaviour of creative firms
   - role of new industries in stimulating demand

---

8. Industry and housing linkages
   - mixed-use activity in live-work, work-live, loft conversions
   - dislocation associated with property market behaviour

Figure 9.1 The recombinant economy of the new inner city: synergy, synthesis, and interdependency in production.

corporate complex of the CBD operated at some spatial and functional remove from goods production.

Fifth, the synergies of specialized production in the new inner city are also exhibited in the relationships between industrialization and space, as interpreted in Soja's (2000) acknowledgment of the 'industry-shaping power of space', as a corollary of the 'space-shaping power of industrialization'; in the development of 'territorial-based systems of innovation' described by Kevin Morgan; and in the seductive appeal of aestheticized landscapes for cultural production, elucidated by David Ley. These new industry sites demonstrate the social, as well as economic, reproduction of space (Lefébvre 1974).

Sixth, the synthetic processes specific to new industry formation and development within the inner city are illustrated in the recombinant production systems deployed by firms and enterprises, consisting of, first, localized production systems situated within industrial districts, incorporating agglomeration economies, as well as positioning within more extended metropolitan and regional production chains; and, second, the rapidly expanding use of the Internet and digital transmission for sourcing, including the transmission of key inputs over space, as well as international staff recruitment for high-skill positions. As an example, Radical Entertainment in Vancouver, a large video game producer for Los Angeles publishing companies, sub-contracts some of its drawing work to Chinese firms. In the case of film production, Neil Coe (2001) characterizes Vancouver as a 'neo-Marshallian satellite' industrial district, reflecting the external exercise of control and decision-making, and the specialization of labour in both Hollywood and 'runaway' sites.

Seventh, the recombinant economy of the inner city is typified by the intimacy of relations between production and consumption, reflecting the marked amenity-seeking behaviour of firms in the creative and knowledge-intensive sectors, evoked in Graeme Evans' work on Clerkenwell; and the role of new industry firms and labour in increasing localized demand for services and products.

Finally, our illustration of the recombinant economy of the new inner city is demonstrated in the complex connections between new industries and local housing markets. There are important complements, as exemplified by the inner city live-works favoured by some creative workers, illustrating a spatial nexus of specialized production and lifestyles. But this relationship is also characterized by conflict, as the dislocative effects of upscale housing in revalorized inner city districts demonstrate.

The new inner city and its theoretical signifiers

An appreciation of new industry formation and its spatial, social, and cultural dimensions suggests the possibility of a more powerful and nuanced contribution to theoretical conjecture and synthesis, informed by the four case studies and the reference cases cited in this volume. These experiences can then be situated within a larger theoretical construct which also takes in leading features of the social reconstruction of the metropolitan core. Here I want to build upon preliminary exercises in theoretical conjecture, derived principally from the Vancouver experience (Hutton 2004a; 2004b), entailing conditional proposals for new typologies of urban change. In these initial efforts the constraints posed by the abbreviated nature of restructuring episodes since the late 1990s were acknowledged, as were the inclusion of cities with quite differentiated story-lines. These problems remain, but can perhaps be addressed in part by embedding a measure of robustness in scenario-building and typologies of change.

While care must be taken not to overstate the significance of the study findings, nor to exaggerate the role of inner city new industry formation in the overall transformation of the city, we can advance the idea that in important respects the nature of change in the urban core implies a marked departure from the defining attributes of postindustrialism, circa 1965–1990. These six signifying shifts can be summarized as follows.

1 Structures of the metropolitan core economy

The basic contours of the postindustrial city economy comprised the following: first, the collapse of Fordist manufacturing and ancillary industries within the inner city; and second, the rapid expansion of specialized services, notably high-value intermediate services concentrated in the CBD's corporate offices: the largest, densest, and most complex industrial agglomeration in the modern history of the city. The postwar takeoff period for the CBD office complex occurred in the late 1950s and 1960s, described in Jean Gottmann's (1961) seminal work on 'Megalopolis' situated on the Northeastern seaboard of America, accelerated by new divisions of production labour favouring service occupations in the 1970s, and the propulsive effects of globalization, deregulation, and privatization over the 1980s, described in Chapter 2.

The basic contours of this period of urban transformation persist, but we can now also acknowledge new and prospectively redefining trends. There is still growth in office development and employment, but it has a marked spatial dimension, favouring global cities, notably London and New York; US Sunbelt cities such as Atlanta, Miami, Houston, and San Diego; and the primate cities of the growth economies of East and South-East Asia, including Shanghai, Seoul, Taiwan, Singapore, and Kuala Lumpur, among others. But in other regional markets office development has appreciably diminished over the past decade or more, owing to a range of factors which include pressure of competition (producing downsizing and reconcentration of corporate control in higher-order centres), oversupply of building stock, land supply constraints for new building, and policy factors, including development quotas and pricing instruments included in growth management strategies.

The reassertion of production within the inner city represents a marker of change, and to these new industry formations we can add a profusion of other economic activities inserted with the core, including galleries, exhibition spaces, 'cultural quarters' (both officially designated and spontaneous), professional sports stadia, specialty retail activities, boutique hotels and the like. These cumulatively present a profile of economic activity in the twenty-first-century metropolitan core markedly different from that defined by the essentially *monocultural office economy* of the postindustrial city.

2 New divisions of production labour

The comprehensive restructuring of employment within the metropolitan core over the 1970s and 1980s constitutes a defining attribute of the postindustrial city, producing new task specializations and social divisions of labour, and social class reformation. Within the CBD, the rapid expansion of a hierarchical and segmented office labour force in the corporate complex was accompanied by relative (or even absolute) decline in non-office employment, including service categories such as retail and wholesale trade, personal services, and residual quasi-industrial uses such as auto repair.

While many non-office industries and occupations within the metropolitan core experienced decline, though, by far the largest contractions were in manufacturing and allied activities, exemplified by the massive collapse of Fordist production labour in our London case study, and with smaller but significant losses in San Francisco, Singapore, and Vancouver. The urban research literature of the period included bitterly critical polemical treatments contesting the implications and causalities of blue-collar employment contractions, and rejecting the terminology of postindustrialism. But the empirical dimensions of this fundamental restructuring of labour, and its visceral social outcomes, transcend any normal crisis of capitalism, and for many scholars validate the heuristic value of the term.

In the first decade of the twenty-first century, the central area office employment sector remains the dominant element of the core's workforce, and, in most cities, the largest agglomeration within the metropolitan labour force. That said, we can again identify some important shifts which suggest a departure from the employment structures of the postindustrial core. First, in many cities, the last decade and a half or so has seen increasing pressures on segments of the office workforce. Middle managers have been affected by successive rounds of corporate mergers, acquisitions, and buyouts, while even in successful companies the demands of market competition have produced leaner workforces. The office workforce tends to be less hierarchical and segmented in many companies, with fewer managers and an increase in multitasking. Capital substitution has cut heavily into the clerical workforce, with coincident processes of upgrading among the most skilled secretarial staff to IT and para-professional positions, and a corresponding shedding of clerical workers engaged in routinized functions, as well as outsourcing. Again, there is a compelling global-regional dimension: at the apex of the global city system, the most specialized workers within financial and commercial occupations, such as derivative traders and fund managers, enjoy astonishingly high incomes including salaries and bonuses, separated by status and privilege from even the higher ranks of executives and managers of the classical postindustrial era: bonuses for the higher echelons of London traders totalled US$17 billion for 2006–2007. But these stratospheric incomes are largely restricted to the world's financial capitals, notably London and New York, with smaller representations in Chicago, Frankfurt, Tokyo, Hong Kong, and Singapore.

As our case study cities and sites in this volume have demonstrated, the (uneven) growth of production employment has been a feature of the new inner city, representing a departure from the postindustrial city and its conditions of industrial disinvestment and employment decline. The economy of the new inner city has generated creative labour among the cultural industries, a large contingent of artists and designers, and, in some cases (notably London), intermediate service industry workers employed in new office clusters. To these we can add the staff of residual Fordist industries (including food and beverage and garment workers), together with the expanded workforce of new public institutions, consumption industry labour, and entertainment sector workers, to produce a revivified twenty-first-century inner city employment base. This new employment growth in the inner city, coupled with the pressures on the CBD office workforce in many cities, indicates a partial recovery of the employment balance within the metropolitan core so comprehensively skewed by the restructuring of the 1970s and 1980s, suggesting the need for new concepts to account for the emergent spatial divisions of labour in the metropolitan core.

3 The recombinant economy, the new inner city, and the restructuring of the metropolitan core

The asymmetrical urban structure produced by the hyper-specialization of the CBD, and industrial disinvestment and decline within the CBD fringe and inner city (Hall 1997), has been supplanted by a reconfigured, more complex, and dynamic metropolitan core, shaped by new industry formation, the social reconstruction of the central city, and the insistent relayering of capital in the metropolis.

What we are now seeing in the core is a partial recovery of the functional diversity, social density, and spaces of consumption and spectacle that defined the central city before the onset of hyperspecialization in the 1970s and 1980s. The twenty-first-century core includes the emergence of new spaces of production, consumption, housing, and spectacle, producing in many instances greater polarization and dislocation, as well as enhanced urban vitality. In the instructive London case study, for example, the space-shaping effects of economic change include: (a) two global financial-corporate spaces, in the form of the City of London and Canary Wharf; (b) major inner city office subcentres, at Paddington Basin and Hammersmith; (c) the cultural production and creative industry formations in Soho, Shoreditch, Islington, Bermondsey, and elsewhere; (d) major 'outlier' office complexes, such as the BNP-Paribas centre in Marylebone, and the proposed Shard/London Bridge Tower for Southwark, designed by Renzo Piano; and (e) the incipient new global space of the 2012 Olympic site at Stratford. These examples effectively demonstrate the role of production in the reformation of space in the core that represents a redefining break with the spatial configuration of the postindustrial city, as depicted in Peter Hall's model of the multifunctional core (Chapter 3).

Our other case study cities demonstrate similar tendencies, at different scales and shaped by local contingencies. In Singapore, the CBD office complex remains the dominant feature of the urban core. But the emergence of Chinatown and Little India as sites of cultural production represents an enlargement of the core's space-economy, while the current plan to increase the inner city's share of Singapore's population from 3 to 7 per cent, and consideration of the adaptive reuse of older office buildings for residential conversion, suggests an expanded role for housing in the reconstruction of the spaces of the core. The San Francisco case demonstrates a more extensive (and deeply problematic) experience of spatial restructuring, with the redevelopment of obsolescent industrial space and marginal housing for the cultural, leisure, and convention activities of the Yerba Buena Center, the tumultuous industrial restructuring experiences in SOMA and Mission, and the long-running saga of Mission Bay each contributing to a substantially reconfigured metropolitan core.

The Vancouver story-line includes an increasing experience of dislocation, as the high-externality redevelopment of the core encroaches upon low-income housing, and as new gentrifiers (and new industries) locate in or adjacent to marginal communities. But Vancouver represents an example of the local state assuming a more direct role in reshaping the spatial structure of the core, principally through the strategic reallocation of space articulated in the seminal *Central Area Plan* (1991). This strategy consolidated the corporate office complex within a reduced CBD, encouraged the social reconstruction of the core via new zoning for high-rise residential communities, and enabled the formation of new industrial districts and sites on the CBD fringe and inner city districts.

We can identify several layers of spatial restructuring in Vancouver's metropolitan core associated with shifts in the economy of production. First, the spatial scope of the territory has been enlarged from the definition of a 'Central Area' in the 1991 plan, limited by Main Street on the east, to now include a terrain extending to Clark Drive, which takes in the New Economy site of False Creek Flats, the mixed production zone along Hastings Street, and a major artists district in the Strathcona–Grandview Woodlands area, constituting a *reterritorialization* of the core. Second, the proliferation of new industry sites within the CBD fringe and inner city, comprising primary production precincts and technical support firms within integrated production networks and ensembles, has inserted a significant measure of functional diversity into the spatial structure of the core. Third, the expansion of skilled workers in the core beyond the CBD to encompass the wider territories of the CBD fringe and inner city has produced a reconfigured spatial division of labour within the City (Figure 8.13), shaping a new core area space-economy which presents a marked contrast to the highly polarized pattern of the postindustrial period privileging the CBD.

It might be tempting to cast these new spaces as expressions of postmodernity, in acknowledgement of the (relative) weakening of the hegemony of the CBD in some cases, the new mosaics of land use characterized by complexity and heterogeneity, and the recovery of diversity, each of which stand in contrast to the sharply bifurcated spatial structures of the postindustrial city. The new spatial, social, and technical divisions of labour in the core also present a departure from the formalized, hierarchical, and highly segmented office workforce of the 1970s and 1980s.

There are, however, clear limits to the postmodern interpretation. The Dear and Flusty (1998) version of postmodern urbanism postulated an amorphous, chaotic, decentred urban structure, with apparently random distributions of diverse (and often conflictual) land use and social groups. But the new industry formations depicted in this volume follow discernible logics of location, shaped by the core's spatiality and built form, agglomeration economies, property markets, sociocultural factors, and policy measures. The latter include land use planning, urban design, and regulation which impose a measure of order on development. Complexity and conflict there is in abundance, as demonstrated in the contested interface between the Cartesian spaces of Vancouver's core area, shaped by land use plans, urban design guidelines, and formal megaproject site plans, and the Hogarthian realm of chronic deprivation and disorder in the Downtown Eastside. But the new industry formations depicted in this volume do not in the main reflect arbitrary dispersion but rather coherent principles of location and development which may be construed in compelling conceptual terms, rather than momentary and discursive interpretations of postmodern urbanism.

4 New industries and the reconstruction of urban form

The growth of the CBD high-rise office complex, with its projection of power and control, and its high visual impact, generated new urban imageries and narratives in the postindustrial city. The high-rise office of course has an earlier provenance, dating to the first towers built in Chicago in the 1880s, and then progressing to the iconic Empire State and Chrysler Buildings in Manhattan in the 1930s. By contrast, many of the office towers constructed in the 1970s and 1980s were built according to an almost Fordist specification, functioning as 'office factories' for the expanding producer services labour, and aggressively marketed in the speculative commercial sector to recoup costs of land acquisition and construction. But with the onset of a new round of globalization and financial deregulation in the 1980s, a trend toward 'signature' office towers, imbued with high design values, was established in major business centres. As Maria Kaika has observed, the landmark office tower is an essential feature of the iconicity of global cities, from London and New York, to Shanghai and Hong Kong: the most exuberant expression of global competition and the struggle for market primacy (Kaika and Thielen 2005).

The office towers of the CBD cast a shadow, figuratively and functionally, on the mixed use and nondescript spaces of the CBD fringe, relegated to quasi-industrial and low-value service operations. But we can identify alternative narratives which find expression in the city's built form, associated in part with the sequence of new industry formations which have been situated within the postindustrial terrains of the inner city. The adaptation of the postindustrial built environment for residential use has been in progress for some decades, as vividly elucidated by Sharon Zukin in her influential treatise on *Loft Living* (1989).

These conversions form an important part of the social reconstruction of the metropolitan core. But the formation of new industry ensembles among the textured spaces of the inner city offers another story-line, a physical consequence of the reassertion of production in the inner city over the past two decades.

In this new instalment of industrial urbanism, the appropriation of former factory and warehouse space for new economic purposes reflects certain values, both functional and symbolic. As Thomas Markus has asserted in his volume on *Buildings and Power* (1994), historic building types may embody integrity of design and meaning – a historic truth that transcends their moment of production. The quality of building design, materials, and construction enables part of this historic building stock to be redeployed for new industries, a century and a half or more after construction, as we saw in the London and Singapore case studies. The adaptive reuse of heritage structures for new production purposes contributes to the revalorization of the postindustrial landscape.

Where new industry formation and the social reconstruction of the metropolitan core come together is in new forms of adaptive reuse which endeavour to combine working and residential space, including work-live and live-work studios, now features of many cities where city planners are encouraging mixed use precincts. The ratio between 'work' and 'live' activities in these units is variable and in some cases contested, and, as we saw in the San Francisco case, has often taken the form of *faux* live-works constructed according to industrial, rather than more stringent residential, building standards. But taken together with the adaptive reuse of structures for innovative production purposes, the development of cultural institutions, stadia and public places, and new residential landscapes in the form of condominiums, the comprehensive reconstruction of the inner city has produced imageries of the city in the twenty-first century which present marked contrasts to those of the classic postindustrial city.

5 New industries, cultural labour, and social class reformation

The social class implications of new industry formation represent perhaps the most contested aspect of the most recent rounds of restructuring in the city, both on normative and theoretical grounds. A central debate concerns the status and positioning of an emergent 'creative class' within the social structure of the metropolis. At the heart of the issue is whether the 'creatives' constitute a semi-autonomous cohort, or, alternatively, can be subsumed within the new middle-class construct deployed by urban geographers and urban sociologists. As in other similar debates, the question of 'difference' comes into play as a fulcrum of theoretical contestation, involving a diverse range of social signifiers including occupation, income, identity, behavioural issues, and housing preferences, as well as the importance of scale which must be considered in any proposal of conceptual innovation or synthesis.

As a means of situating the issue in a deeper historical setting we can go back further to Dickens' polemics on social class inequality in nineteenth-century London, and to the seminal theory of Marx on structures of class, socioeconomic fault lines, and radical political implications articulated in *The Communist Manifesto*. What these foundational theories have in common is first, an articulation of the social class consequences of industrial restructuring; second, the problematic features of social class change, emphasizing structural differences of power, affiliation and identity, and their polarizing effects on society; and, thirdly, a specific urban context for evaluating the complexities of social class reformation. Bell's model of the postindustrial society stressed new social class and cultural divisions without treating space in a particular way, and an important contribution of urban and social geographers was to specify an inner city context when studying the manifestations of the new middle class in its most salient domain. The inner city has been the defining *habitus* of the 'working class' (or proletariat) as well as the postindustrial society, the new middle class, and the 'creative class'.

Where it may be possible to contribute modestly to the ongoing discourse on social class implications of industrial restructuring is in suggesting some insights derived from the inner city case studies and sites incorporated within this volume as potential entrées to more incisive and systematic study. First, it can be persuasively argued that the cohorts of cultural workers in the urban core embody attributes of Bell's 'postindustrial society', in the sense of meeting the axial principle of the supremacy of scientific knowledge, broadly construed. Education, training, and knowledge represent core attributes of Bell's 1973 forecast of an ascendant postindustrial society and the contemporary cultural class. Extending the argument, Bell's pessimistic interpretation of cultural outcomes of postindustrialism, with his elitist disdain for the populist configuration of cultural values, offers another reference point for investigating the social class outcomes of new industry formation in the inner city. Where a divide might be posited between Bell's postindustrial society and the contemporary creative class lies in an assumption of functional separation between service-type occupations and the production of goods: many creative workers (for example, in industrial design) conduct literally 'hands-on' fabrication of physical goods.

If one accepts the very generous definitions of the dimensions of the creative class proposed by Richard Florida and his followers, there is a case for structuring the creatives as a significant occupational cohort and (urban) class. The problems here (mirroring the earlier debates concerning postindustrialism) concern the different analytical vantage points offered by adopting industrial or occupational definitions, difficulties in reconciling time-series data sets using different definitions of principal categories, and the almost intractable problem of sorting out which workers are 'creative' or 'non-creative' within accepted industrial categories (e.g. in advertising).

Addressing definitions of 'difference' between influential constructions of the new middle class and cultural workers might be more fruitful (or at least more interesting). The idea of the new middle class was always elastic enough to accommodate a fairly broad range of actors, distinguished in fundamental ways from the old industrial elites from whom they differed in terms of education, occupation, identity, behaviour, political affiliation, and housing preferences, among other attributes. There was room enough to include artists, designers, and entrepreneurs; the latter group augmented by the increasing numbers of international immigrants to metropolitan cities over the 1980s and 1990s. That said, the core of the new middle class comprised the upper-tier occupations of the services class, managers and professionals working principally in offices, chiefly in business, but also including government and other public agencies and institutions. Urban scholars emphasized the importance of the central city as place of residence, as well as work. This preference was expressed in the purchase of downtown apartments and condominiums, willingness to experiment in co-operative housing arrangements, and engagement in the community and cultural life of the central city. Gentrification occurred primarily within working-class neighbourhoods, vulnerable to the encroachment of more affluent members of the new middle class because of income losses accruing from the collapse of Fordist manufacturing, buttressing a socio-spatial articulation between social upgrading processes and the terrains of industrial disinvestments and decline.

Cultural workers tend to be younger than the professional and managerial members of the new middle class, and in many cases lack the incomes required to support a central residential location, particularly in a context of insistent property inflation driven by 'supergentrifiers' (described by Tim Butler and Loretta Lees 2006), investors, and speculators. This is evidently the case in London, San Francisco, and Vancouver, among our sample cities, where property values in former industrial districts within the inner city are approaching (or in the Vancouver case, exceeding) those of the long-established prestige residential areas. (Singapore, with only 3 per cent of its population living in the central area, presents an exception, although the current plan calls for a rapid increase in the downtown-inner city residential population.) Interviews I conducted with cultural workers in London and Vancouver underscore the difficulty of the (mostly younger) cultural workers gaining access to central city housing, with long commutes common among occupations below senior professional and managerial levels, reinforcing the idea that the new middle class is more about occupational status and incomes than about industrial positioning.

6 The recombinant economy and new policy discourses

At one level the emergence of new industries within the metropolitan core is associated with the idea of a major 'cultural turn' in policy terms, enunciated in state and local plans and programs in support of creative and design-based industries (Evans 2001). A principal reference point is the rhetorical flourish concerning a putative creative class and its centrality to urban development and community regeneration. This discourse has attracted the attention of some of the leading scholars writing on urban-regional development, including (to illustrate) Richard Florida as the most public advocate, Allen Scott as a more sceptical observer, and Jamie Peck as an outright rejectionist, at least with respect to the more euphoric claims of the 'creatives'.

We can return to this debate, but the immediacy of the cultural 'buzz' shouldn't obscure other policy questions arising from the new industry experience within the inner city, which reflect perhaps more fundamental shifts in policy discourses and city planning models for this crucial terrain of the metropolis. The evolving policy responses to episodes of industrial innovation and restructuring in the metropolitan core since the 1970s embody intrinsic significance, but also reflect larger benchmarks of change in the construction of policy priorities and models, and here our four sample cities and sites represent particularly instructive exemplars. In London, the collapse of inner city manufacturing created theoretical disjuncture and a deep urban policy crisis in developmental terms, while the negative externalities associated with the rapid growth of the central office complex, notably commuting, produced increasingly stringent development control measures enacted both by the central and local state. The Greater London Council (GLC), under the leadership of Ken Livingstone in his initial political persona, was supportive of traditional manufacturing and labour and reflexively oppositional to the money men of the City, and attempted to support the former and suppress the latter, until the abolition of the GLC in 1986. The growth of central London's office sector was construed as a regional issue as well as a local planning problem, and Labour governments introduced controls through the operation of Office Development Permits (ODPs) and the Location of Offices Bureau (LOB). Inner London boroughs were also under the control of hard left councillors, and added their weight to the efforts of the GLC and central government to restrict commercial development in central London.

The postindustrial policy environment in San Francisco and Vancouver also reflected concerns about the rampant growth of the central office sector. The regional plan for Metro Vancouver attempted, without much success, to promote alternative office locations within designated Regional Town Centres (RTCs), as the strategy lacked either market inducements or compliance measures. A more robust planning response in San Francisco took the form of Proposition 'M', approved by voters in the municipal elections of 1986, incorporating annual limits on office development, as well as higher development charges designed to offset the public costs of office development.

But new development conditions and changes in political control combined to signal a new era in policies for the core by the latter years of the 1980s (see Harvey's influential essay, 1989). This policy viewpoint emphasized the cardinal economic values of the central bastions of the service economy – professional employment, steady incomes, high-value output and sales, and tax revenues – rather than the 'nuisance value' (negative externalities such as congestion, displacement, and view blockage). The new policy values also included activating the redevelopment potential of the postindustrial landscapes of the inner city. Policy choices for the metropolitan core tended to favour 'change' rather than 'retention', the liberating growth tonic of the market over the hidebound regulatory instincts of government, and global visions over the restrictive confines of the 'local'. The new role of the state, in its local and central agencies, was to remove impediments to restructuring, and indeed to 'incentivize' experimentation, and innovation, particularly in the strategic terrains of the metropolitan core, both in the local and national interest.

Two of our sample cities, Singapore and Vancouver, experienced similar shifts in policy orientation during the mid-1980s, in response to changes in regional circumstances. In Vancouver, a severe downturn in the early 1980s driven by a commodity shock in British Columbia's resource periphery led to the City abandoning a strategic growth management process in favour of an ambitious new development plan in 1983, which emphasized the growth potential of service industries. In 1986, a global exposition on transportation (Expo 86) was convened on the former industrial lands of False Creek North, recreating this terrain as a site of spectacle and globalization. City Council ruled out the possibility of reindustrializing the False Creek North site, but public and private sector investments for the exposition included new amenities which proved attractive to artists, designers, and creative industries in the adjacent heritage districts. In Singapore, a short but sharp recession in 1985 exposed the limitations of a vocation underpinned by regional entrepôt trade, and as in the earlier Vancouver case led to a new development strategy which placed greater value on exportable services.

But as important as these policy initiatives were in signalling a more developmental (as opposed to regulatory) emphasis for cities, the London case study presents the model par excellence of signifying policy shifts in the postindustrial era, involving the larger state as well as local government and agencies. The abolition of the GLC in 1986 and a series of parliamentary Acts promulgated over the 1980s limiting the policy powers of local government effectively forestalled any major policy effort to slow the rate of industrial decline in London's inner city. The long period of industrial (and overall metropolitan) decline had run its course by the early 1990s, and a 'new inner city' was shaped by new policy approaches as well as by the market and sociocultural influences. The obsolescent landscapes of inner London included major development sites, including, first, the new global financial spaces of Canary Wharf, imposed upon the East London docklands by Michael Heseltine and the Reichmann brothers over the strident objections of the local authorities; second, a resurgent cultural economy comprising Europe's largest artists community, new institutions, and creative industry sites overlaying the traditional industrial terrains of Hackney, Southwark, and Islington; and, third, a pervasive residential conversion of much of the old warehouse stock of East London, a process described in Chris Hamnett's recent work. The local boroughs, many of which had vigorously resisted the wave of deindustrialization and the successor commercial industries, now for the most part accepted the new realities of the global metropolis, promoting cultural development as an engine of local community regeneration, and acquiescing to the demands of the commercial and property sectors for space in the old industrial territories of East London. The Mayor of London, Ken Livingstone, who had led the struggle against postindustrialism in the 1980s, now embraced London's global status and the commercial imperatives incumbent with this vocation, vigorously supporting high-rise offices and greater commercial density as a means of generating revenues for housing and enhancing London's sustainable development. There is still a policy vocabulary at the local level which incorporates terminology for 'production', and the 'local' (LDA 2005), but these terms occupy a subordinate status within a new lexicon which privileges globalization and the market.

There is, of course, debate concerning the neo-liberal tendencies of the state, both at central and local government levels, and the policy responses to new industry formation addressed in this volume may have modest implications for this larger discourse. The creative industries concentrated within inner city districts of the metropolis have been incorporated into the relentless construction of competitive advantage by public agencies and business advocates, played out in London, New York, Toronto, Singapore, and a host of smaller places. Local planning systems have been enlisted in this enterprise, recast as development services, in intent if not name. The cultural economy is certainly 'real' enough, as this study has demonstrated, even if the notion of an autonomous creative class seems eminently contestable. The establishment of policies and programs to more fully develop the cultural potential of the city, both as a community development measure and as an instrument of local/regional development policy, is entirely reasonable.

But in the more ebullient and universalistic claims of the cultural economy advocates we can readily detect an element of fantasy, reminiscent of the long parade of panaceas which have included experiments in special enterprise zones, tax holidays and fiscal incentives, domed stadia, science parks, and the rest. The bandwagon effect of this latest magic policy bullet is likely to yield a similarly uneven distribution of costs and benefits.

Research directions and opportunities

This volume, essentially an interim statement of observations from a program of field study over the past decade, constitutes a modest contribution to what is now a burgeoning genre of scholarship on the reassertion of industrial production within the inner city. In some ways the state of research mirrors the early stage of investigation of gentrification, as in the search for more compelling explanations of causality and effect, in questions of universal impacts as against the incidence of exceptionalism, in the role of local property markets versus class and social actors in generative processes of change, and in the quest for potentially effective policy intervention points.

A number of research issues are implied within the five case study chapters of the book. There is scope for more intensive and systematic investigation of specialized labour formation within the individual 'new industrial districts' of the city, identified by Susan Christopherson and Michael Storper, elaborated by Ann Markusen, and now surely ripe for further study. Here Graeme Evans' recasting of the cultural quarters phenomenon within a more historically informed sequence of industrialization processes offers a useful beginning to this exercise.

Second, there is substantial research potential represented by the social class implications of this most recent phase of industrial urbanism, apart from the contemporary furore over the claims of Richard Florida and his acolytes. This new research could perhaps more profitably relate new occupational shifts to broader and more established processes of social upgrading in the metropolitan core, to the behaviour of property markets, and to specific residential forms of tenure and style, such as live-work studios.

There is a particularly pressing need for more incisive research on new industry sites within the Asia-Pacific. The larger world cities in the region, such as Tokyo, Seoul, and Shanghai, are now well advanced in processes of polycentric development, with second and third metropolitan subcentres incorporating ensembles of cultural industries as well as IT, telecomms, and R&D installations. Detailed case study work here would be most welcome, as well as some comparative work on the generative processes and causalities of new industry formation between these cities and exemplary cases in the 'West'.

Finally, we are perhaps approaching a vantage point upon which a more robust retheorization process which takes in the restructuring episodes of the last two decades might be feasible. Here Allen Scott's acknowledgement of the promiscuous neologisms of change, advanced over twenty years from 'sunrise industries' to 'cultural industries', with no fewer than five intervening candidate descriptors, captures the essence of this dilemma. The idea of a 'cultural inflection' of production among advanced societies has a certain but perhaps limited appeal, while Scott's own proposal of a 'cognitive cultural economy' as a contemporary form of post-Fordism may – or may not – catch on among scholars, and within the media and public imagination. My own laboured working through the dimensions of recent change in the metropolitan core, set out in the discussion of theoretical signifiers in this concluding chapter, has led me to abandon the twentieth-century idea of the *post-industrial city* as a model for the contemporary city, as the contrasts between the conditions (economy, structure, labour, urban form, and social class) of the classic postindustrial era and those of the present, I believe, are too pronounced to sustain this terminology. In Asian cities, furthermore, the post-industrial template does not fit the conditions of urbanism in many cases, as Henry Yeung and George Lin have argued (2003). The next step in urban theorization might be to enjoin industrial urbanists and scholars committed to investigating the sociology of the city to jointly undertake projects on transformative change in the metropolis at the urban system level.

Appendix A
Research model and methodology

The basic research model for this volume entailed a blend of theory, empirical analysis generated largely by case studies, normative engagement, and, finally, conceptual synthesis. Research methodologies for the multiperspectival research project included an ongoing literature review drawn principally from economic geography, urban geography, and urban studies; theoretical critique; key informant dialogue; and an extensive process of field work designed to elicit empirical data for the study. The program of field work was conducted principally in key new industry sites in London, Singapore, San Francisco, and Vancouver, with supplementary site visits to other cities which included Amsterdam, Cologne, Florence, and Seattle. (See Appendix B for a schedule of fieldwork.)

This research model was augmented by other methods of obtaining critical feedback and supplementary knowledge, including conference presentations, the production and exchange of working papers and draft chapters, supervision of Master's and Doctoral theses, and graduate and senior undergraduate teaching. The book is therefore a single-authored monograph in the modern social science tradition, but one which has drawn heavily on the work, ideas, and contributions of many colleagues and students.

What follows is a more detailed articulation of research methods and techniques deployed in this project.

1 Literature review and synthesis

Over the course of the study, books and articles drawn from a range of social science and humanities disciplines were assembled for review, including titles from the economic geography, urban geography, urban studies, and city and community planning literatures, together with a smaller selection of scholarship in sociology, urban history, and political science. In the end, an inventory of about 500 titles was compiled.

Much of this body of research related to the processes and outcomes of industrial restructuring over the past four decades, synthesized in order to place the more recent episodes of innovation and change in a temporal context of transformation.

2 Primary documentary sources

For each of the principal case studies an inventory of local planning and policy documents was assembled, dealing principally with urban structure and land use, planning at the community and district level, and industrial or economic development. Here the overarching purpose was to identify the policy and institutional forces shaping (by first intent or incidentally) the nature and distribution of economic activity, notably for production-related industries and firms. These include for the purposes of illustration an extended sequence of industrial land planning policy documents for SOMA in San Francisco, generated in a strategic exercise in balancing the interests and needs of 'traditional' PDR (production, distribution and repair) activity with newer industries and upscale housing from the early 1990s, as well as the intense planning activity associated with the location of industry and employment-generating land uses in Vancouver, initiated in the core area growth management program which commenced in 1980 and effectively abandoned four years later in the depths of the city's worst recession since the Great Depression of the 1930s. Heritage policy materials were also key to understanding the role of the state in the preservation of the built environment for adaptive re-use, specifically new creative industries and cultural institutions, for example, in Telok Ayer (Singapore Chinatown Historic Sub-area), Yaletown (municipal heritage policy guidelines for the warehouses and loading bays), and Bermondsey Street (Bermondsey Street Conservation Area [CA] planning guidelines, London Borough of Southwark).

A number of central government documents were also obtained, notably the seminal Cultural Industries mapping study published by the British Department of Sport, Recreation and Culture (2001), and cultural industries and new economy industry reports released by central Singapore government departments.

3 Fieldwork and site visits

An extensive program of fieldwork lies at the heart of the research program undertaken for this study. This program was framed within different temporal periods, reflecting the length of research tenure or experience in each case (see Appendix B). But the core of the work for this volume specifically was conducted over the period 1998–2007, thus taking in multiple phases of innovation and restructuring in each case.

The program of fieldwork encompassed a range of complementary research techniques, designed to draw out the empirical richness of new industry experiences in the inner city setting, and to depict changes in the nature of industries and firms over time. These techniques included the following:

(a) *Initial observation.* The first stage of research in each of the sites took the form of observation, to develop a qualitative appreciation of the nature of each site, to scope the industries situated *in situ*, to examine the 'relational assets' of each site (built environment, institutions, consumption activity, and housing), and to determine for later more intensive research surveys district and/or site boundaries.

(b) *Key informant interviews*. In most sites, interviews with select key informants were undertaken in order to achieve an operational level of understanding of processes, trends, and issues, as well as other important sources of knowledge and data for the sites. These key informants were typically academics or policy planners, although in some cases included representatives of firms, non-governmental organizations (NGOs), and community-based organizations (CBOs), such as the South of Market Foundation in San Francisco, and the Clerkenwell Green Association in London.

(c) *Mapping exercises*. Spatiality, the reproduction of urban space, and place-making are all central to our understanding of the location, nature, and oper-ation of new industries, as depicted in this volume. Accordingly, mapping exercises constituted a core field research technique for the study in each of the cities and sites, and a complement to the description and analysis pre-sented in the text of each case study chapter. Mapping offered a means of graphically illustrating the narratives of new industry location and restructur-ing in the new economy of the inner city.

The maps prepared for this volume comprised the following types: (i) maps showing city and site boundaries, including in some cases historic continu-ities (such as a map showing the basic congruence between the contemporary City Fringe boundary in London with the old East London industrial dis-tricts circa 1840–1960, as well as changes in district/site boundaries, as for the planning definition of the Vancouver Central Area and metropolitan core, and changes in the definition of Yaletown; (ii) maps showing the dis-tinctive spatiality of inner city districts and sites of the new economy, includ-ing principal arterials, smaller roads, landmarks, and institutions, such as the maps of Shoreditch and Clerkenwell (Chapter 5) and Telok Ayer (Chapter 6); and (iii) maps showing distributions of firms within selected industry groups, notably the time series maps of firm representations in Telok Ayer 2000–2006, and in South Park 2000–2003.

(d) *Interview program*. In each city case study area a program of semi-structured interviews was conducted. (See Appendix B for schedule.) A set of questions and issues for consideration was developed, administered to a sampling of firms in creative industries in each site, designed to elicit data regarding (i) basic enterprise information (date of establishment, employee numbers, and occupations); (ii) creative processes and key production technologies; (iii) perception of the area as operating environment; (iv) networks of sup-pliers and clients; (v) relationships with public agencies e.g. regarding regula-tion, development policies, and the like; (vi) perceptions regarding the future of the industry and the area. As a methodological note, I (and my graduate student RAs working in the Vancouver sites) experienced a gener-ally more difficult interviewing environment relative to earlier survey work on advanced services in Vancouver circa 1985–1995: firms were on the whole more reluctant to commit to scheduled interviews in this current project phase, with likely reasons including a more difficult business environment which exerted greater time pressures on staff; and, second, leaner workforces relative to the earlier interview phases, in which it seemed in the larger firms especially a staff person was designated as available for interviews and perhaps other non-production related activity. In response to this more challenging interview environment, I often suggested a meeting later in the working day, or during a coffee or lunch break, rather than disrupting the rhythms of the business day. I also frequently augmented the program of scheduled interviews with numerous informal conversations with individuals working in each site, including, for example, discussions initiated in cafés and coffee shops, in parks or other open spaces, or in 'passing by' workplaces at street level. These conversations typically included discussion of work experiences, industry trends, the nature of the local area as working environment, and housing and journey-to-work issues. While of course the quality of information was by no means as complete or systematic as that generated in the scheduled interviews, these conversations yielded a rich harvest of information, insights, and experiences, greatly enhancing an understanding of the workings of the new economy of the 'new inner city'.

4 Census and other statistical data sources

Data on employment in each of the four cities were obtained from Census surveys, as well as from supplementary sources.

In the London case, employment data were also obtained from the City Corporation of London and associated consultants. Employment data for Singapore included various government ministry sources, including branches of the Economic Development Board and the Ministry of Trade. Data for the San Francisco case study included the City of San Francisco Planning Department and the American Community Survey. Sources for the Vancouver case study included Census data, as well as data derived from the Vancouver City Planning Department, notably from the current Metropolitan Core Economy, Jobs, and Land Use exercise.

5 Conference presentations and workshops

Interim findings for this urban research project were reported in conference sessions, notably annual meetings of the Association of American Geographers. Valuable feedback on research issues, particularly on the experiences of change in land use and enterprise structure in the case study cities and sites, was obtained and incorporated into successive chapter drafts. Special sessions addressing issues of new industry formation and restructuring were organized at a number of meetings. Conferences, symposia, and seminars also provided opportunities to discuss research observations and insights with colleagues. Multiple workshops with City of Vancouver Planning staff were convened to share observations on the Vancouver new industry sites.

6 Journal articles

A series of papers submitted to journals circa 1998–2006 on experiences of new industry formation within inner city sites supported the concept of a larger monograph project in various ways: first, in developing a theoretical platform for the research enterprise; second, in generating commentary and suggestions from colleagues; and third, in sharpening ideas via the peer referee process. The articles in this series included Hutton (2000, 2004a, 2004b, 2006).

7 Graduate teaching and supervision

I am pleased to acknowledge the great value of graduate teaching and thesis supervision in the development of this book's conceptual foundation. Planning 592 'Structural Change and the City', which I offer every spring in the UBC School of Community & Regional Planning, and which attracts Masters and Doctoral students from Geography and other disciplines as well as Planning students, invariably provides a stimulating forum for discussion and debate. At its best, graduate teaching exemplifies the spirit of the collaborative production of knowledge, and I gratefully acknowledge the many rich insights and suggestions contributed by my students. In particular, graduate students have been instrumental in suggesting ways to enrich the analysis beyond the stock techniques many of us are burdened with, and to enlarge the realm of themes and motifs associated with urban change beyond the more obvious choices. Finally, I have benefited from the ideas, experiences, and energies of my thesis students working in this field, starting with Mary Shaughnessy's Master's thesis on design industries in Vancouver (1988), and Shauna Brail's thesis on Yaletown and Victory Square in Vancouver (1994), followed by Naomi Pope's thesis comparing new media spaces in Belltown (Seattle) and Yaletown (2002), Kevin Eng's thesis on the New Economy site of False Creek Flats in Vancouver (2003), Jennifer Johnston's thesis work on the planning implications of the 24-hour city (2004), and Helen Cain's thesis on cultural planning in the city (2005).

8 Newspapers and other media sources

The new economy of the inner city, in its various manifestations observed over the past decade, has (for better and for worse) attracted a great deal of media attention, associated with both the 'buzz' of dazzling success and social cachet, as well as the depths of despair following periodic (and occasionally terminal) crashes. I have Kris Olds to thank (if that's the right word) for the idea of doggedly collecting piles of newspaper reports on new economy developments in each of the principal case study cities and sites. These newspaper accounts do offer a kind of 'running commentary' both on the high points and more gruesome episodes of industrial experimentation and innovation, as well as sometimes useful counterpoint to academic outputs, which tend to lag events by a year or two in the best of circumstances. My collection of media and related papers includes a formidable stack of real estate and property market reports, which have enabled me to keep abreast of the (generally) spiralling inner city residential, commercial and industrial price points over the last decade in each of the cities in the sample.

9 Membership in related collaborative research projects

Over the last five years or so I've participated in a number of strategic research projects which have fed, in one way or other, into the investigative processes for this volume. These have included, principally, a SSHRC national project on 'multilevel governance in Canada', directed by Bob Young of the University of Western Ontario; a second SSHRC MCRI project on 'social dynamics of innovation and creativity in the city-region', with David Wolfe and Meric Gertler of the University of Toronto, as Principal Investigators; a collaborative project on 'trajectories of the new economy: an international investigation of inner city regeneration and dislocation', undertaken for a special theme issue of *Urban Studies*; and a project on transformative change in the Canadian metropolis and urban system, focusing on employment growth and labour market change, co-managed by Larry Bourne and me. Each of these projects has generated new ideas, and has introduced me to new colleagues who have contributed in some way (without, of course, bearing any responsibility) toward the research exercise culminating in this volume.

Appendix B
Site selection and fieldwork schedule

The cities for the case studies – London, Singapore, San Francisco, and Vancouver – were selected on the basis of multiple criteria, including theoretical significance in key domains, notably economic geography, urban geography, and urban studies; saliency as site(s) of new industry formation; and my accumulated research experience and knowledge levels.

In each case the approach was to position new industry experiences within both a developmental context, emphasizing the sequence of restructuring experiences and industrial change since the 1960s, as well as within a larger spatial setting, notably the evolving space-economy of the metropolis. Each case study also incorporated associations between larger political shifts of the state and their transmission to (or resistance from) local policy systems and planning regimes. At the intensely localized scale of the case study sites, attention was paid to the framing issues of space and place, including the shaping influence of major arterials, local street patterns, and open spaces, as well as characteristics of built form and the existence of iconic buildings and landmarks.

These common attributes impart a level of systematic treatment to the cases under investigation. But the cities were not treated equally, in the manner of Janet Abu-Lughod's incisive study of New York, Chicago, and Los Angeles, or of Saskia Sassen's classic global cities monograph on New York, London, and Tokyo. The contrasts in scale among the four cities represented in this volume are manifestly greater than for the trilogy of cities in either the Abu-Lughod or Sassen volumes. It is also the case that I took different routes to field research in each city, having undertaken graduate study on London initially in the 1970s, in the last years of the Labour Government before the advent of Margaret Thatcher, then again during the 1980s, when the dual trajectories of industrial decline and a globalizing financial and commercial sector were at their maximum degree of divergence. In Vancouver, I enjoyed a sustained period of opportunity for conducting policy research on the City's development characteristics and restructuring episodes while employed in the Department of Finance, a sojourn which included several large-scale surveys of service industries and manufacturing, with outputs including a series of articles with David Ley and Craig Davis. My interest in San Francisco dated to 1993, when I began a series of conversations with Dr Amit Ghosh, Director of the City Planning Department. Themes for our meetings and discussions included the trajectories of restructuring in the City and the Bay area, impacts on industrial mix and employment, and implications for land use decisions, particularly in the mixed use/industrial zones south of Market Street. I was also interested in comparisons between San Francisco and Vancouver in the domain of growth management for the central office district, which emphasized development control, quotas, and pricing in the former, and reallocations of land resources and rezoning in the latter.

Selection of sites for special study and schedule of fieldwork

For the purposes of this book the provenance of a systematic research program was the late 1990s for each city, with site selection undertaken as follows.

London: initial site visits and observations in 1998; return visits for fieldwork 1999, 2000, 2002, 2003, 2004, 2006 (two visits)

For the London site, selection process and fieldwork, the advice of colleagues was crucial. Andy Pratt of LSE suggested Hoxton as a rich terrain of study for creative industries and cultural institutions, and generously shared his deep knowledge of the area. Andy Thornley of LSE recommended adding Clerkenwell to my repertoire of sites for study, including a description of key features of the district in London's economy. In 2002, I elected to add a third site, the Bermondsey Street Conservation Area, owing to the location of a number of key firms and institutions in the area, generating opportunities to explore the connections between heritage conservation and the formation of new production industries.

Singapore: initial site visit in 1999; return visits in 2000 (two visits), 2003, and 2006

The Singapore case demonstrated the uses of serendipity in new industry research in the inner city. While attending a conference at NUS in 1999, I happened to walk through Chinatown and was interested to observe distributions of dot.coms and new media firms in Telok Ayer, immediately to the west of the CBD. These distributions of firms exhibited striking similarities in clustering and industrial structure with the New Economy epicentre sites of Yaletown in Vancouver and South Park in SOMA, in each of which I had begun intensive study. On a six-week study leave in 2000 with my family, I undertook a site-mapping exercise, as well as photography and an initial set of interviews both with selected firms and URA officials. I made a second visit for more interviews in December of the same year while attending the Global Economic Geography Conference at NUS. I returned in January of 2003 to undertake a mapping exercise and another panel of interviews in Telok Ayer, and then again in December 2006 for a further mapping of changes in the distribution of firms.

San Francisco: initial visit in 1993 to discuss land use and industrial planning with San Francisco City Planning; return visits in 1995 and 1997; systematic program of site visits to South Park and SOMA 1999, 2000, 2001, 2003, 2005, 2007

Study of change in San Francisco's economy started with a meeting with Dr Amit Ghosh, Director of Planning (Policy) in 1993, with reference to planning for industrial change in the South of Market Area (SOMA). In 1997, I returned to the San Francisco Planning Department and met with Catherine Bauman (who was leading a study of San Francisco's multimedia industry) and Stephen Shotland, who proved to be a remarkably insightful and helpful interlocutor over subsequent visits. Stephen suggested I might want to conduct a special study in the South Park area, emerging as the epicentre of SOMA's New Economy, an idea which led to the series of mapping exercises and interviews between 1999 and 2005, with a final site visit during the Association of American Geographers annual meeting in San Francisco in April 2007.

Notes

1 The reassertion of production in the inner city

1 For the purposes of this study, the zonal structure of the metropolitan core comprised the following elements, reflecting foundational models of urban structure: (1) the *central business district* (CBD), dominated by the high-rise corporate office complex, the largest single agglomeration in most advanced city-regions, but including a substantial platform of retail, personal services, and other consumption activity; (2) the *CBD fringe* or '*frame*' (as some American scholars describe it), for much of the twentieth century a zone of diverse and somewhat functionally promiscuous low-value services and quasi-industrial activities; and (3) the *inner city*, a larger zone incorporating 'industry' (manufacturing, craft production, warehousing, and distribution) as well as older housing, comprising in many cities a mix of occupations and social classes, but with an emphasis on working-class residential communities and households, some of which had persisted in form, building types, and ethnicity since the early nineteenth century. The precise configurations of these inner city zones, within which both the ravages of restructuring in the latter decades of the twentieth century, and the reassertion of production since the early 1990s, took place, vary from place to place. These will be specified in each of the principal case studies, presented in Chapters 4–8, inclusive.

2 Susan Christopherson has underscored the psychic and intellectual legacies of post-industrialism as a socioeconomic process, in observing that '[m]any contemporary planners and planning educators came of age intellectually during the 1970s and 1980s, when U.S. manufacturing was transformed by firm decisions to cut costs by changing production methods or transferring production to lower wage workers and locations' (Christopherson 2003: 487).

3 Rates of office development tended to level off in many cities in the 1990s, owing in part to the overbuilding of the previous decade, and to the substitution of capital for labour in clerical employment – one of the largest occupational groups in the labour force of most cities, among other factors; but the impact has been uneven across urban systems, a theme to be picked up in the city case studies.

4 In the abstract to their well-known essay on 'Postmodern Urbanism', in the *Annals of the Association of American Geographers* (1998: 50), Michael Dear and Stephen Flusty justifiably assert that 'models of urban structure are scarce', but in the following decade few have accepted their challenge to develop new models derived from tenets of postmodernism, suggesting that the utility of the latter as a framing concept for the study of urban space has been at least temporarily exhausted.

5 See, for example, Henry Yeung and George Lin's influential paper on the theoretical distinctiveness of economic development in Asia, 'Theorizing Economic Geographies of Asia' (2003).

6 Ideology and fundamental political change can influence the power of industry to reshape land use and urban structure; see Chapter 3 in this respect.

7 These land use conflicts are especially important in cities such as London and Vancouver, in which the social reconstruction of the inner city constitutes in part a counterbalance to new enterprise and employment formation, as explicated in Chapters 4 and 5 and Chapter 8, respectively.

8 To illustrate the range of spatial linkage patterns typifying inner city behaviour, firms (for example, graphic design firms, printers, and Internet service firms) situated within new industry formations within the inner city may provide specialized inputs to corporations in the CBD, as is common practice in the London 'City Fringe'; others (such as computer software firms) perform as subcontractors to larger enterprises in the suburbs or ex-urbs (observed in the San Francisco Bay Area); while others are more remote offshoots of industrial complexes situated in other regions, exemplified by the neo-Marshallian 'satellite' film production in Vancouver, described by Neil Coe (see Chapter 3).

9 Critics have observed a tendency on the part of the more ebullient advocates of the creative city and class to universalize the potential of culture-led urban-led development, echoing the initial euphoria concerning the technology-driven 'New Economy' of the late 1990s, while ignoring the social costs, uneven benefits and risks associated with public commitments in this domain. For a particularly trenchant critique of the 'creatives', see Jamie Peck, 'Struggling with the Creative Class' (2005).

2 Process: geographies of production in the central city

1 Prominent examples include Doreen Massey and Richard Meegan's trenchant 'Industrial Restructuring Versus the Cities' (1980), published in the year following the election of the Thatcher Conservative Government, and *The Anatomy of Job Loss* (1982); and, in America, Barry Bluestone and Bennett Harrison's *The Deindustrialization of America: Plant Closing, Community Abandonment, and the Dismantling of Basic Industry* (1982), published in the early years of the Reagan presidency.

2 While this volume is chiefly concerned with the emergence of new production industries, the role of consumption as a major element of the metropolitan economy must be fully acknowledged, both in the general sense of the purchasing power and expenditures which characterize the economy of the large metropolis (see Glaeser *et al.* 2001), as well as the more recent juxtaposition of specialized production, creative workers, and high-amenity consumption in the cultural economy of the city.

3 Many inner city industrial firms and enterprises specialized in the production of food and beverages for local consumption, while others were engaged in tailoring and garment production. With household income growth and higher discretionary spending possibilities, too, certain inner city industries catered to an expanding range of specialty, high-value product demand, such as jewellery, handcrafts, and custom furniture-making.

4 Contractions in the bargaining power of trade unions and organized labour generally, coupled with attractive tax arrangements and other fiscal inducements, in time brought regions of the First World (back) into the new international division of production labour, evidenced by the German and Japanese car manufacturing and assembly plants in the south-eastern US states, and Japanese auto plants in Britain.

5 In some cases too, the tenure of heavy industries in the central and inner city was threatened by growing public disaffection with the negative externalities generated by some of these firms, concerns increasingly inserted into public policy discourses, and eventually into planning for land use change and rezoning. David Ley's classic portrayal of the Vancouver case in the *Annals* of the AAG (1980) represents one of the earliest treatments in this domain, while Hutton's (2004b) article in *Urban Studies* revisits this case and its exemplary impact on planning for Vancouver's metropolitan core.

6 Influential treatments of the growth of specialized services in the urban core include Jean Gottmann, *Megalopolis: The Urbanized Northeastern Seaboard* (1961); John

Goddard, 'Changing Office Location Patterns within Central London', (1967); John Goddard, 'Office Linkages and Location: A Study of Communications and Spatial Patterns in Central London' (1973); Peter Daniels, *Office Location: and Regional Study* (1975); Gunter Gad, 'Face-to-face Linkages and Office Decentralization Potentials: A Study of Toronto' (1979), Thomas Stanback, *Understanding the Service Economy: Employment, Productivity, Location* (1979); and Peter Daniels, *Service Industries: A Geographical Appraisal* (1985).

7 The insertion of major Asia-Pacific metropolitan centres into global city competition has in a sense reintroduced manufacturing within global and world city discourses, as many of these (including Tokyo, Seoul, and Shanghai) sustain very large industrial production sectors within the extended city-region.

8 See, for example, D.W. Jorgenson, 'Information Technology and the U.S. Economy', *American Economic Review* 91 (2001) 1–32, while a more sceptical view was expressed in D.C. Mowery, 'U.S. Postwar Technology Policies and the Creation of New Industries' (2001).

9 In an incisive comparison of Vancouver's cultural industries with those of Italian urban societies, Pier Luigi Sacco has observed that the former has an extraordinarily robust (if in some ways imperfectly articulated) cultural economy, while the latter have been 'coasting' on the foundational achievements *non pareil* of the Renaissance (Sacco *et al.* 2007).

10 The lack of a compelling theoretical construct to replace (or in a major way refurbish) the foundational theories of the 1970s and 1980s testifies to the evident difficulties of the task, despite a large battalion of scholars engaged in this exercise. It seems clear that the 'short cycle' restructuring episodes of the past fifteen years or so can be implicated as central to the problematic nature of this conceptual task: in his keynote address to the Vancouver meeting of the Pacific Rim Regional Science Association in May 2007, Allen Scott identified no fewer than seven descriptors advanced to acknowledge 'new' economic trajectories since the introduction of post-Fordism, including the abject term 'sunrise industries', through to the 'New Economy' of the late 1990s, and more recently to the over-hyped 'creative economy'.

11 The somewhat ambivalent positioning of artists in the socioeconomic transformation of the inner city was vividly captured in two sessions of the annual meeting of the Association of American Geographers in March of 2006 in Chicago: an early morning session led by Dennis Grammenos, Tom Slater, and other urban geographers identified the artist as villain, the inevitable forerunner of gentrification; while in a session immediately following, economic geographers, including Ann Markusen, celebrated the artist as the catalyst of the cultural and economic regeneration of the core.

12 Scott's model of intra-metropolitan industrial location articulated in the *Urban Studies* article of 1982 (updated and elaborated in Scott 1988) is now a quarter of a century old, and has been exposed to the influences of telecommunications, extensive outsourcing, and the ever more distended new international division of labour in advanced production systems, but still holds up well in explaining the basic contours of the metropolitan space-economy.

13 Meric Gertler and David Wolfe at the University of Toronto are leading a collaborative research project which endeavours to explore this theme in greater depth, in the 'Social Dynamics of Innovation and Creativity in the City-Region', supported by the Social Sciences and Humanities Research Council of Canada. (http://www.utoronto.ca/isrn/web_files/bibliography.htm).

3 Place: the revival of inner city industrial districts

1 Lifang Chiang offers a useful perspective on the beneficial roles played by manufacturing and blue-collar labour in 'new economy' regions, by virtue of innovation spillovers and economic and employment multipliers, as reported in 'A Reconsideration of "Old

Economy" Industries within "New Economy" Regions', in *Geography Compass* (in press).

2 The larger San Spirito area in which the Oltrarno is situated was formed at the end of the twelfth century, when Florence was divided into four 'quarters': Santa Maria Novella, San Giovanni, Santa Croce, and San Spirito.

3 Some of the leading industries in Florence have a more recent provenance, as in the case of the shoe industry, the development of which was initially stimulated by League of Nations economic sanctions against Italy, and then another growth phase in the 1960s 'when export demand for Italian products soared, especially in the United States' (Istituto Nazionale per il Commercio Estero 1988: 81).

4 See Chapter 5, for a discussion of traditions of artisanal production in Clerkenwell, together with the contemporary market pressures imposed upon these activities.

5 Council for Mutual Economic Cooperation.

6 Here Turner describes the transition of a precinct of the Ancient Quarter which formerly produced cooking oils, but had recently become engaged in the shoe trade; while another specializing in sedge matting was now trading in plastic, rope, and canvas goods.

7 Coe acknowledges that exchange rate differentials between the US and Canada have been crucial to sustaining runaway film production in Vancouver (and other Canadian centres, notably Toronto), and for much of the period of growth the value of the Canadian dollar occupied a position in the US$0.65–0.75 range. With the Canadian dollar now (June 2007) exceeding US$0.90, and perhaps approaching equivalency by the fall of this year, the comparative economics of runaway film production may change, although this may be partially offset by the highly developed skills of Vancouver-based production crews, and perhaps by tax benefits and other incentives.

8 In the conclusion to his essay in *Regional Studies*, Harrison states:

> It may well be that without the assumption of strong embedding, the 'pure' district model is unstable, after all. As concrete empirical research on the districts proceeds in the years ahead, we may discover that such strong embedding is simply not sustainable under the onslaught of competitive pressures from larger, more powerful, more distant and impersonal economic forces.
>
> (1992: 479)

There is some indication that Harrison's tentative forecast has been vindicated, as witness Italy's increasingly severe economic crisis, associated in part with the decline of high-value fabrication industries under pressure from 'larger, more powerful, more distant' competitors such as China.

9 For a more detailed account of the history of Leipzig's media industry, see H. Bathelt and J. Boggs (2003), 'Towards a Reconceptualization of Regional Development Paths: Is Leipzig's Media Cluster a Continuation or a Rupture with the Past?'

10 The loss of office space in Manhattan accruing from the 9/11 attacks roughly equals the total downtown office floor space inventory in Vancouver (Canada's third largest city, see Chapter 8 in this volume) downtown.

11 Indergaard presumably means San Francisco, which did experience a sharper fall; see Chapter 7 of this volume for the San Francisco New Economy experience in SOMA (South of Market Area).

12 The lingering recessionary effects of the post-9/11 attacks on the economy of Lower Manhattan included a protracted downturn in retail and personal service sales, a doubling or even tripling of office vacancy rates in some areas of the district, and closures of garment factories, described in 'Narrative no. 1' in our vignettes of the Manhattan industrial district. See C. Jones, 'Lower Manhattan Still on the Rebound after 9/11' (2005: 5A).

13 Shanghai's inner city cultural/industrial redevelopment started with one 'spontaneous' artists' and artisanal district, Suzhou Creek, about a decade ago, and now boasts no

fewer than thirty officially sanctioned 'creative clusters' or sites (Sheng Zhong, PhD candidate, University of British Columbia, personal communication, September 2006).

14 Singapore's experience of inner city new economy formation is described and evaluated in Chapter 6.

4 Restructuring narratives in the global metropolis

1 The notion of 'legitimacy' in this contested domain itself raises conceptual and definitional issues, but the saliency of London's claim rests on the following historical and developmental attributes: its increasingly hegemonic role as global centre of trade and empire in the colonial era; the strength of its claim to 'world city' status (incorporating measures of metropolitan scale, growth rates, specialized banking and financial activities, corporate control functions, and industrial power) according to Hall's (1966) original formulation; and London's pre-eminence (with New York) as a 'global city' as conceived by Friedmann and Wolff (1982), Sassen ([1991] 2001), Taylor (2004) and others. No other city meets each of these criteria.

2 These processes and events have yielded some of the classic texts on the development of cities, communities, and city-regions, with commensurate influence on discourses of urbanism and urban studies, notably Ruth Glass's original research on gentrification, Peter Hall's work on urban 'containment' as a cornerstone concept of British postwar planning, and Saskia Sassen's influential writing on London as a global city.

3 For both the Thatcher and Blair Governments, the undertaking of major redevelopment projects in East London were in part designed as expressions of innovation and visionary thinking in the interests of revitalization, the former by drawing on the power of markets, capital, and globalization to reconstruct derelict space in the inner city, and the latter by demonstrating the merits of private-public projects as a demonstration of the 'Third Way' approach to governance and leadership in a millennial context.

4 The much-publicized London Central Area congestion charge represents an example of pricing as means of resource allocation. See also GLA Economics *Annual Report 2006* (pp. 17–20) for a discussion of pricing and the levy of charges in order to 'incentivize' public behaviour.

5 These include novels, such as Monica Dickens' sagas of postwar life in London, and, more recently Monica Ali's account of Bengali immigrant life in *Brick Lane*, and Ian McEwen's *Saturday*, in which a successful London professional experiences an unexpected and violent encounter with London's underworld.

6 Ed Soja has in this connection advocated for studies acknowledging the 'industry-shaping power of spatiality', as a corollary research task of conventional treatments of the role of industry in reshaping space (Soja 2000: 166).

7 The Chicago Mercantile Exchange (CME) has recently purchased the Chicago Board of Trade (CBOT) for US$8 billion, a venture designed to create the world's largest financial marketplace in the exchange-trading of derivatives, with (should the deal gain approval of shareholders and regulators) a market value exceeding the New York Stock Exchange (NYSE) (*Economist* 2006: 100).

8 As Doel and Beaverstock have observed, the New York and London financial sectors were largely insulated from the effects of the 1997 'Asian crisis', which further depressed Tokyo's banking, financing, and corporate sectors, extending what Yuko Aoyama has described as the 'lost decade' of the 1990s for Tokyo and Japan as a whole (personal communication).

9 A report published by the property firm Development Securities has suggested that the quantity of office space in the City of London owned by foreign firms increased from 28 per cent in 2001 to 45 per cent in 2005, demonstrating London's 'continuing unique appeal', but also increasing London's exposure to the potential withdrawal of funds ensuing from external shocks (Development Securities 2006: FP5).

10 Tokyo retains a large manufacturing and industrial sector, relative to London and other

global cities within the 'Atlantic sphere', as do other major Asian metropolitan cities such as Seoul, consistent with the earlier criteria for defining world cities.

11 The EU accounts for about 40 per cent of business service exports from London, while European clients absorb about one-third of London's turnover in international banking (City of London Corporation 2004: 2).

12 In the case of Paris, the principal corporate complex of La Défense, four kilometres down the Seine from Central Paris, has been undertaken in the classic *grands projets* style of French planning over the past four decades, including massive public investment as well as private expenditure, incorporating comprehensive transportation and infrastructural planning, as well as commercial offices and institutions. In contrast, the Zuidas (southern ring) project for the Randstat region in Holland, situated on the southern perimeter of Amsterdam, has proceeded more incrementally, with initial Amsterdam City Planning studies undertaken for the site in 1992, and development accelerated by the decision of ABN/AMBRO to locate its international headquarters to Zuidas in 1994 (in defiance of the Amsterdam government's wishes). An agreement between the city and the state to expand infrastructure provision to Zuidas was followed by the development of a World Trade Center operation, and the corporate offices of the ING Bank. The current plans entail a substantial new residential component, equalling almost one half of the projected floor-space, and an ambitious cultural and recreational amenity element (Salet and Majoor 2005).

13 See Kenneth Powell's *New London Architecture* (2005) for an appraisal of the capital's new built form in most major land use classes, while Maria Kaika and Korinna Thielen offer a critical appraisal of London's new 'iconic buildings' in a recent special theme issue of *City:* (2005) volume 10, no. 1.

14 *The Weekly Telegraph* (Wednesday July 7–Tuesday July 13, 2004) published a map of England on its front page, with the headline 'London Takes Over the Country', with 'London' encompassing a territory extending from Grimsby in the north-east to Bristol in the south-west – effectively one-half the national territory within the capital's sphere of influence.

15 The *Creative Industries Mapping Document 2001* published by the Department for Culture, Media and Sport acknowledged that London and the South East region 'continue to be major magnets for the creative industries' (DCMS 2001: 13), with leading positions in such creative industries as advertising, architecture, design, designer fashion, film, performing arts, publishing, software, television, and media, among others.

16 A number of heavy industries established at some distance from London's core, notably shipbuilding, engineering and chemicals at Blackwall and Millwall, and in Battersea in west London.

17 Martin's complacent view of the stability of London's manufacturing economy at mid-century was widely shared. Saskia Sassen cites the comprehensive review of employment change in London conducted by Buck, Gordon and Young (1986), which 'established that in the postwar decades the city's manufacturing sector was, with a few exceptions, structurally sound, paying relatively good wages overall, with considerable inputs of skilled and craft work and relatively high levels of specialization' (Sassen [1991] 2001: 210).

18 Whether the Thatcher Government pursued policies in a deliberate effort to run down the London (and UK) traditional manufacturing sector, in the interests of advancing the economy to more modern industries and activities, or was merely indifferent to the fate of older industries, workers and constituent communities, may be open to question, but the vehemence of polemical discourse and debate bears eloquent testimony to the depths of social dislocation and deprivation associated with industrial restructuring in late twentieth-century England.

19 With the accession of Eastern European countries to the EU comes a new cohort of international immigrants, many of whom settle in London, according to a recent article on Britain's 'new working class'. In particular, the number of Poles emigrating to

Britain has been estimated at between 600,000 and 1 million, 'unquestionably the largest number to enter Britain in such a short time – there's been nothing like it in the last 300 years, probably never', according to John Salt, Professor of Migration Research at University College London (quoted in Saunders 2006: F3).

20 Francis Sheppard in his history of London reminds us that the hegemony of London's higher-order service class dates not from the industrial restructuring of the late twentieth century, but at least as far back as the mid-nineteenth century: '[t]he members of the élite who ran this metropolitan civilization belonged overwhelmingly to the service sector of the national economy . . . in such citadels of the élite as the clubs of Pall Mall there were very few industrialists among the members' (1998: 309). As Sheppard observes, 'London's service employment included many domestic, transport and distribution workers, but also incorporated rapid occupational growth among the professions, banking and finance, Government and defence' (ibid.).

21 Sheppard quotes Charles Booth's estimate of the incidence of poverty in London's East End in 1889 as exceeding 40 per cent (ibid.: 293).

22 The production of glossy brochures and magazines promoting property in London constitutes a minor industry in itself. One of the leaders in the field is *The London Property News*, which typically features ebullient leaders such as 'Property selling like hotcakes' (June 2002), or 'House prices up 22.7%' (October 2002).

5 London's inner city in the New Economy

1 See, for example, Hamnett (1991), 'The Blind Men and the Elephant: Toward a Theory of Gentrification', *Transactions of the Institute of British Geographers* 16: 173–189.

2 A number of conversations with Circus Space participants disclosed a relatively large number of individuals seeking a distinctive fitness regime, as opposed to aspirant circus performers.

3 Soja's injunction for urban scholars to acknowledge the 'industry-shaping power of spatiality' as well as the 'space-shaping power of industrialization' (Soja 2000: 166) seems especially germane to studies of new industry formation within the inner city.

4 Helbrecht's work underscores the need to appreciate the importance of the 'concrete' physical reality of urban landscapes in attracting creative enterprises, as well as the symbolic representations of places (heritage, style, memory) often cited as crucial to new industry formation.

5 For information on mission and programs, see the Foundation (19–22 Charlotte Road, London EC2A 3SG) web-site: www.princes-foundation.org

6 Among the casualties was a button shop on Rivington Street, a remnant of the district's former vocation as site of garment production and tailoring.

7 See, for example, C. Hamnett (2001) 'Social Segregation and Social Polarization'.

8 A manager of the Delfina Trust, for example, spoke of the rapidly inflating residential price points of Bermondsey, a hitherto unfashionable part of London, which made it increasingly difficult to attract and retain staff. She added that a mitigating factor was the Jubilee Line, which enabled access between Bermondsey and less costly housing in north-west London.

9 The Zandra Rhodes project was Legoretta's first in Europe, with a second now underway in Bilbao. Kenneth Powell suggests that the use of bright orange and pink as external colouring was prima facie 'the antithesis of good taste', but also asserts that '[t]his development could be on the edge of downtown Los Angeles but looks good right where it is' (Powell 2005: 62), demonstrating the transnational mobility of design values.

10 A sign attached to the wall of the bakery exhorts delivery truck drivers to be mindful of the residential neighbours in this southern, rapidly gentrifying, part of Bermondsey Street.

11 While not precisely a 'chain', Britannia Row 2 was descended from the original Britannia Row project, located in Islington north of the Pentonville Road.

6 Inscriptions of restructuring in the developmental state

1 Singapore's sharp mid-1980s recession, its origins in the economic problems of the larger regional hinterland, and the policy response privileging specialized services, international gateway functions, and technology-intensive manufacturing, was mirrored by the experiences of Vancouver, on the opposite littoral of the Pacific; see Chapter 8 for an elaboration of this story.

2 A release (2006) of the Singapore Economic Development Board cited 2,000 Chinese companies domiciled in Singapore, attracted by Singapore's commitment to functioning as 'China's Internationalisation Springboard'. The report ('New Geographies', EDB: 2006) quotes Ms Mary Ma, Senior Vice-President of Lenovo, a Chinese supply chain control centre for personal computing products, as follows: 'Locating our Asia Pacific, and Global Supply Chain, Headquarters in Singapore will allow us to further our mission of delivering innovative, high-quality products and world-class service to our customers, while also benefiting from the operational efficiencies this location provides.'

3 This program of slum clearance has an earlier provenance, in the work of the colonial Singapore Improvement Trust (SIT) following the Second World War. In 1948, the SIT demolished 102 dwellings and shops, 'only the first installment of a program for dealing with all such slum properties' (*The Work of the Singapore Improvement Trust*, 1948: 3–4, cited in Clancey 2004: 39).

4 While many of those rehoused in the new estates were likely satisfied with the quality of new premises, the SIT slum clearance and eviction of 'squatters' created resistance among displaced populations. Clancey notes that organized resistance (seen, for example, in the 'Attap [= palm leaf] Dweller's Association') took place 'against a larger backdrop of anti-colonial protests, strikes and riots, which the authorities were attempting to contain through a gradually-expanding electoral process' (ibid.: 39).

5 The alienation and sense of dislocation experienced by some of the former residents having occasion to revisit the conservation areas are captured in the testimony of Tang Wai Yin. Tang, who had spent the first twelve years of his life in No. 26, Ann Siang Road, part of Telok Ayer, visited his former shophouse dwelling for the first time in a decade, and had this to say about his experience:

> I shared the third floor with 16 other people and various creepy crawlies. Living there meant having mice skitter across the ancient wooden floorboards and cockroaches flying overhead. A good flushing system meant a strong hand hauling a large pail of water. The walls had huge cracks extending from floor to ceiling and someone's arrival was always heralded by the creaking of the stairs. That may not seem like fun, but those were some of the happiest years. It was in that house where, as a five-year old, I watched fascinated as my brother fed his fat pet gecko which lived in the cracks behind our wooden kitchen cabinet . . . [the] building had dignity, despite its age and condition: now it doesn't. With a sweep of the paintbrush all that has changed . . . It has become a house that loudly proclaims its ugliness, shouting its lack of taste to all who walk past and are attacked by the sledgehammer effects of its colour.
>
> (*The New Paper*, 23 April 1994: 42)

6 The initial outlay for restoration was S$3 million, to cover the cost of skilled artisanal workers from China as well as material expenditures.

7 'Saffron Hill' also exists (as a street) in our Clerkenwell site (Chapter 5), just off the Farringdon Road, but the Singapore entity seems likely to have a stronger claim of regional authenticity.

8 This idea of the 'recombinant economy' of the contemporary inner city will be elaborated in Chapter 9, the conclusion to this volume.

7 The New Economy and its dislocations in San Francisco's South of Market Area

1 Despite its swings of economic fortune, San Francisco consistently ranks as the favourite American destination for international tourists, based on its unique 'package' of high urban style and consumption amenities, the proximate attractions of Marin, Sonoma and Napa Counties, and a Mediterranean climate which facilitates year-round visitations.

2 The shift in economic fortunes between San Francisco and Los Angeles follows in part Sydney's primacy over Melbourne, and Toronto's supplanting of Montréal as Canada's dominant business centre, shaped by a mix of market factors, migration and population growth, locational attributes, and (especially in the Canadian example) political factors. In scalar terms, California's population and economy are larger than that of Canada or Australia, and indeed the Golden State has been interpreted as a 'nation' as well as a constituent state of the American Union. As an expression of this cultural separateness, a woman staffing the California history society in SOMA told me that 'People come to California to get away from the U.S.'

3 San Francisco is both a county and a city for administrative and statistical purposes, with a Board of Supervisors rather than a City Council, reflecting its County status.

4 Earnings (wages and salaries) for workers in San Francisco are typically 30 per cent higher than the national average, another measure of the high levels of professionalization within the City's labour force.

5 Hartman notes that 58 hotels and 80 lodging houses were constructed in the South of Market by 1907, juxtaposed within a proliferation of new manufacturing industries, wholesalers, and warehousing (2002: 58).

6 This prejudicial dismissal was untrue, as numerous studies demonstrated the length of tenure and relatively positive social conditions for many of the area's residents.

7 NEMIZ is a planning acronym for 'North East Mission Industrial Zone', an area which experienced its own New Economy phase of new media development, artistic production, and associated consumption, as elucidated by Peter Cohen.

8 As Parker and Pascual observed:

> For builders, live-work was a popular type of development. Considered by planners to be commercial rather than residential development, the building codes for live-work did not require creation of open space, payment of school tax fees, notification of neighbours, conditional review, or parking development. In addition, live-work development could be built higher than any other type of development, aesthetic concerns were minimal, and such development was permitted in industrial areas of the city with few existing residents and thus little risk of organized resistance. Finally, banks allowed developers to finance live-work construction with residential loans rather than commercial ones, and it is much easier to secure a residential construction loan.
>
> (1999: 58)

9 The South End Warehouse recalls the layers of historical description pertaining to the larger area *south* of Market Street. The 'South End' term used in the area's heyday as warehousing and transportation district has been effectively supplanted by 'South Beach' and 'South Park', the latter including elements of the former South End as well as the much tighter designation of the area bounded by Second and Third Streets and Bryant and Brannan.

10 In about twenty visits to South Park over six years I never witnessed any overt conflict between the groups of homeless blacks congregating on the western margins of the

Park and other users, and overall the social tone in the area was palpably one of toler-ance. That said, their presence was a harsh reminder of the gaps in welfare and lifestyle between groups co-existing in South Park and, indeed, the larger City. The inclusive-ness that seemed to encompass a mix of workers and residents did not evidently extend to the homeless.

11 Larissa Sand confirmed that her employees enjoyed the amenity of the park, 5 metres from the entrance to the workshop, and observed that the park spaces functioned as a site for sharing knowledge, including work opportunities.

12 Fuseproject is now located on Second Street, not far from the South Park enclave.

13 On a return visit to South Park in April 2007, I discovered that Jumbo Shrimp had moved to larger premises in South Park, close to the long-established ISDA fashion design and retail business, affirming a commitment to South Park as 'special place' amid the mosaic of economic spaces in the city.

8 New industry formation and the transformation of Vancouver's metropolitan core

1 Vancouver has not experienced a significant downturn since the deep recessionary period 1982–1984, the latter precipitated by a commodity price shock affecting British Columbia's resource economy, and speedily transmitted to Vancouver *via* the multiple networks (processing, finance, services, transportation) linking 'core' and 'periphery' in a staples economy setting. During the recession of the early 1980s, the worst since the Great Depression of the 1930s, unemployment in Vancouver approached 14 per cent, exacerbated by a inflationary credit cycle that saw interest rates for borrowing exceed 20 per cent.

2 For a comparative analysis of industrial restructuring and employment change within Canada's metropolitan cities, see W.J. Coffey, *The Evolution of Canada's Metropolitan Economies* (1994), and (a later work by the same author) *Employment Growth and Change in the Canadian Urban System, 1971–94* (1996).

3 The anomalous persistence of agriculture as a key primary industry in Metro Vancouver may be attributed in part to the high quality of alluvial soils prevalent in the Fraser River Delta, the role of the provincial Agricultural Land Commission (ALC) whose mandate is to protect prime farmland in the region against the encroachment of urbanization and industrialization, the force of local (municipal) zoning and land use policies favour-ing high-output agriculture, and the immigration of South Asians with an agricultural background, prominent in the region's farming industries both as owners and labourers.

4 Metro Vancouver's production economy specializes in small- and medium-sized manu-facturing, including garment industries and food and beverage production, as well as technology-intensive manufacturing in telecommunications, rather than large Fordist plants and labour which have borne the brunt of restructuring processes in Canadian cities since the 1960s.

5 While Metro Vancouver's space-economy is shaped in part by multinucleation, the progress of many of the designated Regional Town Centres (RTCs) has been con-strained by relatively weak municipal policy commitments, with respect to zoning and land use, infrastructure provision, and amenity provision, so the metropolitan area is characterized in large part by dispersion and diffusion. Employment growth in the auto-oriented business and office parks has consistently exceeded that for the desig-nated RTCs, hampering the region's quest for reduced auto-dependency in the inter-ests of sustainable development values.

6 A comprehensive inventory of downloadable research papers prepared by Daniel Hiebert and other members of the 'Research on Immigration and Integration' Centre of Excellence at the University of British Columbia is accessible at: http://www.geog.ubc.ca/metropolis/atlas

7 See, for example, papers on suburban immigration experience in New York, Los Angeles, Silicon Valley, Sydney, Toronto, and Vancouver in Wei Li (ed.), *From Urban Enclave to Ethnic Suburb: New Asian Communities in Pacific Rim Countries* (2006).

8 Following Council's approval of the *Central Area Plan* in 1991, the City Planning Department undertook local planning studies for individual districts of the core, including the Downtown South, Victory Square, and the Downtown Eastside.

9 In advance of Vancouver's international exposition in 1986 a number of single-room occupancy hotels (SROs) were converted to hotels to cater to Expo 86's visitors, displacing long-standing, low-income residents in the process. The Government of British Columbia has purchased ten SRO hotels with a view to providing security of residential tenure in advance of the City's next 'hallmark event' (after Olds 2001), the 2010 Olympics.

10 The depth and complexity of socioeconomic problems in the DTES have given rise to the formation of a tri-level program, 'The Vancouver Agreement', which endeavours to combine the resources and regulatory powers of the federal, provincial and local governments to enhance possibilities of housing, education and training, economic development, and substance abuse treatment for residents of the area.

11 While housing affordability and the problems of security of residential tenure persist in Vancouver's marginal communities especially, David Ley acknowledges the achievements in providing social and non-market housing in the City over the past thirty years, the result of commitments shaped by a 'culture of morality'.

12 A review of the competition for land in Vancouver's downtown between residential and office uses noted that 'the ability to sell condos for $350 a square foot and more was trumping office space that rents for $25 a square foot per year'. Frances Bula, 'Downtowns at Crossroads', *The Vancouver Sun*, Wednesday, 29 March 2006: B2 and B3. Bula's interviews with real estate agents and property managers also suggested that 'conversions of use from office to residential reinforced this tendency favouring housing in the downtown, while the Vancouver office market destabilized as other owners toyed with the idea of converting to residential' (ibid.).

13 The accelerating globalization processes of the last two decades of the twentieth century, combined with Vancouver's peripheral location, and the boom-and-bust nature of British Columbia's staple economy, led to the stripping away of much of Vancouver's head office capacity, while the takeover of the Bank of British Columbia by the Hong Kong & Shanghai Banking Corporation represented both a highly symbolic event and the end of an institutional experiment in local capital formation. More recently, a Statistics Canada report observed that British Columbia suffered the largest drop in head office employees (over 3,100) of any Canadian province in the period 1999–2002, and further suggested that 'Calgary established itself as the leading head office centre in Western Canada, supplanting Vancouver' ('Space Cowboys', *The Vancouver Sun*, Saturday, 13 November 2004).

14 For a more comprehensive discussion of these processes and their theoretical reference points see Thomas Hutton, 'Post-industrialism, Post-modernism, and the Reproduction of Vancouver's Central Area: Retheorising the 21st Century Metropolis', *Urban Studies* 41 (2004), pp. 1953–1982, reprinted in Bruce Stiftel, Vanessa Watson and Henri Acselrad (eds), *Dialogues in Urban and Regional Planning*, London: Routledge (2007), pp. 25–68.

15 Several houses have sold in the $750,000–1,000,000 range in the last year, underscoring the steep gradient of inflation in what has historically been Vancouver's poorest neighbourhood by income.

References

Abu-Lughod, J. (1999) *New York, Chicago, Los Angeles: America's Global Cities.* Minneapolis: University of Minnesota Press.

Ackroyd, P. (2000) *London: The Biography.* London: Chatto & Windus, pp. 461–474.

Amin, A. (1998) 'Globalization and Regional Development: A Relational Perspective', *Competition and Change* 3: 145–165.

Arena, R. and Quéré, M. (eds) (2004) *The Economics of Alfred Marshall: Revisiting Marshall's Legacy.* New York: Palgrave Macmillan.

Barnes, T. (2001) 'Retheorizing Economic Geography: From the Quantitative Revolution to the "Cultural Turn" ', *Annals of the Association of American Geographers* 91: 546–565.

Barnes, T., Edgington, D., Denike, K. and McGee, T.G. (1992) 'Vancouver, the Province, and the Pacific Rim', in G. Wynn and T. Oke, *Vancouver and its Region.* Vancouver: UBC Press.

Bathelt, H. (2005) 'Cluster Relations in the Media Industry: Exploring the "Distanced Neighbour" Paradox in Leipzig', *Regional Studies* 39: 105–128.

Bathelt, H. and Boggs, J. (2003) 'Toward a Reconceptualization of Regional Development Paths: Is Leipzig's Media Cluster a Continuation or a Rupture with the Past?' *Economic Geography* 79: 265–293.

Bathelt, H. and Glückler, J. (2003) 'Toward a Relational Economic Geography', *Journal of Economic Geography* 3: 117–144.

Baum, S. (1999) 'Social Transformations in the Global City: Singapore', *Urban Studies* 36: 1095–1118.

Becattini, G. (1989) 'Sectors and/or Districts: Some Remarks on the Conceptual Foundations of Industrial Economics', in E. Goodman, J. Bamford and P. Saynor (eds), *Small Firms and Industrial Districts in Italy.* New York and London: Routledge.

Becattini, G. (1990) 'The Marshallian Industrial District as a Socio-Economic Notion', in F. Pyke and G. Becattini (eds) *Industrial Districts and Inter-Firm Cooperation in Italy.* Geneva: International Institute for Labour Studies.

Bell, D. (1973) *The Coming of Postindustrial Society: A Venture in Social Forecasting.* New York: Basic Books.

Bell, D. and Jayne, M. (2004) 'Conceptualizing the City of Quarters', in D. Bell and M. Jayne (eds), *City of Quarters: Urban Villages in the Contemporary City.* Aldershot: Ashgate.

Bellandi, M. (2004) 'Some Remarks on Marshallian External Economics and Industrial Tendencies', in R. Arena and M. Quéré (eds) *The Economics of Alfred Marshall: Revisiting Marshall's Legacy.* New York: Palgrave Macmillan.

Bentele, G., Liebert T. and Polifke M. (2003) *Medienstandort Leipzig IV: Studie zum Cluster Medien/Kommunikationstechnik/IT 2002* [Leipzig as a Location of the Media Industry IV: A Study of the Media/Communication/Information Technology Cluster 2002]. Leipzig: Mimeo (available at http://www.leipzig.de/de/business/zukunfit/ Medienstudie_IV.pdf) (accessed on 26 August 2004).

Beyers, W. (2000) 'Cyberspace or Human Space: Whither Cities in the Age of Telecommunications?', in J. Wheeler, Y. Aoyama, and B. Warf (eds), *Cities in the Telecommunications Age*. London: Routledge.

Bianchini, F. and Ghilardi, P. (2004) 'The Culture of Neighbourhoods: A European Perspective', in D. Bell and M. Jayne (eds), *City of Quarters: Urban Villages in the Contemporary City*. Aldershot: Ashgate.

Blomley, N. (2004) *Unsettling the City: Urban Land and the Politics of Property*. New York and London: Routledge.

Bluestone, B. and Harrison, B. (1982) *The Deindustrialization of America: Plant Closing, Community Abandonment, and the Dismantling of Basic Industry*. New York: Basic Books.

Boggs, J. (2001) 'Path Dependency and Agglomeration in the German Book Publishing Industry', paper presented to the Annual Meeting of the Association of American Geographers, New York.

Boschma, R. (2005) 'Proximity and Innovation: A Critical Assessment', *Regional Studies* 39: 61–74.

Bourne, L.S. (ed.) (1982) *Internal Structure of the Metropolis: Readings on Urban Form, Growth and Policy*, 2nd edn. Oxford: Oxford University Press.

Brail, S. (1994) 'Creative Services in Vancouver: a Case Study of Yaletown and Victory Square', MA thesis, University of British Columbia.

Braudel, F. (1982) *Civilization and Capitalism 15th–18th Century:* Vol. II, *The Wheels of Commerce*. London: William Collins & Sons.

Brechin, G. (1999) *Imperial San Francisco: Urban Power, Earthly Ruin*. Berkeley, CA: University of California Press.

Bridge, G. (2001) 'Estate Agents as Interpreters of Economic and Cultural Capital: The Gentrification Premium in the Sydney Housing Market', *International Journal of Urban and Regional Research* 25: 87–101.

Buck, N., Gordon, K. and Young, J. (1986) *The London Employment Problem*. Oxford: Clarendon Press.

Butler, T. (2003) 'Living in the Bubble: Gentrification and its "Others" in North London', *Urban Studies* 40: 2469–2486.

Butler, T. and Lees, L. (2006) 'Supergentrification in Barnsbury, London: Globalization and Gentrifying Global Elites at the Neighbourhood Level', *Transactions of the Institute of British Geographers* 31: 467–487.

Caldari, K. (2004) 'Review of Richard Arena and Michel Quéré (eds) *The Economics of Alfred Marshall: Revisiting Marshall's Legacy*. Economic History Series. September 16, 2004, URL: http://www.eh.net/bookreviews/library/0842.shtml

Camagni, R. (ed.) (1991) *Innovation Networks: Spatial Perspectives*. London: Belhaven.

Capello, R. and Faggian, A. (2005) 'Collective Learning and Relational Capital in Local Innovation Processes', *Regional Studies* 39: 75–87.

Cherry, B. and Pevsner, N. (1983) *The Buildings of England. London 2: South*. London: Penguin Books.

Cherry, B. and Pevsner, N. (1998) *The Buildings of England. London 4: North*. London: Penguin Books.

Chiang, L. (2007) 'A Reconsideration of "Old Economy" Industries within "New Economy" Regions', *Geography Compass* (in press).

Christopherson, S. (2003) 'Review of J. Cowie and Joseph Heathcott (eds), *Beyond the Ruins: The Meanings of Deindustrialization.* (Ithaca, NY, Cornell University Press 2003)', *Journal of the American Planning Association* 70: 487.

City Fringe Partnership (2005) *Analysing the Creative Sector in the City Fringe.* London: TBR Economics.

City Fringe Regeneration Audit Group (1997) *An Overview of the City Fringe Economy: Report of the Audit Group, with Strategic Policy Unit.* London: London Borough of Hackney.

City of London Corporation (2004) *Understanding the Print and Publishing Sectors in the City Fringe.* London: Prepared by Burns Owens Partnership, Parker Associates Economics and Strategy Limited, and Experian Business Strategies.

City of London Corporation (2005) *London's Place in the UK Economy, 2005–06.* London: Oxford Economic Forecasting.

City of San Francisco (1987) *Industrial Employment Bulletin.* San Francisco: City of San Francisco Planning Department.

City of San Francisco (1997) *Multimedia in San Francisco.* San Francisco: City of San Francisco Planning Department.

City of San Francisco (1999) *Zoning Options for Industrial Land: Industrial Protection Zones and Mixed-Use Areas.* San Francisco: City of San Francisco Planning Department (8 April).

City of Vancouver (1982a) *The Core Employment Trend.* Coreplan Background Report (June). Vancouver: City of Vancouver Planning Department.

City of Vancouver (1982b) 'Industry on the North Shore of False Creek', in *Quarterly Review* (October) Vancouver: City of Vancouver Planning Department.

City of Vancouver (1991) *Central Area Plan.* Vancouver: City of Vancouver Planning Department (December).

City of Vancouver (1998) *High-Tech Industry in the Urban Context: A Discussion Paper.* Vancouver: City of Vancouver Planning Department (September).

City of Vancouver (1999) *Council Report: False Creek Flats 1–3 Zone.* Vancouver: City Planning Department, City Manager's Office, General Manager of Community Services.

City of Vancouver (2005) 'Metropolitan Core Jobs, Economy and Land Use' (staff bulletin). Vancouver: City of Vancouver Planning Department, Central Area Division.

Clancey, G. (2004) 'Toward a Spatial History of Emergency: Notes from Singapore', in R. Bishop, J. Phillips and W.-W. Yeo, *Beyond Description: Singapore Space Historicity.* London: Routledge, p. 39.

Coe, N. (2001) 'A Hybrid Agglomeration? The Development of a Satellite-Marshallian Industrial District in Vancouver's Film Industry', *Urban Studies* 38: 1753–1775.

Coffey, W.J. (1994) *The Evolution of Canada's Metropolitan Economies.* Montreal: Institute for Research on Public Policy.

Coffey, W.J. (1996) *Employment Growth and Change in the Canadian Urban System, 1971–94,* Working Paper No. W/02, Ottawa: Canadian Policy Research Networks.

Cohen, P. (1998) 'A Transforming San Francisco Industrial Landscape', *Pacific (Journal of the Association of Pacific Coast Geographers)*, Fall: 1: 712.

Cooke, P. and Morgan, K. (1998) *The Associational Economy: Firms, Regions and Innovation.* Oxford: Oxford University Press.

Costa, P. (2004) 'Milieu Effects and Sustainable Development in a Cultural Quarter: the

"Bairro Alto–Chiado" Area in Lisbon', paper presented to the Annual Meeting of the Association of American Geographers, Philadelphia, PA (March).

Crampton, G. and Evans, A. (1992) 'The Economy of an Agglomeration: The Case of London', *Urban Studies* 29: 259–271.

Cybriwsky, R., Ley, D. and Western, S. (1986) 'The Political and Social Construction of Revitalized Neighbourhoods: Society Hill, Philadelphia, and False Creek, Vancouver', in N. Smith and P. Williams (eds), *Gentrification of the City*. Boston, MA: Allen & Unwin.

Daniels, P. (1975) *Office Location: An Urban and Regional Study*. London: Bell.

Daniels, P. (1985) *Service Industries: A Geographical Appraisal*. London: Methuen.

Davis, H.C. and Hutton, T.A. (1991) 'Producer Service Exports from the Vancouver Metropolitan Region', *Canadian Journal of Regional Science* XIV: 371–389.

Dear, M. and Flusty, S. (1998) 'Postmodern Urbanism', *Annals of the Association of American Geographers* 88: 50–72.

Dennis, R. (1978) 'The Decline of Manufacturing Employment in Greater London', *Urban Studies* 15: 63–73.

Department for Culture, Media and Sport (2001) *Creative Industries Mapping Document*, London: DCMS and the Creative Industries Task Force.

Development Securities (2006) 'Who Owns the City?', cited in the *National Post*, Toronto: June 5 2006: p. FP5.

Doel, M. and Beaverstock, J. (1999) 'London and New York's Response to the Asian Financial Crisis', paper presented to the 95th Annual Meeting of the Association of American Geographers, Honolulu, HA: 23–27 March.

Drennan, M. (1987) *The Performance of Metropolitan Area Industries*. New York: Federal Reserve Bank of New York.

Driver, S. and Gillespie, A. (1993) 'Information and Communication Technologies and the Geography of Magazine Print Publishing', *Regional Studies* 27: 53–64.

Economist, The (2006) 'Chicago Bulls', *The Economist*, October 21–27: 100.

Elliott, P. (2005) 'Intra-Metropolitan Agglomerations of Producer Services Firms: The Case of Graphic Design Firms in Metropolitan Melbourne, 1981 to 2001', Master of Planning and Design thesis, University of Melbourne, Faculty of Architecture, Building and Planning.

Erie, S.P. (2004) *Globalizing L.A.: Trade, Infrastructure and Regional Development*. Stanford, CA: Stanford University Press.

Evans, G. (2001) *Cultural Planning: An Urban Renaissance?* London: Routledge.

Evans, G. (2003) 'Hard Branding the Cultural City: From Prado to Prada', *International Journal of Urban and Regional Research* 27: 417–440.

Evans, G. (2004) 'Cultural Industry Quarters: From Pre-Industrial to Post-Industrial Production', in D. Bell and M. Jayne (eds), *City of Quarters: Urban Villages in the Contemporary City*. Aldershot: Ashgate.

Fainstein, S. and Harloe, M. (2000) 'Ups and Downs in the Global City: London and New York at the Millennium', in G. Bridge and S. Watson (eds) *A Companion to the City*. Oxford: Blackwell.

Fainstein, S., Harloe, M. and Gordon, I. (eds) (1992). *Divided Cities: New York and London in the Contemporary World*. Oxford: Blackwell.

Florence, P.S. (1948) *Investment, Location and Size of Plant*. Cambridge: Cambridge University Press.

Florida, R. (2002) *The Rise of the Creative Class: And How It's Transforming Work, Leisure, Community and Everyday Life*. New York: Basic Books.

Foord, J., Evans, G. *et al.* (2005) *Creative Spaces: Phase 1 Report*. London: London Development Agency. www.creativelondon.org.uk

Fost, D. (2006) 'Web 2.0 Has a Local Address: South Park, the Neighborhood that Fostered the dot.com Boom Is Back', *San Francisco Chronicle*, 16 April.

Fothergill, S. and Gudgin, G. (1982) *Unequal Growth: Urban and Regional Employment Change in the UK*. London: Heinemann.

Fox, C. (ed.) (1992) *London: World City 1800–1840*. New Haven, CT: Yale University Press.

Friedmann, J. and Wolff, K. (1982) 'World City Formation: An Agenda for Research and Action', *International Journal of Urban and Regional Research* 6: 309–344.

Fröbel, F., Heinrichs, J. and Kreye, O. (1980) *The New International Division of Labour*. Cambridge: Cambridge University Press.

Gad, G. (1979) 'Face-to-face Linkages and Office Decentralization Potentials: A Study of Toronto', in P.W. Daniels, *Spatial Patterns of Office Growth and Location*. London: Wiley.

Gerber, J. (2004) 'The Shrinking and Fading Garment Center', *The New York Times*, 23 August 2004: B1 + B6.

Gertler, M. (1995) ' "Being There": Proximity, Organization and Culture in the Development and Adoption of Advanced Manufacturing Technologies', *Economic Geography* 71: 1–26.

Gertler, M. (2003) 'Tacit Knowledge and the Economic Geography of Context, or the Undefinable Tacitness of (Being) There', *Journal of Economic Geography* 3: 79–99.

Glaeser, E.L, Kolko, J. and Saiz, A. (2001) 'Consumer City', *Journal of Economic Geography* 1: 27–50.

Glass, R. (1963) 'Introduction' to *London: Aspects of Change*. London: Centre for Urban Studies, reprinted in R. Glass (ed.) (1989) *Clichés of Urban Doom*. Oxford: Blackwell.

Goddard, J. (1967) 'Changing Office Location Patterns within Central London', *Urban Studies* 4: 276–284.

Goddard, J. (1973) 'Office Linkages and Location: a Study of Communications and Spatial Patterns in Central London', *Progress in Planning* 1: 109–232.

Gottmann, J. (1961) *Megalopolis: The Urbanized Northeastern Seaboard*. New York: The Twentieth Century Fund.

Gottmann, J. (1970) 'Urban Centality and the Interweaving of Quaternary Activities', *Ekistics* 29: 322–331.

Grabher, G. (2001) 'Ecologies of Creativity: The Village, the Group, and the Heterarchic Organisation of the British Advertising Industry', *Environment and Planning A* 33: 351–374.

Grabher, G. (2002) 'Cool Projects, Boring Institutions: Temporary Collaboration in Social Context', *Regional Studies* 36: 205–214.

Graham, S. and Marvin, S. (2001) *Splintering Urbanism: Networked Infrastructures, Technological Mobilities, and the Urban Condition*. London: Routledge.

Granovetter, M. (1985) 'Economic Action and Social Structure: The Problem of Embeddedness', *American Journal of Sociology* 91: 481–510.

Greater London Authority (2006) *GLA Economics Annual Report 2006*, London: GLA Economics.

Gripaios, P. (1977) 'Industrial Decline in London: An Examination of its Causes', *Urban Studies* 14: 181–189.

Hackworth, J. and Smith, N. (2001) 'The Changing State of Gentrification', *Tidschrift voor Economische en Sociale Geografie* 92: 464–477.

Haig, R.M. (1927) *Major Economic Factors in Metropolitan Growth and Arrangement.* New York: Regional Plan of New York and its Environs.

Hall, C. (2006) 'Urban Entrepreneurship, Corporate Interests and Sports Mega-events: The Thin Policies of Competitiveness within the Hard Outcomes of Neoliberalism', *The Sociological Review* 54: 59–70.

Hall, P.G. (1960) 'The Location of the Clothing Trades in London, 1861–1951', *Transactions and Papers of the Institute of British Geographers*, Publication No. 28.

Hall, P.G. (1962a) *The Industries of London since 1861.* London: Hutchinson.

Hall, P.G. (1962b) 'The East London Footwear Industry: An Industrial Quarter in Decline', *East London Papers* 5: 3–21.

Hall, P.G. (1966) *The World Cities.* London: Weidenfeld & Nicolson.

Hall, P.G. (1997) 'The Future of the Metropolis and its Form', *Regional Studies* 31; reprinted in *Regional Studies* 40th Anniversary Classic Papers Supplement 41: S137–S146.

Hall, P.G. (1998) *Cities in Civilisation.* London: Weidenfeld & Nicolson.

Hall, P.G. (2000) 'Creative Cities and Economic Development', *Urban Studies* 37: 639–651.

Hall, P.G. (2006) 'The Polycentric City', PowerPoint presentation, The Bartlett School, University College London.

Hall, P.G., Thomas, R., Gracey, H. and Drewett, R. (1973) *The Containment of Urban England*, vol. 2, *The Planning System: Objectives, Operations, Impacts.* London: George Allen & Unwin/Sage Publications.

Hamnett, C. (1991) 'The Blind Men and the Elephant: Toward a Theory of Gentrification', *Transactions of the Institute of British Geographers* 16: 173–189.

Hamnett, C. (1994) 'Socio-economic Change in London: Professionalisation not Polarisation', *Built Environment* 20: 192–203.

Hamnett, C. (2001) 'Social Segregation and Social Polarization', in R. Paddison (ed.) *Handbook of Urban Studies.* London: Sage.

Hamnett, C. (2003) *Unequal City: London in the Global Arena.* London: Routledge.

Hamnett, C. (2006) 'Loft Conversion in London: From Industrial to Postindustrial', Kings College London, Department of Geography.

Hamnett, C. and Whitelegg, A. (in press) 'Loft Living in London: An Analysis of City Centre Gentrification', *Urban Studies.*

Hardwick, W. (1974) *Vancouver.* Don Mills, ON: Collier-Macmillan.

Harrison, B. (1992) 'Industrial Districts: Old Wine in New Bottles?', *Regional Studies* 26: 469–483.

Hartman, C. with S. Carnochan (2002) *City for Sale: The Transformation of San Francisco.* Berkeley, CA: University of California Press.

Harvey, D. (1989) 'From Managerialism to Entrepreneurialism: Transformation in Urban Governance in Late Capitalism', *Geografiska Annaler Series B-Human Geography* 88B: 145–158.

Harvey, D. (2003) *Paris: Capital of Modernity.* London and New York: Routledge.

Helbrecht, I. (2004) 'Bare Geographies in Knowledge Societies – Creative Cities as Text and Piece of Art: Two Eyes, One Vision', *Built Environment* 30: 191–200.

Hiebert, D. (2005) 'Migration and the Demographic Transformation of Canadian Cities: The Social Geography of Canada's Major Metropolitan Centres in 2017. Vancouver: Research on Immigration and Integration in the Metropolis, University of British Columbia, Working Paper No. 05–14.

Ho, K.C. (1994) 'Industrial Restructuring, the Singapore City-State, and the Regional Division of Labour', *Environment and Planning A* 26: 33–51.

Ho, K.C. (2005) 'Service Industries and Occupational Change; Implications for Identity, Citizenship and Politics', in P.W. Daniels, T.A. Hutton, and K.C. Ho, *Service Industries and Asia-Pacific Cities: New Development Trajectories.* London and New York: Routledge.

Ho, K.C. (2007) 'The Neighbourhood in the Creative Economy', Working Paper, Department of Sociology, National University of Singapore.

Hoover, E.M. and Vernon, R. (1959) *Anatomy of a Metropolis.* Cambridge, MA: Harvard University Press.

Hubbard, P. (1996) 'Urban Design and City Regeneration: Social Representation of Entrepreneurial Landscapes', *Urban Studies* 35: 1441–1461.

Hutton, T.A. (1997) 'The Innisian Core–Periphery Revisited: Vancouver's Changing Relationships with British Columbia's Staple Economy', *BC Studies* 113(Spring): 69–100.

Hutton, T.A. (2000) 'Reconstructed Production Landscapes in the Postmodern City: Applied Design and Creative Services in the Metropolitan Core', *Urban Geography* 21: 285–317.

Hutton, T.A. (2004a) 'The New Economy of the Inner City', *Cities* 21: 89–108.

Hutton, T.A. (2004b) 'Post-industrialism, Post-modernism, and the Reproduction of Vancouver's Metropolitan Core: Retheorising the 21st Century City', *Urban Studies* 41: 1953–1982.

Hutton, T.A. (2004c) 'Service Industries, Globalization, and Urban Restructuring within the Asia-Pacific: New Development Trajectories and Planning Responses', *Progress in Planning* 61: 1–74.

Hutton, T. A. (2006) 'Spatiality, Built Form and Creative Industry Development in the Inner City', *Environment and Planning A* 38: 1819–1841.

Hutton, T.A. and Ley, D.F. (1987) 'Location, Linkages and Labor: The Corporate Office Complex in a Medium Size City, Vancouver', *Economic Geography* 63: 126–141.

Indergaard, M. (2004) *Silicon Alley: The Rise and Fall of a New Media District.* London and New York: Routledge.

Indergaard, M. (forthcoming) 'What to Make of New York's New Economy? The Politics of the Creative Field'.

Istituto Nazionale per il Commercio Estero (1988) *The Italian Model: Industrial Organisation and Production.* Rome: INCE and Centro di Economia e Politica Industriale, University of Bologna.

Jacobs, J. ([1961], 1984) *The Death and Life of Great American Cities: The Failure of Modern Town Planning.* London: Peregrine Books.

Jacobs, J.M. (1996) *Edge of Empire: Postcolonialism and the City,* London and New York: Routledge.

Jones, C. (2005) 'Lower Manhattan Still on the Rebound after 9/11', *USA Today* 14 October: 5A.

Jorgenson, D.W. (2001) 'Information Technology and the U.S. Economy', *American Economic Review* 91: 1–32.

Judt, T. (2005) *Postwar: A History of Europe since 1945.* London: Penguin.

Kaika, M. and Thielen, K. (2005) 'Form Follows Power: A Genealogy of Urban Shrines', *City* 10: 59–69.

Keeble, D. (1978) 'Industrial Decline in the Inner City and Conurbation', *Transactions of the Institute of British Geographers* 3 (ns): 101–114.

Keltie, L. (2006) 'The Contribution of Managed Workspace to Urban Regeneration',

MSc thesis, Faculty of the Built Environment, Bartlett School, University College London.

Krugman, P. (1994) 'The Myth of Asia's Miracle', *Foreign Affairs* 73: 62–78.

Lampard, E.E. (1955) 'The History of Cities in Economically Advanced Areas', *Economic Development and Cultural Change*, 3: 81–136.

Leaf, M. (2005) 'The Bazaar and the Normal: Informalization and Tertiarization in Urban Asia' in P. Daniels, K.C. Ho and T.A. Hutton, *Service Industries and Asia-Pacific Cities: New Development Trajectories*. London and New York: Routledge.

Lefèbvre, H. (1974) *Production de l'espace*, Paris: Anthropos; published in English as *The Production of Space*, trans. D. Nicholson-Smith (1991) Oxford: Blackwell.

Ley, D.F. (1980) 'Liberal Ideology and the Post-Industrial City', *Annals of the Association of American Geographers* 70: 238–258.

Ley, D.F (1994) 'The Downtown Eastside: "One Hundred Years of Struggle"', in S. Hassan and D. Ley (eds) *Neighbourhood Organizations and the Welfare State*. Toronto: University of Toronto Press.

Ley, D.F. (1996) *The New Middle Class and the Remaking of the Central City*. Oxford: Oxford University Press.

Ley, D.F. (2003) 'Artists, Aestheticization and the Field of Gentrification', *Urban Studies* 40: 2527–2544.

Li, W. (ed.) (2006) *From Urban Enclave to Ethnic Suburb: New Asian Communities in Pacific Rim Countries*. Honolulu: University of Hawai'i Press.

Lloyd, R. (2006) *Neo-Bohemia: Art and Commerce in the Postindustrial City*. New York and Abingdon: Routledge.

Lomas, G. (1975) *The Inner City*. London: London Council of Social Services.

London Development Authority (2005) *Production Industries Strategy for London*. London: LDA and the Office of the Mayor of London.

Lorenz, E. (1988) 'Neither Friends nor Strangers: Informal Networks of Subcontracting in French Industry', in D. Gambetta (ed.) *Trust: Making and Breaking Co-operative Relations*. Oxford and New York: Basil Blackwell.

Lorenz, E. (1989) 'The Search for Flexibility: Subcontracting Networks in French and British Engineering', in P. Hirst and J. Zeitlin (eds) *Reversing Industrial Decline?* Oxford: Berg.

Markus, T. (1994) *Buildings and Power: Freedom and Control in the Origin of Modern Building Types*. London: Routledge.

Markusen, A. (1996) 'Sticky Places in Slippery Spaces: A Typology of Industrial Districts', *Economic Geography* 72: 293–313.

Markusen, A. and Schrock, G. (2004) 'The Artistic Dividend: Urban Artistic Specialization and Economic Development Implications', presentation to the North American Regional Science Council's annual meeting, Seattle, WA: October.

Marshall, A. ([1890] 1972) *Principles of Economics*. London: The Macmillan Press.

Marshall, A. (1927) *Industry and Trade: A Study of Industrial Technique and Business Organization; and Their Influences on the Conditions of Various Classes and Nations*, 3rd edn. London: Macmillan.

Martin, J.E. (1964) 'The Industrial Geography of Greater London', in R. Clayton (ed.) *The Geography of Greater London*. London: George Philip and Son Limited, pp 111–142.

Massey, D. (1984) *Spatial Divisions of Labour: Social Structures and the Geography of Production*. London: Macmillan.

Massey, D. and Meegan, R. (1980) 'Industrial Restructuring Versus the Cities', in A. Evans and D. Eversley (eds) *The Inner City: Employment and Industry*. London: Heinemann.

Massey, D. and Meegan, R. (1982) *The Anatomy of Job Loss*. London: Methuen.

Mitchell, D. (2003) *The Right to the City: Social Justice and the Fight for Public Space*. New York: Guilford Press.

Molotch, H. (1996) 'LA as Design Product', in A. Scott and E. Soja (eds), *The City: Los Angeles and Urban Theory at the End of the Twentieth Century*. Berkeley, CA: University of California Press.

Morgan, K.J. (2004) 'The Exaggerated Death of Geography: Learning, Proximity and Territorial Innovation Systems', *Journal of Economic Geography* 4: 3–21.

Morgan, K.J. (2006) *Devolution and Development: Territorial Justice and the North-South Divide*, Cardiff: School of City and Regional Planning, Cardiff University.

Mowery, D.C. (2001) 'U.S. Postwar Technology Policies and the Creation of New Industries', in *Creativity, Innovation and Job Creation*. Paris: OECD.

Mugerauer, R. (2000) 'Milieu Preferences among High-Technology Companies', in J. Wheeler, Y. Aoyama and B. Warf (eds), *Cities in the Telecommunications Age*. London: Routledge.

Nicholson, B.M., Brinkley, I. and Evans, A. (1981) 'The Role of the Inner City in the Development of Manufacturing Industry', *Urban Studies* 18: 57–71.

Norcliffe, G. and Eberts, D. (1999) 'The New Artisan and Metropolitan Space: The Computer Animation Industry in Toronto', in J.-M. Fontan, J.-L. Klein and D.-G. Tremblay (eds) *Entre la métropolisation et la village global: Les scènes territoriales de la reconversion*. Québec: Presses de l'Université du Québec.

Office of the Deputy Prime Minister (2006) *State of the English Cities*, vols 1 and 2. London: The Stationery Office.

Olds, K. (2001) *Globalization and Urban Change: Capital, Culture, and Pacific Rim Megaprojects*. Oxford: Oxford University Press.

Olson, D. (1982) *Town Planning in London: The Eighteenth and Nineteenth Centuries*. New Haven, CT: Yale University Press.

Ott, T. (2001) 'From Concentration to De-concentration – Migration Patterns in the Post-socialist City', *Cities* 18: 403–412.

Oxford Economic Forecasting (2006) *London's Place in the UK Economy, 2005–06*. London: City of London Corporation.

Papin, P. (2001) *Histoire de Hanoi*. Paris: Fayard.

Parker, C. and Pascual, A. (1999) 'A Voice that Could Not Be Ignored: Community GIS and Gentrification Battles in San Francisco.', in D. Weiner, W.J. Craig and T.M. Harris (eds), *Community Participation and Geographical Information Systems*. London: Taylor & Francis.

Peck, J. (2001) 'Neoliberalizing States: Thin Policies/Hard Outcomes', *Progress in Human Geography* 253: 445–455.

Peck, J. (2005) 'Struggling with the Creative Class', *International Journal of Urban and Regional Research* 29: 740–770.

Perroux, F. (1955) 'Note sur la Notion de "Pôle de Croissance" ', *Economic Appliqué*, Jan.–June: 307–320.

Perry, M., Kong, L. and Yeoh, B. (1997) *Singapore: A Developmental City-State*. Chichester: Wiley.

Phillips, J. (2005) 'The Future of the Past: Archiving Singapore', in M Crinson (ed.), *Urban Memory: History and Amnesia in the Modern City*. London and New York: Routledge.

Piore, M.J. and Sabel, C.F. (1984) *The Second Industrial Divide: Possibilities for Prosperity*. New York: Basic Books.

Polanyi, K. (1944) *The Great Transformation*. New York: Holt, Rinehart.

Pope, N. (2002) 'A Tale of Two Towns: Conflations of Global and Local in the New Media Spaces of Belltown and Yaletown', MA thesis, School of Community and Regional Planning, University of British Columbia.

Post, A. (1999) *Fringe Benefits: Property Market Competition and Industrial Innovation in London's City Fringe*. London: London School of Economics and Political Science, Urban and Metropolitan Research.

Powell, K. (2005) *New London Architecture*. London and New York: Merrell Publishers.

Power, D. and Scott, A.J. (eds) (2004) *Cultural Industries and the Production of Culture*. London and New York: Routledge.

Pratt, A.C. (1997) 'The Cultural Industries Production System: A Case Study of Employment Change in Britain', *Environment and Planning A* 29: 1953–1974.

Pratt, A.C. (2000) 'New Media, the New Economy, and New Spaces', *Geoforum* 31: 425–436.

Pratt, A.C. (2007) 'Urban Regeneration: From the Arts "Feel Good" Factor to the Cultural Economy: A Case Study of Hoxton, London', Working Paper, London: Department of Geography, London School of Economics and Political Science.

Pratt, A.C. and Jarvis, H. (2002) 'Creative Destruction: The Struggle for Work-Life Balance in San Francisco's "New Knowledge Economy" Milieu', paper presented at the 'Cities and Regions in the 21st Century' conference, Newcastle, UK, Centre for Urban and Regional Development, 17–18 September.

Punter, J. (2003) *The Vancouver Achievement: Urban Planning and Design*. Vancouver: UBC Press.

Rantisi, N. (2002) 'The Competitive Foundations of Localized Learning and Innovation: The Case of Women's Garment Production in New York City', *Economic Geography* 78: 441–462.

Rennie Marketing Systems (2005) 'Woodwards: the Intellectual Property', Marketing Broadsheet, Vancouver.

Rigg, J. (1997) *Southeast Asia: The Human Landscape of Modernization and Development*. New York and London: Routledge.

Sacco, P.L. with Del Bianco, E. and Williams, R. (2007) *The Power of the Arts in Vancouver: Creating a Great City*. Vancouver: VanCity Capital.

Salet, W. and Majoor, S. (2005) *Amsterdam Zuidas: European Space*. Rotterdam: 010 Publications.

Sandercock. L. (2005) 'An Activist Civic Agenda in Vancouver', *Harvard Design Magazine* 22 (March): 36–43.

Sassen, S. ([1991] 2001) *The Global City: New York, London, Tokyo*, 2nd edn. Princeton, NJ: Princeton University Press.

Sassen, S. (2006) 'Chicago's Deep Economic History: Its Specialized Advantage in the Global Network', in R. Greene, M. Bouman, and D. Grammenos (eds), *Chicago's Geographies: Metropolis for the 21st Century*. Chicago: Association of American Geographers.

Saunders, D. (2006) 'Britain's New Working Class Speaks Polish', *The Globe and Mail* (Toronto), 23 September: F3.

Saxenian, A. (1991) 'The Origins and Dynamics of Production Networks in Silicon Valley', *Research Policy* 20: 423–437.

Schön, D., Sanyal, B. and Mitchell W.J. (1998) *High Technology and Low Income Communities*. Cambridge, MA: The MIT Press.

Scitovsky, T. (1963) 'Two Concepts of External Economies', reprinted in A.N. Agarwala and S.P. Singh (eds), *The Economics of Underdevelopment*. Oxford and New York: Oxford University Press.

Scott, A.J. (1982) 'Locational Patterns and Dynamics of Industrial Activity in the Modern Metropolis: a Review Essay', *Urban Studies* 19: 111–142.

Scott, A.J. (1988) *Metropolis: From the Division of Labor to Urban Form*. Berkeley, CA: The University of California Press.

Scott, A.J. (1997) 'The Cultural Economy of the City', *International Journal of Urban and Regional Research* 21: 323–339.

Scott, A.J. (2000) *The Cultural Economy of Cities*. London: Sage.

Scott, A.J. (2003) 'Cultural-Products Industries and Urban Economic Development: Prospects for Growth and Market Contestation in Global Context', Working Paper, Center for Globalization and Policy Research, University of California at Los Angeles.

Scott, A.J. (2006) 'Entrepreneurship, Innovation and Industrial Development: Geography and the Creative Field Revisited', *Small Business Economics* 26: 1–24.

Scott, A.J. (2007) 'Exploring the Creative City Paradigm', Keynote Address to the Biennial Meeting of the Pacific Regional Science Council, Vancouver, 6 May.

Sheppard, F. (1998) *London: A History*. Oxford: Oxford University Press.

Simmie, J. (1985) 'The Spatial Division of Labour in London', *International Journal of Urban and Regional Research* 9: 556–568.

Sjoberg, G. (1965) 'Cities in Developing and Industrial Societies: A Cross-cultural Analysis', in P.M. Hauser and L.F. Schnore (eds), *The Study of Urbanization*. New York: Wiley.

Smith, H. (2000) 'Where Worlds Collide: Social Polarisation at the Community Level in Vancouver's Gastown/Downtown Eastside', PhD thesis, Department of Geography, University of British Columbia.

Smith, M.P. (2001) *Transnational Urbanism: Locating Globalization*. Oxford: Blackwell.

Soja, E. (2000) *Postmetropolis: Critical Studies of Cities and Regions*. Oxford: Blackwell.

Solnit, R. with Schwartzenberg, S. (2000) *Hollow City: The Siege of San Francisco and the Crisis of American Urbanism*. London and New York: Verso.

Spate, O.H.K. (1938) 'Geographical Aspects of the Industrial Evolution of London till 1850', *Geographical Journal* XCII.

Stanback, T. (1979) *Understanding the Service Economy: Employment, Productivity, Location*. Baltimore, MD: Johns Hopkins University Press.

Starr, K. (1996) *Endangered Dreams: The Great Depression in California*. New York and Oxford: Oxford University Press.

Storper, M. (1997) *The Regional World: Territorial Development in a Global Economy*. New York: The Guilford Press.

Storper, M. and Christopherson, S. (1987) 'Flexible Specialization and Regional Industrial Agglomeration', *Annals of the Association of American Geographers* 77: 104–117.

Storper, M. and Walker, R. (1989) *The Capitalist Imperative: Territory, Technology, and Industrial Growth*. Oxford: Blackwell.

The Straits Times (2003) 'Singapore Faces New Competitors in Services', 24 January: 1.

Surborg, B. (2006) 'Advanced Services, the New Economy and the Built Environment in Hanoi', *Cities* 23: 239–249.

Tames, R. (1999) *Clerkenwell and Finsbury Past*. London: Historical Publications.

Tan, E. (2006) 'An Appreciation of the Social Constructs in the Making of the Local Shophouse', *Journal of Southeast Asian Architecture* 8: 41–51.

Taylor, P.J. (2004) 'Leading World Cities', in *Global and World Cities*. Loughborough: Loughborough University.

The Arts Business Ltd (1997) *The City Fringe Multimedia Study: Final Report*. London: The Arts Business Ltd.

Thornley, A. (ed.) (1992) *The Crisis of London*. London: Routledge.

Thornley, A. (2000) 'Dome Alone: London's Millennium Project and the Strategic Planning Deficit', *International Journal of Urban and Regional Research* 24: 689–699.

Thrift N. and Olds, K. (1996) 'Refiguring the Economic in Economic Geography', *Progress in Human Geography* 20: 311–337.

Thrift, N. and Olds, K. (2005) 'Assembling the "Global Schoolhouse" in Pacific Asia: The Case of Singapore', in P.W. Daniels, K.C. Ho, and T.A. Hutton (eds), *Service Industries and Asia-Pacific Cities: New Development Trajectories*. London and New York: Routledge.

Toh, M.H., Choo, A., and Ho, T. (2006) *Economic Contributions of Singapore's Creative Industries*. Singapore: Economics Division, Ministry of Trade and Industry, and Creative Industries Strategy Group, Ministry of Information, Communication and the Arts.

Turner, S. (2006) 'Hanoi's Ancient Quarter: "Traditional Traders" in a Rapidly Transforming City', thesis, McGill University, Department of Geography, Montreal.

Webber, M.J. (1984) *Explanation, Prediction and Planning: The Lowry Model*. London: Pion.

Weber, A. ([1899] 1963) *The Growth of Cities in the Nineteenth Century*. Ithaca, New York: Cornell University Press.

Wise, M.J. (1949) 'On the Evolution of the Jewellery and Gun Quarters in Birmingham', *Transactions of the Institute of British Geographers* 15: 57–72.

Wise, M.J. (1956) 'The Role of London in the Industrial Geography of Great Britain', *Geography* XLI (November).

Wong, K.W. and Bunnell, T. (2006) ' "New Economy" Discourse and Space in Singapore: A Case Study of One-North', *Environment and Planning A* 38: 69–84.

Wu, F. and Yeh, A.G.-O. (1999) 'Urban Spatial Structure in a Transitional Economy: The Case of Guangzhou, China', *Journal of the American Planning Association* 65: 377–394.

Yeoh, B. and Huang, S. (1996) 'The Conservation–Redevelopment Dilemma in Singapore', *Cities* 13: 411–422.

Yeoh, B. and Kong, L. (1994) 'Reading Landscape Meanings: State Constructions and Lived Experience in Singapore's Chinatown', *Habitat International* 18: 17–35.

Yeoh, B. and Kong, L. (eds) (1995) *Portraits of Places: History, Community and Identity in Singapore*. Singapore: Times Editions.

Yeung, H. (2005) 'Rethinking Relational Geography', *Transactions of the Institute of British Geographers* 30: 37–51.

Yeung, H. and Lin, G.C.-S. (2003) 'Theorizing Economic Geographies of Asia', *Economic Geography* 79: 107–128.

Young, M. and Willmott, P. (1957) *Family and Kinship in East London*. London: Routledge.

Yue, A. (2006) 'Cultural Governance and Creative Industries in Singapore', *International Journal of Cultural Policy* 12: 17–33.

Zukin, S. (1989) *Loft Living: Culture and Capital in Urban Change*. New Brunswick, NJ: Rutgers University Press.

Zukin, S. (1995) *The Cultures of Cities*. Oxford: Blackwell.

Zukin, S. (1998) 'Urban Lifestyles: Diversity and Standardisation in Spaces of Consumption', *Urban Studies* 35: 825–839.

Index